Launching into Commercial Space

Innovations in Space Travel

By
Joseph N. Pelton
and
Peter Marshall

Published by
American Institute of Aeronautics and Astronautics, Inc.
1801 Alexander Bell Drive, Reston, VA 20191-4344

Library of Congress Cataloging-in-Publication Data
Pelton, Joseph N.
  Launching into commercial space : innovations in space travel / Joseph N. Pelton, Peter Marshall.
     pages cm -- (Library of flight series)
  Includes bibliographical references and index.
  ISBN 978-1-62410-258-5 (Print) -- ISBN 978-1-62410-241-7 (Kindle)
  1. Space tourism. 2. Outer space--Civilian use.
  I. Marshall, Peter (Peter P.) II. Title.
  TL794.7.P45 2013
  338.4'791--dc23   2013045236

Copyright © 2015 by the American Institute of Aeronautics and Astronautics, Inc. All rights reserved. Printed in the United States of America. No part of this publication may be reproduced, distributed, or transmitted, in any form or by any means, or stored in a database or retrieval system, without the prior written permission of the publisher.
  Data and information appearing in this book are for informational purposes only. AIAA is not responsible for any injury or damage resulting from use or reliance, nor does AIAA warrant that use or reliance will be free from privately owned rights.

CONTENTS

Prologue v

Chapter 1: The Spectacular Debut of Commercial Space Travel 1

Chapter 2: Building a Business Case 29

Chapter 3: The First "Citizen Astronauts" 53

Chapter 4: The Gigabuck Space Entrepreneurs 73

Chapter 5: The Billion-Dollar Corporations and the Other Leading Players 117

Chapter 6: The Challengers and the "Wannabes" in the United States 137

Chapter 7: The International Scene 161

Chapter 8: The Spaceport Stampede 191

Chapter 9: How Safe Is Private Space Travel? 225

Chapter 10: Recap of Other Key Issues and Potential Show Stoppers 257

Chapter 11: Spaceplane Systems and Their Strategic Application 283

Chapter 12: Hypersonic Transport: The Golden Goose of Commercial Space? 307

Chapter 13: The Top Ten Things to Know about the Future of Commercial Space 333

Appendix A: Inventory of Private Space Companies around the World (Past and Present) 373

Appendix B: Membership List of the Commercial Spaceflight Federation (CSF) (As of April 2015) 387

Appendix C: Chronology of Events in the Evolution of Commercial
  Spaceflight 389
Appendix D: Recent Commercial Space Launches Licensed by
the U.S. FAA 401

Author Biographies 403

PROLOGUE

The Space Shuttle Is Grounded—And a New Era Begins

Few people are aware that the 2012 grounding of the Space Shuttle came exactly 11 years *after* the date that the Challenger Accident Review Commission of 1986 had recommended. But the truth is, the three-decades-old, obsolete technology of the Space Shuttle made it a dangerous craft in many ways. Three miles of wiring could not be adequately serviced. The foam insulation problem was never completely fixed. The solid fuel motor did not allow shutdown for astronaut escape, and so on. It was indeed well past the time for the shuttle to go. In many ways, the shuttle and the International Space Station (ISS) ventures—as interesting and exciting as they have been—stood in the path of progress. Now is the time for a new Commercial Space Age. Now is the time for NASA and the other space agencies to allow innovation from private space commerce to blossom and soar into tomorrow.

Fig. P.1 The remaining space shuttles were piggybacked to their final resting places at museums in New York, Washington D.C., Florida, and Los Angeles. *(Courtesy of NASA.)*

Now that the shuttle is finally grounded and new commercial space vehicles are under development, a new trail to the future can be blazed. As the door closes on the remarkable technology of the 1970s, the amazingly innovative commercial "Space Billionaires" of the 21st century are opening a new door. Certainly, we still have a long way to go. That new door to the future is now just barely ajar. With the right space vision, however, for both the public and commercial space sectors, we can swing this new door wide open and more rapidly accelerate into the "non-earth world" of tomorrow. Please journey with us through this door.

Just who is in this gang of space entrepreneurs who are trying to wedge open brand new opportunities in space? Well, they include Paul Allen of Microsoft fame; Jeff Bezos of Amazon.com; Sir Richard Branson of The Virgin Group; Elon Musk, the founder of PayPal and Tesla Motors; Robert Bigelow of Budget Suites of America; and even John Carmack, who became a billionaire by creating "Doom" and other computer games.

In these pages, we will profile these exciting space billionaires. But first, let's find out what is coming to a spaceport near you very soon. The opening up of space—not just for astronauts but also for a new band of space zealots—will be an exciting part of the new commercial space future. During the next few years, the number of humans who have flown into space and usher in the age of "Citizen Astronauts" will double, as Arthur C. Clarke foretold in 2001 (the year, not his famous movie).

New sub-orbital flights and the emergence of "Citizen Astronauts" are but the start of a new commercial space revolution. Thus, we will also introduce the space entrepreneurs who are developing real spacecraft that can actually fly into orbit and deploy private space stations. This requires a bold new approach to space rocketry, and this is very much underway in these new commercial rocket development sites.

The new commercial space era has the potential to revolutionize many other things as well. With the innovations now in the pipeline, we may see the evolution of hypersonic transport that allows us to span the world in about three hours. We may even see radical new technology that will allow us to achieve low cost to orbit or even ways to colonize the moon for a fraction of the cost that would have been required if NASA or other space agencies were to mount these efforts using the old-fashioned ways that were first developed in the 20th century.

Hold onto your hats—or space helmets. We are in for an exciting new ride into space, in more ways than one.

CHAPTER 1

The Spectacular Debut of Commercial Space Travel

"I believe that the Golden Age of space travel is still ahead of us. Before the current decade is out, fee-paying passengers will be experiencing sub-orbital flights aboard privately funded passenger vehicles, built by a new generation of engineer-entrepreneurs with an unstoppable passion for space. We are seeing the emergence of a new breed of 'Citizen Astronauts' and private space enterprise." – Sir Arthur Clarke, 2001

Sir Richard Branson's Virgin Galactic spaceplane, after years of delay, is now on the verge of flying fare-paying passengers to space. If the SpaceShipTwo (SS2) flights prove successful, this could lead to new era in space adventures. This, in turn, could ultimately lead to hypersonic travel by commercial spaceplanes, which will allow transcontinental travel across the oceans in as few as three hours.

Before this can happen, however, this fledgling spaceplane travel industry will have to overcome the setback of the fatal test flight crash that took place on October 31, 2014. This accident, which ended in the death of co-pilot Michael Alsbury, was triggered by the still-unexplained deployment of the so-called feathering device. This was supposed to slow the spaceplane on reentry, not plunge it into a death spiral as it left the carrier vehicle. In fact this device was not to be deployed unless speeds in excess of 1600 kilometers/hour were reached. The SS2 pilot, Pete Siebold, fortunately parachuted to the ground.

Although Siebold suffered major injuries, he still survived. His unlucky mate, Michael Alsbury, gave his life to prove the viability of spaceplane flights into outer space. Test flights have resumed, and once the feathering braking device and its deployment gears are fixed properly, the SS2 should be close to being able to go into service. [1]

Commercial space travel and the so-called NewSpace industries are, however, much more than just space tourism. We are in fact seeing the blossoming of a host of new enterprises in the NewSpace arena. This includes the debut of quite a few new, private space ventures. These encompass resupply missions to the International Space Station (ISS) (already a reality) and now plans for private space stations and private flights to orbit—and even to land on—the moon (and possibly on Mars). We are also seeing the birth of a number of other so-called "protozone industries." These are new developments such as robotic freighters that fly above commercial air space; dark-sky stations for research and staging to orbit via ion thrusters (for small payloads); and high-altitude platforms for telecommunications, police and military surveillance, and remote sensing.

No matter how you look at the future, it is happening at an accelerating rate. Once again, another of Arthur C. Clarke's remarkable predictions is becoming reality. The ever narrower and amorphous line between science fiction and science fact is vaporizing into nothingness like the atmosphere that disappears at the Von Karman line.

So are you ready to buy a ticket to fly off into space? Or maybe do something even more exciting? In the next decade, this new future will offer much more than just a several-hours-long flight to view the earth from the vantage point of dark sky and to experience weightlessness for a few minutes. How about spending a few days aboard the ISS and viewing the Big Blue Marble from several hundred miles up in space? In the coming years, you might be able to take a flight to a private inflatable space hotel for a week or two. This is the current business plan of Robert Bigelow, a Budget Suites of America executive. He plans to deploy his very own space hotel and research facility in low earth orbit (LEO) in the very near future. Or perhaps you might even aspire to take a trip around the moon or sign up with the Golden Spike Corporation to actually land on the moon. This is possible, according to the Golden Spike people, and for a bargain price of less than a billion dollars. The current price tag set by Golden Spike for two people to go to the moon is actually $1.5 billion. Yes, that is a billion with a "b."

No way, José! At least that is what most people would say if you asked them right now. A trip with Virgin Galactic will set you back a couple hundred thousand dollars! And indeed the next tranche of travelers will have to pony up $250,000 due to inflationary price increases. The cost of even more adventurous and exotic flights is of course in the millions.

But, hold on. The price of so-called space tourism will in time be coming down—and likely sharply so. We have reached the threshold of an exciting new era in which private enterprise is taking the lead in human spaceflight. These innovative companies have a remarkable new space agenda, and entrepreneurial NewSpace pioneers have the ambition to approach flight into the cosmos in unique ways. A remarkably diverse set of organizations now seeks to follow on from the pioneering achievements that were once led by NASA and other state-funded agencies in Russia, Europe, and China over the past half century. Much of this new activity arises from an unexpected but very important source—the space billionaires.

A Remarkable Bevy of Space Billionaires

This book is about many things. But a good deal of the story is about the space billionaires who, for better or worse, are ushering us into this new commercial age of spaceflight. South African Elon Musk of SpaceX (who just got a cool billion dollars in funding from Google) and Branson, with his Virgin Galactic enterprise, have already taken the first bold steps to turn their hopes and dreams into reality. In May 2012, SpaceX, after a series of tests and setbacks, succeeded with the historic, first private launch of a spacecraft to carry supplies to the ISS. Musk, who made his first billion as the co-founder of PayPal, spoke of his "indescribable joy" at this achievement. [2] Alan Lindemoyer, manager of NASA's commercial crew and cargo programs, said, "This really is the beginning of a new era in commercial spaceflight." [3]

And in this case at least, NASA, which has often been wrong about the future of NewSpace enterprises, was absolutely right. This remarkable first ISS resupply mission by the Falcon 9 rocket and the Dragon spacecraft was only "a demonstration" flight. But just five months later, on October 8, 2012, the SpaceX Dragon spacecraft was launched again on Falcon 9 for the first "official" delivery mission to the ISS to prove this was not a lucky fluke. The return flight was equally successful, and the Dragon spacecraft splashed down 250 miles off the California coast on October 28. The SpaceX recovery team then transported the spacecraft back to the company's base in McGregor, Texas, for processing and delivering the return cargo back to NASA.

A delighted Musk, then at the ripe old age of 41, said, "This historic mission signifies the restoration of America's ability to deliver and return critical space station cargo. The reliability of SpaceX technology and the strength of our partnership with NASA provide a strong foundation for future missions and achievements to come." By January 2015, there had been six successful missions by the Falcon 9 and the Dragon spacecraft to deliver supplies to the ISS. And Musk has indicated that in a few years he fully intends to be ferrying astronauts to LEO.

Fig. 1.1 History is made as SpaceX Dragon makes its first delivery of supplies by commercial spacecraft to ISS. *(Courtesy of NASA.)*

Back in 2011, Branson took a confident step forward in his ambitious program when he opened his dramatic new quarter-billion-dollar Spaceport America in New Mexico. And it is there that his Virgin Galactic SS2 and its White Knight mother ship have resumed test flights now that the National Transportation Safety Board has concluded its official review of the October 31, 2014, accident. This review confirmed that the premature release of a feathered landing assist mechanism caused the crash. This is important in several ways, but the most important is the confirmation that the crash had nothing to do with the new rocket motors. These new motors substitute the cleaner polyimide plastic fuel for the "dirtier" HTPB rubber fuel originally developed by the Sierra Nevada Corporation. The safety design of the SS2 depends on the use of nitrous oxide (i.e., laughing gas) as the oxidizer for the solid fuel, which can be shut off by the crew. This approach is thus unlike typical solid-fueled rockets, which once ignited can't be shut down. The latest innovation for the SS2, however, is to substitute a new aluminum-based fuel in place of the dirtier artificial rubber fuel, which is considered less efficient

Fig. 1.2 A Virgin Galactic test flight in New Mexico, with SS2 carried by the White Knight. *(Courtesy of Virgin Galactic.)*

Branson insists that the program is proceeding ahead despite the accident and that in the near future, the New Mexico spaceport will see the first actual flights with fare-paying passengers. Branson also claims that very few of the 750 signed up passengers asked for their deposits back after the October 2014 crash, although exact figures have not been released.

Apparently he has not recanted the claim he made at the Farnborough Air Show in England in July 2012 that he and his son Sam and daughter Holly would be on the first flight when it goes up. [4]

So what does all this NewSpace activity by billionaire entrepreneurs mean? Well, for one thing, it means that the grounding of the NASA Space Shuttle was not—as some commentators had suggested—the end of the Space Age. What we are seeing now is instead the start of a remarkable new era of commercial spaceflight. And Musk and Branson are only two of the new space billionaires who will make this happen. Actually, a whole bevy of them is pushing new innovations around the globe.

Fig. 1.3 The Space Shuttle heading for museum display. *(Courtesy of NASA.)*

Boldly leading the way are various intrepid entrepreneurs who are prepared to take the risk of demonstrating that traveling into space is not just for highly trained astronauts and cosmonauts but also for a growing number of would-be "Citizen Astronauts." For this new breed of space explorer, all that is needed is about a quarter million dollars in spending cash, the ability to stand up to 6 to 8Gs in fitness testing, and the willingness to take the risk. This risks includes signing a waiver to hold the U.S. government and the company concerned harmless.

And who are these space billionaires? They include Paul Allen (who co-founded Microsoft with Bill Gates), Jeff Bezos (who founded Amazon.com), John Carmack (who made his fortune developing computer games like "Doom" and "Quake"), and Bigelow of Budget Suites of America. A more recent addition to the list is the Moon Express project from the Indian entrepreneur and technology pioneer Naveen Jain, who moved to California in 1983. Beyond those investing their personal fortunes, there are some multi-billion-dollar corporations in the hunt as well. Google, for instance, has established the $30 million Lunar XPRIZE, which we will tell you about shortly. Google, along with Fidelity Investments, has now also provided a billion dollars in backing to Musk's SpaceX business.

Moon exploration is the target for another new space venture, known as Golden Spike. In December 2012, this newly organized company announced plans to use existing technology to take humans back to the moon and thus to follow in the footsteps of Neil Armstrong and the Apollo astronauts. Initially Golden Spike said the company would be able to land a crew of two on the lunar surface by 2020 for only $750 million a pop. It is not clear whether that 2020 date still holds.

In addition, there are many more actors in this unfolding NewSpace drama, and they involve a myriad of commercial space travel initiatives. In fact, in the past decade, some 50 entrepreneurial, private enterprises—large and small, and all around the world—have been involved in trying to develop spaceplanes for sub-orbital flights. Branson's Virgin Galactic is only one of those aiming to provide a brief flight into outer space at prices somewhere in the range of $200,000 to $250,000 per person. The U.S. company XCOR, for instance, is promising a trip that will take you 37 miles up for only $100,000.

These prices may seem steep. But we peons who fly tourist class and who think that tickets to fly should be calculated in hundreds of dollars rather than hundreds of thousands are not the prime target market.

Celebrities, Royals, and Rock Stars

For the past several years, the new space tourism companies have been busy trying to sign up millionaires, celebrities, royals, rock stars, movie idols, and star athletes who have money to burn on the rocket fuel *du jour*. Virgin Galactic reportedly has booked some 750 paying passengers at $200,000 each, although, as mentioned earlier, the company has confirmed it refunded some flight deposit payments after the October 2014 crash.

There has been much speculation as to who will be the first celebrity to go into space. The names of Hollywood actresses Victoria Principal and Sigourney Weaver have been reported, together with "A British Royal" (believed to be Princess Beatrice, Queen Elizabeth II's 25-year-old granddaughter, whose boyfriend works for Branson's Virgin Atlantic company). Moreover, Charles Simonyi, one of the first fare-paying astronauts on the Russian Soyuz spacecraft (and whom we will discuss in more depth in Chapter 3), has worked closely with Bill Gates since the start of Microsoft, and he has suggested that Gates may indeed be among those who will take a ride into space in the not too distant future.

Simonyi and Gates have also both been in the news for giving $20 million and $10 million, respectively, for the construction of the Large Scale Synoptic Telescope (LSST). This gigantic observatory project, with a massive 8.4-meter (27.5-foot) lens, is to be constructed in northern Chile in the Andes. When complete, this synoptic telescope will be able to sweep the entire visible sky every week. It will join the hunt for killer asteroids but will also engage in scientific exploration of the heavens. Its ultra-resolution digital cameras and gigantic lens will allow this Southern Hemisphere optics tool to see back almost to the Big Bang. [5] The one common thread to all the space billionaires is that they are intent on changing the world as we know it—and the heavens too.

The key to the business plans of space adventure companies is to get the passenger volume and their fleet of spaceplanes UP and their prices DOWN. The truth is that times do change. Travel by airplanes used to be expensive, risky, and just for the elite. The space tourism business hopes that it can repeat what happened in the aviation industry. In the first half of the 20th century, what would any sensible person have said to the idea of booking a flight from New York to London for a weekend jaunt? Or how many people would have said, "Oh, yes. One day we will be buying handheld electronic computers for a few hundred bucks. And by the way, just tell me what an electronic computer actually is and exactly how ordinary people use it"?

Back in the 1940s, if you had asked people about their views on lasers, transistors, satellites, bioengineering, and the cloning of animals, their reaction would have been much the same. They would have simply stared at you and wondered what asylum had let you loose on an unsuspecting world. Manned rockets to land on the moon? The Internet? Even spandex? All these things and more would have been seen as sheer science fiction just a few decades ago.

In these fast-changing times, science fiction becomes harder and harder to write, simply because technology keeps accelerating us to more and more improbable levels. Today, that change comes at us all too fast! It has created a crazy world where genetic engineering and space travel now co-exist with subsistence farmers living a near–Stone Age life in remote villages. We still have nomads living off the land in the African and Asian deserts while we are contemplating clean nuclear fusion energy systems, telescopes that can see back to the creation of the universe, and hypersonic spaceplanes that could fly from London to Sydney in about 180 minutes. We are suspended somewhere between future shock and time compression in a world where the rate of invention just keeps speeding up.

Signing Your Life Away

We used to say, "The sky's the limit," but that is no longer true. Now we can all sign up to become future "Citizen Astronauts" (a term coined by Arthur C. Clarke). But there are a few catches. If you want to become a "Citizen Astronaut," you will have to sign some heavy-duty waivers, which will certainly be daunting in almost every way. They will say in legalese that we know that a sub-orbital flight—going up over 60 miles or so into outer space on a commercial spaceplane, and then down—is a very risky, experimental flight. You will need to sign documents that say something like, "I know that I may be killed and that my estate will have no grounds for collecting damages because I have waived all rights to liability claims against the company or the government." You would have to be a raving lunatic to sign such a document and not understand that commercial spaceflights are in their very early days and that the risks are still quite great. We have only to read the obituary of Michael Alsbury to know that spaceplanes lack the safety of commercial jets.

After waiving all your rights to seek damages and after shelling out something in the ballpark of $200,000 to $250,000, all passengers must go through somewhat rigorous training to prepare for a very sharp pullout on the landing path, which is actually the hardest part. This is because you could be subjected to a force up to six to eight times that of your normal weight. For example, if you weigh 100 kg (220 pounds), experiencing 6 Gs would make you feel as if you weighed 600 kg (1320 pounds). Thus, before the flight you need some training for what will in truth be a somewhat taxing ordeal. And all of this will be to experience three to four minutes of weightlessness and to spend only a very brief time in outer space looking at the Big Blue Marble called earth. Space tourism is clearly not for everyone—at least not yet. But if the space billionaires succeed, the chance for John Q. Public to ride into space may come sooner than you might think.

By Spaceplane from London to Sydney in a Few Hours?

Within another decade or two, as this new business develops, travel through the stratosphere might not even break your piggy bank. Spaceplanes traveling from, say, London to Sydney or from New York to Tokyo in only a few hours may one day become commonplace. Today, rocket engineers are seriously working on new technologies that will enable more and more people to fly into space. Their mission statement is quite simple: to make very high altitude travel much faster, a whole lot cheaper, and not nearly as scary. Spaceplane travel for business executives may become phase two of the space tourism business.

Such a mission statement is key to the success of the new private-enterprise spaceflight companies. No real market will develop until spaceflight becomes not only affordable but also a lot safer than previously. It may never be as safe as riding a passenger jet, but that is the objective of the industry leaders behind the Commercial Spaceflight Federation (CSF). This private organization, which is devoted to commercial spaceflight development, was formed in 2005 as the Personal Spaceflight Federation. It then changed its name from

"Personal" to "Commercial" in 2009 since this more accurately reflected its mission.

Despite the high prices and major risks, this commercial space travel industry will soon become the new fashion. In perhaps a decade—or maybe less—it is expected that thousands of fare-paying passengers will fly up more than 60 to 70 miles (or over 100 to 120 kilometers) into space on board the next-generation spaceplanes. If these entrepreneurial companies succeed in their ambitions to establish a new commercial space business, a trip to outer space will take on new meaning. Within a decade, this new generation of space travelers could far outnumber the 500 or so astronauts and cosmonauts who have gone into space so far. In another two decades, tens of thousands of people (not just trained astronauts or wealthy celebrities) might be able to experience weightlessness and to see the world from the perspective of space—perhaps even as part of a rather routine trip to Japan, China, or Australia.

Ordinary people will start to follow in the footsteps of cosmonaut Yuri Gagarin, astronauts John Glenn and Neil Armstrong, and the other history-making spacefarers who led the way in manned spaceflight. This will be the new era of "Citizen Astronauts" as we enter a new commercial space age. In short, fantasy could become reality—and much sooner than we thought just a decade ago.

Entrepreneurial Talent and Risk-Takers

But who is making this all possible? It is not NASA, not the European Space Agency (ESA), and not the governmental space agency of some other country like China, India, or Japan. No! Rather amazingly, it is those space billionaires who have already made their fortunes in other areas such as computer games, music, hotels, and the Internet. These are new, 21st-century entrepreneurs with "gigabucks," and what they have in common are innovative thoughts about how to create a new future for *Homo sapiens.* They also seem to have both the entrepreneurial talent and the gutsy, "bet-the-company" risk-taking chutzpah to make it so.

Suddenly, the private sector—including not only the bold new start-ups but also some of the established U.S. and European aerospace giants—is jumping into the fray. A crazy quilt band of movers and shakers is now joining this brash effort to create "mass and low cost access to space." These commercial ventures, as described in the following chapters, hope to make civilian access to space happen much faster than a NASA, or another officially constrained governmental space agency, could ever contemplate. In addition, the result might come at a much lower cost. Reportedly, Paul Allen supplied some $20 million and probably much more to finance the Burt Rutan–designed SpaceShipOne. NASA, in contrast, had previously spent hundreds of millions on spaceplanes such as the X-33, X-34, X-35, and X-38 without notable success. When the successful first flight of Rutan's craft to space and back (Flight 15P) landed in 2004, an observer in the Mojave Desert handed the pilot, Michael Melville, a sign proclaiming: "SpaceShipOne—Government Zero." This sign was proudly held aloft by Melville, and the audience cheered. Apollo 11 Astronaut Buzz Aldrin then joined in the congratulations. [6]

Space: Here I Come

This quest for the new, high frontier secured a remarkable endorsement from the famous British physicist and cosmologist Professor Stephen Hawking in 2007. He overcame his severe disabilities with the motor-neuron disease known as Lou Gehrig's Disease to experience weightlessness on a ZERO-G flight. This was on an aircraft designed to fly in high parabolic arcs to achieve on each high arc about 30 to 40 seconds of weightlessness.

Fig. 1.4 Professor Stephen Hawking defies gravity, with Peter Diamandis in foreground. *(Courtesy of ZERO-G Corp.)*

ZERO-G is the company of Peter Diamandis, who was one of the co-founders of the International Space University and the founder of the XPRIZE. A flight on board ZERO-G's aircraft is a much cheaper option for those who want to experience weightlessness but who cannot pony up a quarter of a million dollars.

After Hawking's ride, he said, "Space, here I come." Apparently, he found it exhilarating to escape for just a short while from his wheelchair, and he declared his ambition to make a sub-orbital flight with Branson's new space travel company. Sir Richard has agreed to "fix it" for the professor and to arrange a Virgin Galactic flight for him.

In a more prophetic mood, Hawking also told *The Statesman*, a British publication, "I am hopeful that if we can engage this mass market, the cost of spaceflight will drop and we will be able to gain access to the resources of space and also spread humanity beyond just an earth-based existence. Sooner or later, some disaster may wipe out 'intelligent life' on earth."

Such concerns have led to the creation of groups including the B612 Foundation and the Lifeboat Foundation and serious efforts related to planetary defense and space hazards. Algae and lichen might survive, but future explosions of nuclear or biological weapons or a "killer" meteorite could conceivably extinguish humanity. In the view of Hawking and many other noted scientists, the long-term survival of the human race requires that we spread our "seeds" into space. The question is whether we will be ready for a "non-earth" existence before this type of extinction event occurs.

Orbiting the Moon—The Next Private Space Tourist Feat?
Hawking's ZERO-G flight was arranged by another remarkable entrepreneur, Eric Anderson, who founded Space Adventures, Ltd., headquartered in Vienna, Virginia, with satellite offices in Cape Canaveral, Florida; Moscow, Russia; and Tokyo, Japan. The company offers a wide range of programs as well as the spaceflight mission to the ISS. For example, the company announced that negotiations were underway for the first private human flight to orbit the moon. According to its website, two space tourists, each paying approximately $100 million, would ride with a Russian cosmonaut aboard a modified Soyuz

spacecraft. The Soyuz would get a boost from a docked Russian upper-stage rocket and would fly a boomerang trajectory around the moon. Anderson also masterminded the first fare-paying flights to the ISS. (This bit of space history is described more fully in the next chapter.)

Over the years, Space Adventures has developed quite a few options for those space enthusiasts who don't feel comfortable laying out a few hundred thousand for a vacation. These options start with a trip from either Florida or Las Vegas with the ZERO-G company, aboard the so-called "vomit comet." These flights provide a number of parabolic arcs with short periods of weightlessness on each arc, and some 5000 paying clients had experienced these flights by 2015, at a ticket price of about $5000 per head. For rather more dollars, you can train to be an astronaut at Star City, simulate an entire trip to space, or take a flight on a Soviet Foxbat at supersonic speeds to the true edge of space.

Space Adventures is a company that is always developing new programs, including working with leading universities to carry out new space experiments. The company's marketing skills are also a response to a recent comment by a colleague who remarked that the space tourism business will be highly dependent on what comes next. One key certainly does seem to be having a good answer for potential customers who ask, "What's next?" or "And then what can I do?"

Creating Astronauts by the Thousands

There is a wealth of experience on the Space Adventures advisory board, including the Apollo 11 moonwalker Buzz Aldrin; shuttle astronauts Sam Durrance, Tom Jones, Byron Lichtenberg, Norm Thagard, Kathy Thornton, Pierre Thuot, and Charles Walker; Skylab astronaut Owen Garriott; and Russian cosmonaut Yuri Usachev. The founder and CEO, Anderson, commented, "Ten years ago many people thought that space tourism was in the realm of science fiction. However, through innovative thinking and pushing the boundary of what is possible, Space Adventures sent the first private citizen to the ISS in 2001 on board the Russian Soyuz spacecraft, proving that commercial spaceflight is a reality. In the next 10 years, Space Adventures will fly more people to space than have flown in all of human history. We will create astronauts by the thousands, not by one or two."

Space Adventures has also announced plans to locate a spaceport in the United Arab Emirates. It will be funded by various parties, along with shared investments by Space Adventures and the government of Ras Al-Khaimah. Also, Sheikh Saud Bin Saqr Al Qasimi, along with the U.A.E. Department of Civilian Aviation, has granted clearance to operate sub-orbital spaceflights in their air space.

NanoRacks and Student Space Exhibits

Not everything involving commercial space is for fun and games. The recent flights by the SpaceX Falcon 9 and its Dragon spacecraft carried experiments to the ISS as well as supplies. Some of these are secondary-school experiments competitively selected to be conducted in space. These science experiments are made possible through the partnership between The National Center for Earth and Space Science Education (NCESSE), the Arthur C. Clarke Institute for Space Education, and NanoRacks, which provides very small experimental racks aboard the ISS for micro-experiments. Virtually of these first experiments have come from U.S. students selected from a competition throughout their home cities. Now, through the Arthur C. Clarke Institute for Space Science Education, student experiments are being sought from all partner countries that are involved with the ISS and that can develop and fly space experiments.

Of course, much more than school experiments are involved. Many of the entrepreneurs who are backing the development of commercial space see as much as half of their future revenues coming from experiments by universities, corporations, and even governmental entities. The plans for private platforms in LEO might provide research racks for a wide range of experiments and at a fraction of the cost for conducting these on governmentally launched and operated spacecraft. Currently NASA is talking to Virgin Galactic, SpaceX, Virginia-based Orbital Sciences Corporation (OSC), and others about using their vehicles for both micro-gravity experiments and astronaut training. NASA's interest in buying flights on these new vehicles extends to using them for experiments in earth's upper atmosphere as an analog for the thin atmosphere of Mars, plus a host of other possible tests.

The Next Moneymaker in Space?

Now, as the new breed of space entrepreneurs makes plans for their business enterprises, they are aiming to bring the cost for future space travelers to a level closer to that of other exotic vacations. Several of the start-up companies are pursuing this goal. The first of these, of course, is Virgin Galactic, which has ordered a fleet of the SS2 vehicles. This is the piggyback design of a launcher aircraft and spaceplane that separates and then launches at high altitude. This vehicle underwent extensive flight testing in 2012 and in 2013 in preparation for its maiden fare-paying flight. It is the follow-on version of the XPRIZE winner, SpaceShipOne.

This new generation of spacecraft is being developed, built, and tested by The Spaceship Company (TSC), which was originally a joint venture between Rutan's extraordinary and creative company, Scaled Composites, and Virgin Galactic. In subsequent reorganizations, Scaled Composites was wholly acquired by the aerospace giant Northrop Grumman, and then towards the end of 2013, Virgin Galactic became the 100 percent owner of TSC. According to the announcement by Virgin Galactic,

> "This acquisition, details of which are not being disclosed, marks the successful completion of a long-term strategy and signifies the end of the first phase of TSC's development. During this development phase, TSC completed the build out of manufacturing and assembly facilities in Mojave, CA, established a specialized workforce and transitioned necessary assets from Scaled in order to begin building Virgin Galactic's commercial fleet of WhiteKnightTwo (WK2) carrier aircraft and SpaceShipTwo (SS2)—the manned sub-orbital spacecraft. These vehicles will be utilized for Virgin Galactic's planned spaceline operations, which will be based at Spaceport America in southern New Mexico. The completion of the acquisition came as Virgin Galactic and Scaled began to plan the handover of the SS2 development program to Virgin Galactic, with Scaled remaining fully committed to the final

portion of the WK2 and SS2 test flight programs prior to Virgin Galactic commencing commercial operations." [7]

But TSC has competition, including XCOR and others described later. They are all endeavoring to make the ticket price for space tourism at least two orders of magnitude less, down from the $25 million or more for the Soyuz trip to the ISS to the Virgin Galactic ticket price, which is more than a hundred times less. In time, as the market expands and competition increases, the prices will of course fall.

Going Up and Down Is Much Easier Than Staying Up

Another group of commercial space organizations is attempting something much harder than simply achieving a sub-orbital flight for the purposes of space tourism. Several of the newer commercial space organizations are actually seeking to launch spacecraft into orbit and then to be able to bring at least a capsule safely back down. Some are even seeking to operate a private space station.

Table 1.1 shows the relative speeds and thrust levels needed to travel by airplane, jet, spaceplane, and rocket. A spaceplane ride, in terms of energy requirements, can thus be clearly seen to be more like a jet plane ride than going into earth orbit. The specific energy needed to launch to orbit is almost 25 times greater than that required just to achieve a sub-orbital flight.

Table 1.1 Comparing airplanes, jets, sub-orbital spaceplanes, and rockets to LEO.
(Data courtesy of Prof. Nikolai Tolyarenko, International Space University.)

Comparative Factor	Airplane	Jet	Spaceplane	Rocket to LEO
Velocity (m/sec)	250	500	1600	7800
Height (km)	Up to 10	Up to 20	Up to 120	200+
Specific Energy (Joules/kg)	0.13	0.7	14.5	324

Although there is synergy between the organizations trying to provide space tourism flights and those seeking to launch to orbit, the latter efforts are technically a much different type of effort—truly much harder.

Most space tourism activities are clearly separated from the private space launch initiatives. However, Allen and Rutan have also announced their highly innovative Stratolauncher enterprise, which spans the two sets of activities. The planned Stratolauncher is a carrier aircraft that will carry a payload rocket to 30,000 feet and then launch this large 6800-kilogram (15,000-pound) payload into earth orbit. This will be one of the largest aircraft ever built, and with a wingspan of 117 meters (385 feet) and twin fuselage length of 65 meters (215 feet), it could not fit within a football field. This giant carrier aircraft would have a maximum take-off weight of 545,000 kilograms (1.2 million pounds). It is being built with many components re-used from two Boeing B747-400s purchased by Stratolaunch Systems, including six GE engines. Initially Musk's SpaceX was to have been the third partner in this project, but subsequently SpaceX withdrew and OSC became the substitute partner. [8]

Branson, however, is never going to be "one upped," and thus he has proposed to launch smaller payloads to orbit using the White Knight launcher vehicle and a so-called "Launcher One" rocket. Apparently this capability will play into plans by Branson to be a partner in the Web One project, which intends to provide global Internet connectivity via a mega grid of LEO satellites in partnership with Google and others, as discussed later.

The Commercial Launch to Orbit Initiatives

Musk's SpaceX Corporation, coming from almost out of nowhere, is now a major player in this arena and now even in space-based Internet systems. Musk got a kick-start when he was awarded one of the two contracts in NASA's multi-million-dollar Commercial Orbital Transportation Services (COTS) program. The objective here was to develop a commercial re-supply vehicle for future astronaut and cargo transport to the ISS, as a replacement for the Space Shuttle. To date, he has now accomplished six such flights. The next step is NASA's Commercial Crew Integrated Capability (CCiCap) initiative, which is to lead to commercial human spaceflight services—for both government and commercial customers. There was a spirited competition among

Sierra Nevada, SpaceX, Boeing, and ATK to develop this final phase of a commercial flight capability to the ISS for crew and thus to end NASA's dependence on Russian Soyuz rockets.

As of September 16, 2014, NASA announced $6.8 billion worth of contracts that would go to Boeing and SpaceX to design, build, and test systems for transporting crews to the ISS, with the objective of being able to start actual flights as early as 2017. These contracts include money to develop complete systems. Thus the Boeing and SpaceX contracts provide funding to develop ground operations, launch services, in-orbit operations, re-entry, and landing. It is significant that the Boeing contract, representing the "established" U.S. industry, received a contract valued at $4.2 billion, while the SpaceX contract was for a smaller amount: $2.6 billion. [9]

Enter OSC

The second private company originally named by NASA under the COTS program was Rocketplane Kistler (RpK), headquartered in Oklahoma City. But towards the end of 2007, it was announced that RpK had failed to meet the stringent financial deadlines required under its contract. RpK continued to develop a private launch capability despite its financial difficulties but eventually went into Chapter 7 bankruptcy in October 2011. Meanwhile, after a further review of the various alternatives, NASA had announced in February 2008 that the second company to compete with SpaceX would be OSC.

OSC, a now well-established company, was founded in 1982 by David Thompson, Bruce Ferguson, and Scott Webster. In 1990, OSC successfully carried out eight space missions, highlighted by the initial launch of the Pegasus rocket, the world's first privately developed space launch vehicle. In 2006, OSC conducted its 500th mission since the company's founding. Since the development of the Pegasus, Thompson and company have gone on to develop the Taurus rocket. And then for the COTS program they decided to upgrade the Taurus to the Antares rocket and also to develop the Cygnus spacecraft. Unlike the SpaceX Dragon spacecraft, which returns cargo safely to earth, the Cygnus was designed simply to burn up on re-entry.

Fig. 1.5 OSC's Antares rocket prepares to launch the Cygnus spacecraft with supplies to the ISS. *(Courtesy of NASA.)*

Under OSC's development contract with NASA, OSC is obligated to provide a number of cargo supply missions to the ISS. However, this contract focuses on developing a launch system that can deliver cargo to the ISS but not its return.

OSC successfully launched its liquid-fueled Antares rocket on April 21, 2013, with a dummy Cygnus capsule aboard. The second launch, however, was a disaster for OSC. This rocket, with experiments and ISS supplies on board, blew up shortly after takeoff due to a malfunction of the Russian rocket motors, which had already been planned for retrofit for motors designed and built in the United States. This placed OSC even further behind SpaceX and its Falcon 9 and Dragon capsule resupply systems under the NASA COTS program. Certainly SpaceX performance in this regard contributed to its selection along with Boeing for the crewed vehicle transport system for the ISS.

A New Role for NASA

Time will tell if we have now entered a new era of space exploration where private investment plays an even greater role than before and where governmental space agencies ultimately assume a secondary role. Certainly in the United States, the policy now appears to be a shift of emphasis for NASA to aim beyond routine flights to the ISS and to focus instead on new challenges farther out to explore asteroids, the moon, and Mars. However, effective public and private cooperation will be needed to succeed even in such exotic space ventures.

After the successful launch of the SpaceX mission to deliver supplies to the ISS in May 2012, NASA's Deputy Administrator Lori Garver said that the agency's plan was now to "let private industry do what it does best and let NASA tackle the challenge of pushing the boundary further."

The debate on NASA's future role was a feature of a panel discussion during a conference in Pasadena, California, as long ago as September 2007. During the conference, Rutan insisted that "taxpayer-funded NASA should only fund research and not development." He said that the goal of private space tourism is to reduce the cost of space travel and exploration.

Former NASA Administrator Michael Griffin responded to Rutan's vision in a speech at the same event. "Unlike Rutan, I will continue to think [national] space programs are important," Griffin said. "We [NASA] have here a program which is affordable, sustainable and which can be highly correlated to historical successes and developments from the past." [10] Many, however, would tend to quibble a bit with the former NASA administrator's assessment about how "affordable" NASA development programs have been. The ISS's cost, for instance, has now run to $140 billion, and its completion was many years late and well over budget. Currently NASA and its partners' plans are to splash the ISS down in the ocean rather than redeploy at least parts of it to other space ventures. Currently only DARPA, among U.S. federal agencies, is exploring how to redeploy usable space assets in orbit rather than simply deorbiting them.

Unlike the European, Japanese, Russian, and Chinese space agencies, which continue to concentrate on space applications and missions that involve earth-based activities, NASA is now focused heavily on the cosmos—and especially the moon, Mars, and beyond. Here NASA's research role is much clearer.

In short, critics continue to say that NASA's role should be in research missions to the solar system and beyond and in planetary defense against solar, asteroid, and comet defense rather than commercial development of near earth and the protozone, where commercial ventures can be more cost effective and efficient.

A Clear and Present Danger

Certainly, this much is true: Private initiatives with private test pilots and private astronauts who are insured against risk (and whose families are provided for) add a new dimension of flexibility and risk management that government space agencies do not now have. Neither did they have the experience of flying fare-paying passengers who had signed personal waivers before venturing into space. One must hope that all of the new entrepreneurial space operators know—deep down inside—that failure means ruin for their programs whereas success and continued safe operation mean world acclaim.

Thus, there is a clear and present danger that even one or two of the many organizations trying to develop systems to fly into space will give priority to being the first to fly commercial flights to space and will overlook key safety features and concerns.

Private organizations such as OSC, SpaceX, Scaled Composites, Sierra Nevada, ATK, Masten, Armadillo Aerospace, XCOR, and Bigelow Aerospace clearly have a different set of values and different types of incentives to succeed. The challenge for NASA, in its role as space explorer and developer of new launch systems, is to find new and flexible ways to fuel new intellectual energies, find better technology, and create safer space systems. The question is whether NASA accomplishes these tasks through "conventional" or "innovative" new governmental processes.

The incentives that the FAA, NASA, and Congress have created to stimulate the space tourism business, plus the magnetic allure of the various prizes, all clearly seem to be working to spur new levels of innovation. It is these powerful stimulants that have triggered the large number of established and start-up companies to innovate in the design and launch of new space systems.

Successes—And Early Setbacks

The development of new space technology is not simply a matter of R&D and engineering. The business and marketing aspects are perhaps equally important for all those seeking "a license to orbit."

This 21st-century phenomenon of commercial space development in many ways parallels the early days of the aircraft industry. Just as in the early part of the 20th century, when a plethora of planes was being designed and built, today's world is beset by a diverse group of companies—large, medium, and small—that are designing craft they hope will take hundreds and then thousands of space tourists into the black sky of sub-orbital space. Others aspire to build systems that can go into orbit and even allow space tourists to walk in space or stay at a space hotel, as envisioned by Clarke in *2001: A Space Odyssey*. These various efforts largely share two goals that in many ways are in conflict. The space entrepreneurs want to be the first to launch space tourists into this wondrous and largely unknown environment. Yet they recognize that a space tourism enterprise will succeed only if the flights are safe and passengers return safely from these excursions.

Nevertheless, just as in the early days of flight, and in the first stages of the U.S. and Soviet space programs, the development of private space tourism has not been free of setbacks. In July 2007, there was a stark reminder of the risks and dangers in spaceflight: a fatal accident at Scaled Composites' remote test site at the Mojave Desert facility in California. It occurred during the testing of components for the hybrid rocket engine that would be used on the next-generation rocket plane the company was building for Virgin Galactic. It was reported that "something went wrong" with the tanks of nitrous oxide being used to test SS2's motor and that the blast of pressurized gas "went off like a bomb." Three workers were killed in the explosion and three others severely injured.

Rutan told a press conference that the cause of the explosion was a mystery. "We were doing a test we believed was safe. We don't know why it exploded," he said.

The U.S. Occupational Safety and Hazard Agency (OSHA) in the Department of Labor imposed a fine on Scaled Composites for safety failings in the Mojave accident, and this incident underlined that this is an experimental and dangerous undertaking. This mishap has, on one hand, certainly led to a delay in the original timetable for commercial sub-orbital flights; but it has perhaps helped to make sure that when the commercial launches do occur they will be better engineered and, it is hoped, safer.

The "tragedy in the Mojave" was not the only setback for commercial space development, however. Two unmanned Falcon rockets launched by the SpaceX company exploded during their test flights, but Musk, with true entrepreneurial spirit, made it clear that much had been learned from both launches. OSC also had its failed launch in 2013, when a rocket blew up shortly after leaving the launchpad. Then, of course, there was the fatal crash of Virgin Galactic's SS2 on a test flight in the Mojave Desert on October 31, 2014.

These costly disasters are clear reminders that spaceflight is a hazardous business and that there will inevitably be difficulties to surmount along the way. After the Apollo 1 accident in January 1967, NASA did a complete review of its rocket design and test program from stem to stern. Perhaps propelled by this incident, NASA recovered and memorably succeeded in landing the first astronauts on the moon just 29 months later, in July 1969.

Such setbacks, even human tragedies, lead to rethinking, redesigning, and caution. We can only hope that this means the space tourism business is working out the kinks before real passengers fly.

The 50-Plus New Companies in Constant Brownian Motion

The field of commercial space companies is increasingly international and crowded. During the past decade, more than 50 such enterprises have been involved in developing spaceplanes, lighter-than-air craft that can fly into orbit and operate from spaceports. Most of these firms are in the United States, but a number are in other countries. All of them, big and small, are playing a part in developing the space tourism business. They range from service companies, such as Space Adventures, that can arrange dozens of alternatives for would-be space travelers, to research and technology companies working on new ways to provide access to a sub-orbital space ride to LEO or even beyond.

Appendix A, at the end of the book, summarizes many of these entities, the means by which they are seeking to provide access to space, the names of their craft or spaceport, and their current goals and objectives. In many cases, these companies envision a staged development with expanded capabilities over time.

Space tourism is clearly a highly fluid "industry" reminiscent of the barnstorming days of early aviation 100 years ago. A great deal of shakeout will continue to occur, with some dropping out, going bankrupt, or merging with others. Already over two dozen hopeful commercial space companies have bitten the dust. Meanwhile, other new entities or subsidiaries of established firms have joined the process of developing new, reliable, and affordable access to space.

Maintaining safety, setting reasonable and viable risk management rules and regulations, and undertaking effective independent verification and validation processes for this totally new type of industry are clearly great challenges, and they will require a great deal of balance and judgment. On one hand, there must be the opportunity for innovation and totally new approaches to space travel to obtain new and better solutions, yet at the same time safety controls and inspections must be stringent so as to prevent needless and foolish loss of life.

In the United States, Congress has designated the FAA to be in charge of overseeing this burgeoning new industry, starting with the Commercial Space Launch Act of 2004. Four years later, in 2008, the FAA and NASA formed an Executive Committee to consider aviation safety and airspace systems, efficiency, and environmental compatibility. Then, in June 2012, NASA and the FAA announced an important agreement to align commercial spaceflight requirements between the two agencies, a move welcomed by the CSF. [11]

In February 2012, Congress passed the FAA Authorization Bill, which extended the original eight-year learning period for commercial spaceflight for a further three years, through October 2015. Under this reauthorization, the FAA's Office of Commercial Space Transportation retains all of its current regulatory authority, including the ability to ensure the safety of the "uninvolved public." In addition, the industry continues to have access to U.S. governmental liability protection. It also oversees the "informed consent" regime for spaceflight participants who willingly accept the risk. Other countries have also been considering how such activity is to be regulated, and they are looking very closely at progress made in the United States.

The Way Forward in the Decade Ahead

Many centuries ago, or so the story goes, a would-be astronaut in China sat atop a cluster of rockets powered by gunpowder. Thus, he sought ascent to the heavens. Instead, however, to the horror of the assembled witnesses, he was rather immediately incinerated.

Fortunately, we have come a long way in rocketry and space travel since that time. Careful and prudent safety regulation of this newly emerging "industry" is essential to protect the foolishly innovative and overly daring inventor and entrepreneur. Care is also needed to protect the property and lives of others. And perhaps equally important to the aerospace world is the need for safety regulations and thoughtful licensing activities in order to prevent rash and risky experiments that spoil or retard the development of an entirely new industry and type of human endeavor. In short, the risk is not only to the flight crew and passengers of a dangerous vehicle; in addition, rash and premature efforts to jump start the space tourism business may destroy the initiatives of many who have well-conceived ideas as to how to provide safe access to space.

But above all, we are seeing a tremendous surge of enthusiasm and creativity. Engineers, innovators, and entrepreneurs abound. They all seem to think that they can develop new and better spaceplane technology. Many will fail, but it will only take a handful of the very best to support the growth of a new space tourism industry.

Although the prospect of space tourism seems scary, and to some almost science fiction, the challenge of sub-orbital flight is much less than the challenge of actually going into space. The range of space tourism activities will, if anything, likely continue to expand as the business matures. At the low end will be earth-based simulations of spaceflight, training at Star City, and parabolic flights to give brief periods of weightlessness. At the high end, organizations like Space

Adventures will continue to push the envelope for truly wealthy space explorers seeking to accomplish new firsts. The next space tourism frontier apparently will be a $100 million ride around the moon on a Soyuz vehicle. Or perhaps it will even be the $750-million-a-seat landing on the moon as conceived by the Golden Spike Company or even Mars One or Dennis Tito's "Inspiration Mars" effort to send a manned mission to Mars.

Science fiction just keeps getting more and more difficult to write as the commercial space industry continues to push the boundaries forward.

REFERENCES
[1] http://science.psu.edu/news-and-events/2008-news/.
[2] *Satellite TODAY*, "SpaceX Flight Successful," May 25, 2012.
[3] Dragon Capsule Resupplies International Space Station, http://businesstech.co.za/news/general/13755/spacex-dragon-capsule-opens-new-era/.
[4] "Richard Branson Speaks Out," *Satellite TODAY*, October 29, 2012.
[5] "LSST Receives $30 Million from Charles Simonyi and Bill Gates," Brookhaven National Labs, http://www.bnl.gov/newsroom/news.php?a=1729.
[6] "SpaceShipOne flight 15P," http://en.wikipedia.org/wiki/SpaceShipOne_flight_15P.
[7] "Virgin Galactic Acquires Full Ownership of the Spaceship Company," October 5, 2012, http://www.businesswire.com/news/home/20121005005907/en/Virgin-Galactic-Acquires-Full-Ownership-Spaceship-Company#.VYGQVPlVhHw.
[8] "Orbital Sciences Replaces SpaceX on Stratolaunch Project," Space News, December 3, 2012, http://www.space.com/18747-stratolaunch-orbital-sciences-replaces-spacex.html.
[9] Foust, J., "NASA Selects Boeing and SpaceX for Commercial Crew Contracts," September 16, 2014, http://spacenews.com/41891nasa-selects-boeing-and-spacex-for-commercial-crew-contracts/.
[10] ZDNet Technology News - ZDNet News: September 20, 2007.
[11] *Space Safety Magazine*, June 20, 2012.

CHAPTER 2

Building a Business Case

The new promise of space travel brings with it opportunities—and risks. Three people were killed in Mojave, California, on July 26, 2007, when a 4500-kilogram (10,000-pound) tank of nitrous oxide exploded at the Scaled Composites facility. This resulted in a substantial fine for Scaled Composites by the California Office of OSHA. Then, on October 31, 2014—which was ironically Halloween in the United States—there was the fatal crash during the test flight of SS2. These events tend to make space tourists who are signed up for a flight on a spaceplane a bit skittish. Enticing customers to a new product or service is always a challenge. When the service might prove fatal or ultra-dangerous, the challenge is even greater.

Commercial space travel ventures will not be measured against the old metrics of national glory, nor will they be sustained by generous government budgets and artificial subsidies. Ultimately the normal business measures of profitability, growth potential, and return on capital investment will apply. One might argue that technological breakthroughs or advancement of society should be considered too, but most stockholders demand short-term return above all other considerations. Perhaps that is foolish and short-sighted.

Furthermore, the commercial space travel market is—at least during a rather lengthy start-up period—a "thin" and high-end market. A single Wal-Mart store in a major metropolis will likely clear more revenue than this entire industry for at least some years to come.

Unlike the dot.com online ventures of the 1990s, where investors were willing to look at growth and wait years on profits, this is a much different kettle of fish—largely because of a very low volume of potential customers versus a global market with a potential of billions of consumers. In essence, this is a business that seeks to serve a quite different, much smaller, and more elite group of potential buyers. The number of potential commercial space travelers is hard to quantify at this time. You can bet your boots, however, that far fewer are ready to sign up for a space jaunt than those likely to buy a book or a CD from

Amazon.com or BarnesandNoble.com. And space tourists will definitely not be booking e-tail because they will likely have a host of questions to be answered and a raft of waiver forms to be completed and legally witnessed.

The big question of the day is whether this specialized market is perhaps a thousand times smaller or a million times smaller than the online consumer market. In this realm, like in so many others, size really does matter.

The commercial space business entrepreneurs we profile are well experienced in the harsh world of competition. They have become rich enough to embark on yet another exciting new venture because they have been successful in the past and have plenty of capital to invest. So far, none of them has made public their longer-term business plans and financial projections. But amid all the speculation and guesswork, there are still helpful insights into the realistic costs of these enterprises and the potential size of the emerging markets. Several organizations, including the space agencies, have also attempted to study and project the market for space tourism as a serious business. The bottom line is that no one will know how successful this new business will be until after it has been launched in a serious way. Fortunately for the fledgling industry, some very deep-pocketed billionaires and some very competent technical engineers are involved.

Most people, especially when the offering means putting your life on the line, will take a serious wait-and-see attitude about any such new business. Early adopters are fewer in number when the ante might be eternity.

The Market Studies Are Consistently Overly Optimistic

In 2002 the Futron Corporation prepared the first comprehensive market study of projected growth for the anticipated new space tourism market and then updated it in 2006 with the latest survey information. Just two years before the 2006 update the Burt Rutan and Paul Allen team had claimed the Ansari XPRIZE, and enthusiasm and expectations were high. The thought of commercial travel into space was an inspiring thought. For instance, after the first moon landing in 1969, the head of Cal Tech projected that we might send astronauts to Mars in the 1980s.

In the space arena, early success can often lead to exuberant enthusiasm and overly ambitious market projections.

The Futron study projected that experimental flights would begin as early as 2009. This study, which included the results of many interviews of high-worth individuals, projected a reasonably high uptake in seat bookings for this exciting new business based on wealthy space enthusiasts and risk takers. Futron projected traffic would climb sharply to nearly 16,000 passenger seats a year by 2021. At the time, it was also expected that Congress might pass new legislation, as early as 2012, that would give the FAA authority to start licensing vehicles and perhaps see the start of regular commercial flights for this new industry. None of these projections turned out to be close to the actual events, for a variety of reasons. [1]

A more recent study of the space tourism market was undertaken in 2012, with funding provided by the FAA Office of Space Transportation and Space Florida, a spaceport company. This study was carried out by the Tauri Group, a consulting business that had also conducted a previous study on space tourism business development for the CSF.

The Tauri Group interviewed widely and indeed conducted in-depth conversations with some 120 individuals of high net worth. It also consulted market researchers in the field, delved into a number of open source studies and analogous market studies, and then reviewed relevant government reports and budgets. The Tauri Group then proceeded to develop three market profiles based on a "constrained" market growth model, a baseline market growth projection, and a high-growth and more optimistic growth model. The result was a projection of total market flight forecasts based on the total number of passenger seats associated with space tourism flights covering a 10-year period. The totals ranged from about 2400 total passengers flown on the low end to a projected total of over 13,000 for the high-end forecasts. However, these were for the total number of passenger seats for the entire decade-long study period. These projections were thus less than 10 percent of the Futron forecasts of six years earlier. The Tauri Group also wisely did not indicate the start date for commercial flights but merely gave a "year one, year two, year three" projection, based on whenever these space adventures might begin.

When one compares these projections to the millions of passengers flown on commercial airlines every day, it is hard to forecast this as a major global industry of importance for some time to come. Even if the high-end scenario is achieved, the total revenue based on the Tauri Group forecast for the number of passengers (in Table 2.1) is at most around $2 billion for the entire 10-year market forecast. And if the low-end figures are used (assuming no reduction in the price from $200,000 per flight), then the total revenue would only be around $450 million, or about $45 million per year. The cost of equipment and operations could of course easily exceed that number. [2]

Table 2.1 A Tauri Group study of three growth scenarios for the commercial spaceflight industry.

Market Scenario	Year 1	Year 2	Year 3	Year 4	Year 5	Year 6	Year 7	Year 8	Year 9	Year 10	Total
Constrained	213	226	232	229	239	243	241	247	252	255	2378
Baseline	373	390	405	421	238	451	489	501	517	533	4518
Optimistic	1096	1127	1169	1223	1260	1299	1394	1445	1529	1592	13,134

In short, it seems that the new markets that stem from this activity, such as the launching of experimental payloads, defense-related experiments, and supersonic or hypersonic air travel, may well be the true markets that spin off the real commercial opportunities. It is, in fact, the business of supersonic and hypersonic transportation that seems to be the only future business that could generate millions of passengers and blossom into a highly profitable multi-billion-dollar or, in time, even a trillion-dollar enterprise. Despite the modest traffic projections in the latest market studies, investment in these enterprises continues to mount—perhaps because of the spin-off enterprises that hold major market potential. Or perhaps it is because of governmental incentives and encouragement or even the hubris of the billionaire backers. People who have been able to become billionaires, however, usually have a good business sense and often foresee the markets of the future before others do.

Total Investment of $3 Billion and Rocketing Higher?

The CSF has previously commissioned a number of studies to measure the revenue, investment, and employment levels of this new business sector. The CSF website, www.commercialspaceflight.org, cites a 2009 study that was also conducted by the Tauri Group. This study sought to measure total investment by conducting a survey of 22 companies and accumulating governmental and other data. This study showed that through 2008, the total investment in the commercial human spaceflight business had reached a cumulative total of $1.46 billion. Today this investment is growing to very near $3 billion. At the same time, revenues and deposits for services, hardware, and support services represented a total of only about $800 million, with most of this money accruing to Virgin Galactic, Space Adventures, ZERO-G Corporation, and XCOR. This figure includes the flights of private citizens to the ISS and deposits made on future sub-orbital human spaceflights, which mainly are with Virgin Galactic. [3]

According to the Tauri Group 2009 study, the investment total of $1.46 billion was made up in this manner: about 52 percent from individual and "angel" investors, about 30 percent in private equity, about 15 percent from government and military agencies, and about 4 percent from corporate reinvestment. This study showed that facility space dedicated to commercial spaceflight had increased to 1,180,000 square feet and that employee levels had reached 1186 workers. [4] Although no current hard figures are available, those numbers seem to have doubled as of 2015. Most of the investment is in the United States, with projects such as Spaceport America and other spaceport projects in Florida, Texas, Oklahoma, and California. In addition, the companies noted below are moving ahead with spaceplane developments. However, there is global development as well. European and Asian projects are also beginning to evolve. [4]

Surveys conducted since then by Futron, the CSF, and the International Association for the Advancement of Space Safety variously indicate that globally these numbers for employment, office and plant facilities, and payrolls had more than doubled by the end of 2014. [5] The question is whether such growth can be sustained without an impulse boost to revenues in the next few years. The original Futron study projected a start date of 2009, and for a variety of reasons the real start date is at least six years later.

The CSF also notes that the Tauri Group review of the "state of the industry" in 2008 did not include the activities of companies such as SpaceX, Orbital Sciences, United Launch Alliance (the Boeing-Lockheed joint venture), Airbus (Astrium), or other companies that had expressed interest in commercial human spaceflight.

There were also surveys taken by *Space Future* as far back as the early 1990s. These found that as many as 70 percent of us want to go into space. Almost half of the respondents in one survey said they would pay three months' salary to do so. The problem with such surveys is that the answers were given on the basis of hypothetical play money. Respondents thus were not paying out of a real bank account. Further, no one knows what impact will occur on the potential market once going into space has become rather commonplace, with thousands of people actually having gone there and back. How many will decide to do it again? No one knows for sure. If at a future date one can take a three-hour flight from London to Sydney, with a view of the Big Blue Marble from the dark sky of space thrown in, does the space tourism business die out, rather like pagers being replaced by smart phones?

Some 2011 market studies were sponsored by the FAA Office of Commercial Space Transportation (FAA/AST). In its 2011 report, the FAA/AST identified six commercial spaceflight companies that were developing sub-orbital reusable vehicles that appeared to have viable plans to provide commercial sub-orbital reusable spaceflight vehicles. Only four of these—Armadillo, Blue Origin, Virgin Galactic, and XCOR—were developing vehicles for passenger flights. Masten and UP Aerospace were just to carry cargo and not passengers. Table 2.2 provides the assessment of their progress as of 2011. [6]

Table 2.2 Overview of the most likely early providers of sub-orbital flights.

Company	Main Vehicle	First Year of Test Flights	Operates From	Number of Seats	Comment
Armadillo Aerospace	Hyperion	2014	Spaceport America 2	2	Behind schedule
Blue Origin	New Shepherd	To be decided	West Texas Spaceport	3+	May use SS2 vehicles to start
Masten Space Systems	Xaero	2011	Mojave Air and Spaceport	0	To launch small payloads
UP Aerospace	Spaceloft	2006	Spaceport America	0	To launch small payloads
Virgin Galactic	SS2	2010	Spaceport America	8	
XCOR	Lynx	2012	Mojave Air and Spaceport	2	Pilot and passenger as co-pilot

The FAA/AST assessment was primarily focused on spaceplane companies based in the United States, but there is, of course, a growing amount of activity in Europe. Airbus (Astrium) has developed several prototype spaceplanes. The European Space Agency (ESA) is offering some incentives to encourage private spaceplane developments, whereas Reaction Engines is developing a single stage to orbital vehicle. On the other hand, the European Aviation Safety Agency (EASA) has de-emphasized its efforts to develop safety standards related to spaceplane flights, and its top expert in the field has now left EASA. [7]

Some Clues to Costs and Revenues

A great deal of publicity has surrounded the plans of Virgin Galactic, founded by Virgin Group chairman Sir Richard Branson, the king of hype, who often commands the lead headlines in this field. Branson also makes news with his stratospheric ballooning, the lavish parties on his private Caribbean island, and his spaceplane initiatives with Virgin Galactic. He is certainly known around the world for his bold new ventures. Several start-up dates were forecast by Branson from 2009 onward, but the dates have continuously slipped. Finally, the first powered sub-orbital test flights actually began in 2013, and actual experimental flights with paying customers were planned for 2014. The tragic accident in October 2014, of course, led to new setbacks. The current schedule for Virgin Galactic, XCOR, and others is discussed in the conclusion of this book.

Stephen Attenborough, head of commercial development at Virgin Galactic, has addressed future plans and prices at various forums. He explained that while the initial calculated cost of $200,000 per seat was judged necessary to recover the many millions invested in the venture, over time the aim was to reduce the costs to make space travel affordable for the masses. More recently, however, Virgin Galactic has talked about escalating expenses and inflation and about the need to boost prices to at least $250,000 for those that sign up for flights in 2015. In short, just as in the air travel business, prices will go up before they go down. Indeed the general public skepticism about lower-cost tickets to space is understandably high. The first posted comment on the Internet about a recent interview with Attenborough on the future of space tourism flight was as follows: "Obviously, this 'future' is only for the Filthy Rich to enjoy, while the rest of us toil below, trying live off ramen and work two shifts at Burger King." [8]

Branson himself has said that he expects up to 3000 passengers would travel with Virgin Galactic in the first five years, raising some $600 million in revenue. Eventually he hopes the cost of the flights will fall to around $20,000. No doubt all this is factored into his business plan, but further details are elusive. The truth is that the possibility of $20,000 flights into space on sub-orbital spaceplanes is now decades away. Thus, this hypothetical future price in, say, 2035 dollars is meaningless when adjusted for inflation.

Flights on Virgin Galactic are far from the normal type of commercial aviation flight fully licensed by the FAA in the United States or the EASA in Europe. Instead, the FAA, under the commercial space law as last amended by Congress in 2012, will continue to authorize "experimental flights" on a one-at-a-time basis whereby passengers will sign waiver documents to acknowledge they are undertaking a "high-risk" flight and will hold the U.S. government and Virgin Galactic completely and unconditionally harmless against all liability claims. [8] The current legislation indicates that this experimental licensing will continue until there is an accident involving a passenger. The accident on October 31, 2014, is considered a test flight, and thus the provisions of the law do not come into effect.

The reliance of Virgin Galactic on "Citizen Astronauts" to sustain a viable market clearly has created a few worries for Branson's business planning. However, a further business is now on the horizon for 2016 in the form of a new enterprise named "Launcher One," based on a new rocket that Virgin Galactic is developing in partnership with Planetary Resources. This is the new "asteroid mining" company that is looking for low-cost launches. Several other firms are seeking to deploy low-cost satellite constellations that are also potential clients. Of these the innovative Skybox Imaging enterprise, which has now been acquired by Google, could be a significant Launcher One client as it deploys its next-generation low-cost remote sensing constellation into LEO. Launcher One, a small two-stage rocket, would be launched by the same White Knight carrier vehicle designed for SS2. In this case, however, the White Knight carrier would not be carrying passengers on sub-orbital flights; rather, it would deploy the Launcher One rocket, which would carry payloads of up to 500 pounds (227 kilograms) to LEO. [9]

At least three other partners (Skybox Imaging, GeoOptics, and Spaceflight Inc.) have committed to using this new low-cost launcher, which is projected to cost $10 million per launch. The two leading manufacturers of small satellites—the Sierra Nevada Corporation and the U.K.'s Surrey Space Center (now owned by Airbus-Astrium)—have now both indicated that they would design new spacecraft buses that would match the technical characteristics needed to launch on the Launcher One rocket. The investment firm known as aabar Investments PJS has raised the funds for developing this new rocket system. [10] This puts Virgin Galactic in the enviable position of having a new market for its services with only a modest amount of additional capital at risk and dual use of its carrier vehicle.

Fig. 2.1 The White Knight carrier vehicle and Launcher One rocket—planned for 2017. *(Courtesy of Virgin Galactic.)*

A Shorter Flight at Half the Price

The pricing debate became more interesting when Jeff Greason of XCOR claimed that his Lynx spaceplane would fly up to 37 miles (or nearly 60 kilometers) high for only $100,000 per passenger and that flights would possibly start at about the same time as Virgin Galactic's. Greason certainly changed the competitive dynamics by announcing his rocket-powered vehicle, which is substantially smaller, slower, and less expensive to build than any of those proposed by rivals. Greason indicated that his Mark I spaceplane will cost less than $10 million to build (versus the estimated cost of $50 million for Rutan's SS2). The Lynx Mark I vehicle is intended to carry only a pilot and a single passenger at twice the speed of sound to about 37 miles (60 kilometers) above the earth. The entire outing, which would begin and end at a conventional airport and include about two minutes of sub-orbital zero gravity, would take less than an hour. The initial test flights originate from the Mojave Air and Spaceport, but when operational they will fly from the Midland, Texas, International Airport, where XCOR operations will be permanently based. The plans are to follow this spaceplane with the Lynx Mark II, which would be capable of flights to altitudes of over 100 kilometers, and the Mark III, which could launch small satellites to LEO. [11]

Just as in the case of Virgin Galactic, the plans have slipped significantly. When the Lynx Mark I was announced in 2008, flights were expected to begin in 2010. The Lynx Mark II, with expanded speed and altitude capacity, was supposed to follow 18 to 24 months behind (the exact start dates for flights with passengers will be discussed later). The claim that the Lynx could complete its two-hour mission and then carry out four flights a day also remains in significant doubt. Close to 200 people have signed up for flights on Lynx. With only one passenger flying as "co-pilot," it will take some time to satisfy this backlog demand, unless it is indeed possible to fly four flights a day.

Of course, this is a significantly shorter trip—and only half the ticket price—than that envisioned by Virgin Galactic, which uses a more powerful six-passenger craft, plus pilot and co-pilot, and which is designed to travel at about four times the speed of sound and to fly over 60 miles above the earth. The key may be whether passengers might consider the slower, lower, and cheaper flight to be safer as well. To date, some 200 people have expressed interest in these shorter sub-orbital flights, and so a number of people are taking this option quite seriously.

These and other companies entering the commercial space and space tourism sector are led by serious-minded businesspeople who, without doubt, have their experienced number crunchers in the back office, as well as the engineers and marketers to make it happen. However, the announcements, speeches, and demonstrations are long on optimism and enthusiasm—and short on details of firm launch dates, financial plans, break-even dates, and returns on capital investment. There are few precedents in private industry on a comparable scale, and so the spaceplane business is creating new precedents in business planning. We are literally finding new ways to "launch" a new business.

Hotels in Space Will NOT Be Budget Suites

Bigelow, another billionaire space entrepreneur, has reportedly spent $95 million of his own cash on the development of his inflatable space hotels for tourists. And he stands ready to spend as much again. Speaking at the Space Foundation's National Space Symposium, Bigelow said he expected to offer four-week stays in LEO for high-net-worth individuals. The estimated price he cited would be about $16 million in 2015 dollars. This would cover pre-flight training, and transportation there and back. He kindly offered to any would-be space travelers that they could now place a fully refundable deposit of 10 percent of the cost of the trip. To date, we know of no individual takers.

Bigelow has already deployed his Genesis inflatable structures, and under his NASA Space Act Agreement, he has prepared his GATE 1 and GATE 2 reports. These reports set forth his plans for space habitats in LEO and detail his concepts for habitable structures on the moon's surface and L2 Lagrangian points. [12]

Fig. 2.2 Bigelow spacelab based on inflatable space structures.

Bigelow sees his main target as governments and aerospace firms seeking to do long-term experiments in space. In due course, he says, he will be able to lease space on his orbiting stations to what he calls "Sovereign Clients." These are governments and industries ranging from biotechnology to automotive. They could lease sections of the space station for confidential research work. Bigelow's leasing prices for the station module were first quoted at $88 million to lease a module for a full year, or about a little over $7 million a month, but the price will certainly go up. What is of particular note, however, is that NASA and other space agencies are increasingly working together with new commercial space entities and that while these costs may sound high, they are often an order of magnitude less than the cost of governmental programs.

Fig. 2.3 Bigelow Aerospace inflatable space stations in the company's construction facility in Las Vegas. *(Courtesy of Bigelow Aerospace.)*

Another example of a firm that quoted prices is the now-defunct Canadian Arrow company, which offered reservations for sub-orbital flights priced at $250,000 each, including two weeks of training. The company forecast 2000 "new astronauts" in the first five years of operations of its Silver Dart orbital vehicle, with revenue of $200 million in the fifth year. The company was merged into PlanetSpace of Chicago in 2006.

"If the airline business had been left to the government to develop, it would probably cost $10,000 a pound to fly on an airplane," said Geoff Sheerin, Canadian Arrow's president and CEO. Ironically, the government is still here, but Canadian Arrow is gone. [13]

There Are Other Business Opportunities Too
The commercial space business involves more than travel services to the public. Some will build and service the spaceplanes, and others will provide the spaceports from which the "Citizen Astronauts" will travel. The leading developers of spaceplanes at this point are Rutan and Scaled Composites (now owned by Northrop Grumman); XCOR (developer of

the Lynx Mark I, II, and III); and Virgin Galactic, which has now acquired 100 percent of the SpaceShip Corporation. Others in the hunt include Masten, Armadillo Aerospace, Blue Origin, Reaction Engines, and Bristol Aerospace; we will read more about these companies later. Many of those registered for the Lunar XPRIZE might also be considered viable future spaceplane suppliers.

Two observations, from a business perspective, seem obvious when it comes to the supply of spaceplanes. One straightforward conclusion is that there are far too many potential suppliers at this incipient stage of the space tourism business. The 40 something companies listed in Appendix A have already been winnowed down considerably. There are already more than two dozen failed enterprises, and that number will only increase. Economies of scale in production will drive down the cost of future spaceplanes, and this is likely to be a critical success factor for the production companies. Only a handful of suppliers seem likely to be able to sustain the production of these sophisticated and expensive vehicles.

The second observation is that the expansion of the market for spaceplanes could be a tremendous boost for the potential suppliers. The growth in demand for spaceplanes, from space tourism flights to the supply of craft for executive jet travel, thus becomes a key element of business planning. To date, only a few companies have expressed serious intent to supply supersonic jets for executive and VIP travel; however, the size of this potential market and the ongoing need for executive travel, as opposed to the largely one-time space trips for "Citizen Astronauts," is a major consideration that manufacturers will undoubtedly consider in their future business plans. These decisions are nevertheless far from simple. Lockheed Martin, for example, has estimated that it will cost $2.5 billion to achieve the transition from prototype to large-scale production of a quiet supersonic executive transport.

As we address later, in Chapter 12, the future of supersonic versus hypersonic air flight is cloudy at best. We believe that based on current research and development, there are at least three possible scenarios for the future:

- **Scenario One:** Major R&D projects to develop sonic boom mitigation are currently underway, and then these will proceed to the development of new supersonic transport. These efforts are funded by NASA and ESA and are supported by large and established aircraft manufacturers such as Lockheed Martin, Boeing, Northrop Grumman, and Airbus. These efforts seem poised to develop supersonic vehicles that would fly at speeds of around Mach 2 within 20 to 25 years. These efforts would develop both supersonic executive jets and commercial craft for aviation transport.

- **Scenario Two:** There are conceptual studies, also supported by ESA and NASA, to develop both spaceplane technology as well as hypersonic transport. These efforts also are seeking to develop lower-cost access to LEO. They are reflected in work carried out by groups such as Bristol SpacePlanes, Sierra Nevada, XCOR, and Reaction Engines, which would leap over supersonic planes to develop hypersonic transport that would fly in protospace at flight speeds such as Mach 2 to Mach 6 (and even above). These would be variously sized to support executive jets and commercial aviation. It is not clear the extent to which Boeing, Lockheed Martin, Northrop Grumman, Airbus, and QSST would be involved in these developments, at least not in the early-on systems.

- **Scenario Three:** This would be a combination of Scenarios One and Two. In this case, hypersonic transport would be developed for executive jets and the elite first, but larger commercial companies would deploy supersonic aviation systems as a follow-on. In this case, hypersonic vehicles for regular commercial transport might evolve a number of years later for regular passenger travel.

The bottom line is that market demand—as much as technology—might determine the direction taken in the future and which scenario most closely follows actual events.

The business elements of spaceport development are perhaps even more difficult to project. The economics of these facilities may actually turn on issues of national prestige and the desire to develop new jobs and industries as much as the desire to generate profits from spaceflights. Chapter 8 explains how many countries and state governments are offering tax relief and other incentives to attract a spaceport, together with the new high-tech jobs associated with such operations.

In some business plans, a spaceport is seen as some sort of space museum and advanced technology amusement park as much as a transportation hub. The emir of Ras Al-Khaimah has put up substantial capital toward converting the Ras Al-Khaimah International Airport to become a spaceport in partnership with Space Adventures. It is difficult to say how quickly the projected $265 million in investment might actually be recouped. The same could be said about the Spaceport America facility now being developed in New Mexico to support Virgin Galactic operations; this was built at a cost of a quarter billion dollars—much of which came from a publicly funded bond issue.

The Lesson of Concorde

The Anglo-French Concorde was primarily an inter-governmental project that produced technological advances and huge national prestige, but it never recovered its development costs of several billion dollars—at a time when a billion dollars actually was a huge amount of money.

Fig. 2.4 The Anglo-French Concorde of the 1970s. *(Courtesy of British Air.)*

This aircraft, developed jointly by the British Aircraft Corporation and Aerospatiale of France, became famous as the only commercial supersonic passenger jet, and it provided services for more than 28 years. Two fleets of Concordes were operated—one by British Air and the other by Air France—between January 1976 and October 2003. Their scheduled services linked London and Paris with New York and Washington D.C. in just over three hours' flying time. At times, other routes were operated between Paris and Rio de Janeiro (with a refueling stop in Dakar, West Africa), and these enabled one of the authors to make an urgent business trip from Europe to Brazil in 1983 in under 6 hours instead of 18! There were also a limited number of flights on various routes in the Asia-Pacific region. With fares in the $5000 to $10,000 range for about 100 seats and luxury services, the aircraft were usually fully booked, but even so, the full capital and operating costs could not be recouped. Although it was technologically brilliant, Concorde was a commercial failure compared with, for example, the Boeing 747, which operated in the same long-range commercial market but with a much higher passenger-carrying capacity and longer flight capability.

The United States canceled its own supersonic (SST) program in 1971. Two designs had been submitted: the Lockheed L-2000, which looked like a scaled-up Concorde, and the Boeing 2707, which was intended to be faster, to carry 300 passengers, and to feature a swing-wing design.

The Soviet Union also developed a competitor, the Tupolev TU-144, which actually made its maiden flight before Concorde. The first production aircraft crashed at the Paris Air Show in 1973, but the Konkordski (as it became known) went on to make 102 scheduled flights between 1975 and 1978, when it was grounded after another crash. Ironically, the Concorde and TU-144A combined had an excellent record in terms of fatalities per air miles flown, yet they were also cost-ineffective in terms of seating capacity, and expensive and difficult to operate. These craft were also constrained by regulations, particularly by problems related to sonic booms, and they were still considered "unsafe" because of supersonic speed and high-profile accidents.

Clearly, the historical lesson is obvious. There is a need to develop systems that are soundly conceived, designed, manufactured, and operated—from start to finish. It is a matter of not only a viable technical design and health, safety, and other regulatory concerns such as air pollution in the stratosphere and control of sonic booms, but also cost efficiency and a good fit to market demand. One must implement new vehicles that "work" in terms of overall viability when considered in terms of all aspects: operations, business case, fuel efficiency, and regulatory compatibility.

Financial Viability Was Limited

Factors that limited the financial viability of both supersonic ventures included the limited range (for example, they were unable to fly from Europe to the Far East, Australia, or California without a refueling stop) and objections in various countries to overflying their territory because of the "sonic boom" and environmental pollution. It was even suggested in France and the United Kingdom that part of the American opposition to Concorde on the grounds of noise was in fact encouraged by nationalistic embarrassment and loss of face at not being able to produce a viable competitor. However, other countries, such as Malaysia, also ruled out Concorde's supersonic landings due to sonic boom issues. Somewhat ironically, there is first-hand evidence of the degree of problem as experienced in Malaysia. Back in the 1980s during an early morning layover for one of the authors, the Concorde landed at the Kuala Lumpur airport. A sudden and dramatic explosion rocked the terminal building, and everyone rushed to see if a bomb had exploded because the concrete and steel structure trembled under the force. In fact it had been a sonic boom from a landing Concorde flight. Clearly, noise abatement procedures had not been properly followed. Less solidly built structures in the area were damaged. Sonic booms and their destructive potential are not imaginary problems.

The Concorde was statistically the safest airliner in the world, measured by passenger deaths per distance traveled. But its reputation never fully recovered from the tragic crash of Air France Flight 4590 at Gonesse, near Paris, shortly after allegedly striking debris on the runway during its take-off on July 25, 2000.

There was another problem with the Concorde that was not as well publicized but that was quite serious. This was the damage done to the protective ozone layer at the top of the stratosphere. A thin layer of ozone serves as a protective shield from the deadly radiation that comes from the sun every day. On some days, when space weather is bad and there are solar eruptions, the radiation is particularly lethal. Without this protective layer, there is a very serious danger of mutation to human genes—and, indeed, those of all fauna. In short, without the ozone protection, the human species either dies or becomes a race of mutants. Corporate profits are good; survival of the species is, however, a tad more important. Money and wealth do not count for much if there are no humans around to take advantage of them. It is thus an inconvenient truth—someone has used that phrase before—that we need to look at the so-called "ozone issue" very seriously before large numbers of commercial space travelers take to the skies.

Anyway, there were mounting environmental and ecological pressures on the Concorde to end its flights when the SST was indeed grounded. It is our view that spaceplanes and the damage they might do to the ozone layer is a significant public issue that needs to be addressed before the space tourism business takes off. To date, the FAA has addressed public safety issues, but the concerns about spaceplane sonic booms, the ozone layer, and even the proliferation of carbon-based pollutants must be addressed in a serious way as commercial space enterprises move forward.

The Spaceplane Corporation has shifted from its solid fuel

system while keeping nitrous oxide (i.e., laughing gas) as its oxidizer. Thus, the rocket engine in Rocket Motor 2 is no longer fueled by HTPB (sort of like polyurethane rubber), as was the case with Rocket Motor 1. Instead, it is fueled by more consistently burning (and some believe to be cleaner) polyimide plastic (i.e., nylon) that burns up in Rocket Motor 2. Solid fuels spew particulates into the atmosphere, and so they create more pollution. In the much less dense stratosphere, this pollution is a very serious issue.

If spaceplanes move beyond space tourism to regularly scheduled flights for passengers from point A to point B and if the volume of traffic surges upward, this becomes a major concern. This issue is currently taken more seriously in Europe and within the United Nations than it is in the United States. This concern over environmental effects is why European-backed aviation research, for instance, is giving high priority to electrical propulsion systems for aircraft.

The Market for the "Vomit Comet"

Those looking for evidence of passenger numbers and public interest in spaceflight are encouraged by the success of Space Adventures. But perhaps even more significant currently are the ZERO-G flights for passengers of more modest means who are paying far less money to experience weightlessness. It is not just high-profile customers, such as famed physicist Stephen Hawking, who flew into weightlessness in early 2007. So far, some 5000 people have taken ZERO-G flights for $5000 apiece.

The ZERO-G company, now part of Space Adventures, uses a modified 35-passenger Boeing 727-200, sometimes affectionately referred to as the "vomit comet." Flying from the Shuttle Landing Facility at the Kennedy Space Center in Florida, it provides 90-minute flights similar to those conducted by NASA to train its astronauts. The flights reach altitudes of between 24,000 and 32,000 feet, and passengers experience weightlessness for about 25 to 40 seconds as high arching parabolas are flown. Flights are also possible from Las Vegas.

Space Adventures' Eric Anderson and ZERO-G's Peter Diamandis have cleverly recognized that the dynamic range of the consumer market that exists between those who might fork out around $5000 for a ride on the "vomit comet" and $20 to $35 million for a flight to the ISS is huge. The market size of those who might go into a casino and bet $50 is a lot different from the number of people who might go into a casino and buy it. Thus the Space Adventures' website, for instance, shows a lot of other options that cover the range from the casino bettor to the casino buyer. These options include a ride on the supersonic Russian Foxbat jet plane for around $19,000, or perhaps a week of astronaut training at the Russian Star City including a high "g" ride on the centrifuge to emulate a lift-off to the moon.

In addition to these "low end bets," of course, Space Adventures successfully marketed its flights on the Russian Soyuz spacecraft to the ISS. Seven paying passengers made the trip, at a cost of some $20 to $25 million each, before the availability of places ended (at least for the time being) and when the Soyuz became the only operating link to the ISS.

Space Adventures has also begun to test-market a spacewalk experience and a trip around the moon for even bigger bucks. Clearly, Anderson and company recognize that it is a significant part of the business planning process to see that a wide range of options are offered to broaden market opportunities. Future market developments range from some version of a space amusement park, where people spend perhaps hundreds of dollars, up to a sustained stay in space for tens of millions of dollars.

And Now Fly Me to the Moon—And Mars?

If you think the $20 $25 million to go to the ISS or the $16 million for a month's stay at Bigelow's space hotel is pricey, you have still not reached the upper limit. Space Adventures is seeking to negotiate a deal for a trip around the moon for a very large price tag. However, the championship for chutzpah pricing goes to the newly formed Golden Spike Company, which is offering clients the chance to make a moon landing for the announced price of $750 million a seat, or $1.5 billion for two people.

And then, in February 2013, the first of the original fare-paying astronauts, Dennis Tito, made headlines again by announcing his "Inspiration Mars" project to send a married couple on a "fly by" journey to the Red Planet in 2018, and he is planning to raise between $1 billion and $2 billion to accomplish this remarkable journey!

The Disney Experience

In considering the business case for spaceplane rides, the "risk" or safety environment is likely to be the most critical, and any accident that led to the loss of life would have an immediate and major impact on the sub-orbital space tourism business, especially if there were more than one accident. Futron notes that in 2003, Walt Disney World's Epcot Center unveiled a new space-themed amusement park ride called "Mission: SPACE." The ride provides a simulation of spaceflight from launch to return, with a centrifuge providing the experience of twice the normal pull of gravity on the riders. Though this is lower in terms of G-forces than those experienced on other rides in the park, the centrifugal force is sustained for longer periods throughout the ride, as it would be on a spaceflight. This ride is clearly marked with signs indicating that any potential riders with blood pressure, heart, back, or neck problems; motion sickness; or other conditions that can be aggravated by this adventure should steer clear of the ride. Since the ride's opening, there have been reports of health problems resulting from the experience, including two deaths within months of each other. A sub-orbital flight may likely involve G-forces several times the force of gravity, and so health conditions may still be an important factor to consider. It would seem unwise to entertain any thought of relaxing health and safety standards, at least until we have gained some years of experience. For this reason, we believe we should now heavily discount Futron's more optimistic market projection that the sub-orbital tourism market might actually rise to over $1 billion per year by 2021, if relaxed physical fitness standards were allowed.

Finally, there is the issue of how rapidly the sub-orbital spaceflight business might transition to actual orbital flights, space walks, and access to commercially operated space habitats such as those that Bigelow Aerospace is developing. There are also other questions such as how much basic and applied research money will be available to support the emergence of this new industry. This money will come from governmental sources as well as the various high-flying entrepreneurs (in both senses of the words) who are supporting the development of this industry. For example, ESA has a program to provide finance and business advice to new companies, as described later in the book. At the Paris Air Show. EADS (now trading as Airbus) announced its initiative to fly its spaceplane on commercial flights for space tourism; this announcement has in part come from technical and institutional support provided by ESA.

The Next 15 Years

What is of particular note is that this industry needs a tremendous amount of basic and applied research for it to succeed. These R&D costs are for not only the development of launch systems and cost-effective operations but also for safety-related research, space systems safety design, and life-support suits. This industry, for some considerable time, will need to be heavily subsidized by governmental space agencies, such as NASA's $500 million commitment to support COTS to the ISS, as well as by the entrepreneurial support from commercial leaders such as Branson, Allen, Bezos, Bigelow, Carmack, and Musk.

The real question thus tends to become whether there is simply a large enough market for the sub-orbital tourist business to sustain the needed investment. It may be that this market is only a part—perhaps a small part—of a future space commercialization market that can be sized in the billions of dollars a year and that will be needed to make the necessary transport and safety research investments viable. Key among the various questions, therefore, is whether the commercial space market will at some point include sub-orbital transportation systems. In short, will new spaceplane systems also be used to accommodate business executive travel using sub-orbital trajectories? There is also the important sub-question about the rise of a large number of such flights: whether a large-scale volume of spaceplane flights could have a negative impact on the atmosphere—and especially the stratosphere and the ozone layer. This issue will also be addressed later in the book.

REFERENCES
[1] Futron Corporation, Updated Projected Demand: Sub-Orbital Space Tourism 2006,
http://www.futron.com/upload/wysiwyg/Resources/Whitepapers/Suborbital_Space_Tourism_Revisited_0806.pdf.

[2] Tauri Group, "Suborbital Reusable Vehicles: A Ten Year Forecast of Market Demand," http://www.faa.gov/about/office_org/headquarters_offices/ast/media/Suborbital_Reusable_Vehicles_Report_Full.pdf.
[3] Tauri Group Study of Investment for the Commercial Spaceflight Federation, http://www.commercialspaceflight.org/programs/industry-metrics/.
[4] Ibid.
[5] http://www.futron.com and http://wwww.iaass.org.
[6] U.S. Department of Transportation, Federal Aviation Administration (FAA/AST), "The U.S. Commercial Suborbital Industry: A Space Renaissance in the Making" (2011), https://www.faa.gov/about/office_org/headquarters_offices/ast/media/111460.pdf.
[7] "Evaluation of the European Market Potential for Commercial Spaceflight," http://ec.europa.eu/enterprise/policies/space/files/policy/commercial-suborbital-flights-final-report_en.pdf.
[8] Carrington, D., CNN, "What does a $250,000 ticket to space with Virgin Galactic actually buy you?" August 16, 2013, http://www.cnn.com/2013/08/15/travel/virgin-galactic-250000-ticket-to-space/. Also see Grady, M., "Virgin Galactic Space Launch Expected This Year," *Aviation News*, January 2, 2013.
[9] Klingler, D., "Virgin Galactic Announces New Launch Vehicle," Ars Technica, July 12, 2012, http://arstechnica.com/science/2012/07/virgin-galactic-announces-new-launch-vehicle/.
[10] PR Newswire, "Virgin Galactic Reveals Privately Fund Satellite Launcher and Confirms SpaceShipTwo Poised for Powered Flight," July 11, 2012, http://www.prnewswire.com/news-releases/virgin-galactic-reveals-privately-funded-satellite-launcher-and-confirms-spaceshiptwo-poised-for-powered-flight-162037045.html.
[11] http://www.xcor.com/press/2014/14-09-17_midland_airport_receives_spaceport_license.html. Also see "Midland International Airport Receives Historic Federal Aviation Administration Spaceport License Approval," September 13, 2014, http://www.xcor.com/press/2014/14-09-17_midland_airport_receives_spaceport_license.html.
[12] Gebhart, C. and Grondin, Y.-A., "From Space Station to Moon Base – Bigelow expands on inflatable ambitions," *NASA Spaceflight*, May 30, 2013, http://www.nasaspaceflight.com/2013/05/space-station-moon-base-bigelows-expands-inflatable-ambitions/.
[13] Noyes, K., *TechNewsWorld*, May 17, 2007.

CHAPTER 3

The First "Citizen Astronauts"

Commercial spaceflight is an exciting new field, and with it comes the intriguing "chicken versus the egg" question. Who or what came first? Was it those private citizens who simply wanted to go into space? Was it those who envisioned designing and building new commercial spaceplanes? Or was it those who thought about operating a private commercial spaceplane service for profit. To wit, was it someone who was brash enough to think they could operate spaceplanes that were safe enough to carry passengers willing to pay big bucks for a jaunt into space?

The answer, in fact, seems to be all of above. These would-be-explorers, spaceplane designers and builders, and space entrepreneurs who wanted to operate a spaceplane service not only fed off each other but at times have overlapped. Indeed, Sir Richard Branson, owner of Virgin Galactic, the spaceplane service, has commandeered a seat on the first flight of the spaceplane he personally financed. Virgin Galactic has now also bought out full ownership of the SpaceShip Corporation, which is building the SS2 [1]. And Dennis Tito, who flew into space in 2001 as the first private "Citizen Astronaut," now wants to finance a private mission to fly around Mars.

Anyway this chapter is largely about those "Citizen Astronauts" who were indeed the first to go into space and thereby created a totally new commercial market. In the chapters that follow, we will present details about those who developed the technology and are now trying to create the spaceplane service companies to make it happen.

Those willing to become the first "Citizen Astronauts" have paid some incredibly high costs for what some would call an "ego trip" into space. But ego trip or not, these daring mega-millionaires championed the cause of opening space to a much wider audience and much sooner than almost anyone thought possible. In short, the new aerospace technology, the allure of a profitable new space market, and the excitement of citizens being able to go into space have all congealed together. Peter Diamandis' XPRIZE competition awakened the competitive juices of those who were willing to dream the impossible dream.

The first step occurred in 1994, when Diamandis, one of the founders of the International Space University, read the biography of Charles Lindbergh. He was intrigued by Lindbergh's May 20-21, 1927, solo flight over the Atlantic Ocean but was even more so about Lindbergh's motivation: to win the competitive Orteig Prize. Diamandis, whose motto is "The Meek shall inherit the Earth and the rest of us will go to Mars," immediately said he was going to start an XPRIZE Foundation to spur someone to fly a private spaceplane safely into space. And so he did it in 1998; one of the authors was there at the National Air and Space Museum when he announced it among much fanfare.

It actually took a while to come up with the $10 million in prize money to make the XPRIZE a reality. In fact, all Diamandis ended up doing was convincing the Ansari family enterprise to purchase a million-dollar insurance policy with a $10 million payout, against the long-odds possibility that someone could actually fly a spaceplane safely into "space" twice within a short period of time and do it by the fall of 2004.

The insurance company was told by credible experts that this specific feat could never be accomplished by the 2004 deadline and thus this was an "easy money" policy to write. As has been repeatedly true in American history, the impossible became possible when the spaceplane designed by aerospace guru Burt Rutan flew twice into space and safely landed in late September 2004 and again on October 4, 2004.

As Diamandis was putting the XPRIZE Foundation together and raising the funds via the Ansari family, Dennis Tito was on the cusp of becoming the world's first space tourist. Tito, who heads a financial investment and risk analysis firm that represents over $70 billion in investor funds, linked up with an amazing entrepreneur, Eric Anderson, to make his dream of going into space a reality.

In April 2001, Tito actually ponied up over $20 million to board a Russian Soyuz spacecraft so he could visit the ISS. NASA almost threw a hissy fit over allowing a "civilian" on board the ISS, but actually it worked out fine. Tito was trained to carry out a number of tasks that he actually did quite competently. Since then, a half dozen more "Citizen Astronauts" have slipped the bounds of earth's gravity, including the first female private space tourist, Anousheh Ansari. And now number eight is pending, and number nine is waiting in the wings. The famous "Phantom of the Opera" singer Sarah Brightman is scheduled to fly before the end of 2015, and a Japanese advertising ace is poised to be number nine. These pioneering commercial space adventurers have taken this unique industry to the threshold of a new era of space tourism almost undreamed of even a mere decade earlier.

Space Adventures Made It Happen

Back in 2001, the company that negotiated the deal to put the first millionaires into orbit, in partnership with Russia's Federal Space Agency (Roskosmos), was the Virginia-based Space Adventures. It was formed in 1998 by Eric Anderson, a former NASA employee who thought he could make things happen faster from outside the walls of NASA. Anderson boldly proclaimed his mission: "to open spaceflight and the space frontier to private citizens." Most people dismissed it as a pipe dream. Yet his mission has proven to be quite viable. Along the way he has become a wealthy space industrialist himself. In fact, he has led his innovative company very competently and creatively for more than 15 years.

Anderson was born in Denver, Colorado, in 1974, and he has literally led the way to "invent" the space tourism business. The former aerospace engineer has now sold almost $300 million in thrill-ride tickets and has flown more than 2000 customers on a dozen different aircraft without incident. For $19,000, you can get to see the blackness of space and the curvature of the earth from the cockpit of a MiG-25 Foxbat hurtling at 2.5 times the speed of sound. For $7000, you get ten 30-second bursts of weightlessness on the Ilyushin 76, a cargo plane Russian cosmonauts used for zero-gravity training. Alternatively, a flight with America-based ZERO-G, a Space Adventures affiliate company headed by Diamandis, provides a weightless experience for $4950 plus tax—or for $165,000 you can book a private charter for up to 36 people. ZERO-G claims to have netted $1.5 million in profits on revenues of $15 million in a single year—much of it from corporations like Oracle, American Express, Volkswagen, and Citibank, which offer rides as customer and employee promotions. [2]

Fig. 3.1 Dennis Tito, first of the "Citizen Astronauts."
(Courtesy of Space Adventures.)

Fig. 3.2 Eric Anderson, founder of Space Adventures.
(Courtesy of Space Adventures.)

Going Up—The Fare to the ISS

Space Adventures has 20 full-time employees in its Virginia offices, and Anderson, its founder, owns a large part of it, with several private investors owning the rest. These investors have no doubt seen a NASA/Space Transport Association study that predicts the private space tourism market will reach $10 to $20 billion annually within a few decades. The key phrase here, of course, is "within a few decades."

The opportunity for paying passengers arose in the first place because Russia's space industries experienced hard times after the 1991 Soviet collapse, when once-generous state funding dried up. These industries have survived mostly thanks to launches of foreign commercial satellites and revenue from the so-called "space tourists." More recently, Russia's oil-driven economic boom has led to increases in government spending on the nation's space program, reducing the space agency's dependence on revenue generated by commercial flights.

The arrangements made by Space Adventures with Roskosmos led to those first "Citizen Astronauts," who paid a reported $20 to $25 million each to fly to the ISS and back. But in 2007, the price increased, and in a July 2007 interview with *Space News*, Anderson said, "Actually, it's $30 million now. For the next couple of seats, that's the price."

In April 2008, Russia's space agency chief said it might stop selling seats on its spacecraft to "tourists" because of the planned expansion of the ISS crew from the current three to six or even nine. Anatoly Perminov said the space station's expansion will mean that Russia will have fewer seats available for tourists on its Soyuz spacecraft. "We will continue flying tourists to the ISS in accordance with the existing programs, but we may have problems ... because of a planned increase of the ISS' crew," Perminov said. [3]

Thus, when the NASA Space Shuttle was finally retired from service in 2012, Soyuz became the only remaining link for astronauts and cosmonauts to and from the ISS—and seats for paying passengers became even harder to arrange.

The agency did, however, maintain its commitment to fly astronauts from Malaysia and South Korea as members of the regular missions to the ISS. (The South Korean was the delightful Dr. So-yeon Yi, who sang "Fly Me to the Moon" from the ISS. She is a scientist as well as a keyboard artist and songstress. This amazing female astronaut is both a good friend to and one of the favorite instructors of the many students at the International Space University. She has recently moved to the United States, married, and earned an MBA at the University of California at Berkeley.)

What's Next—Spacewalks and Lunar Missions?

Looking into the future, as usual—and presumably assuming that seats are again available—Space Adventures has quoted $15 million, on top of the $30 million new base price, for those who would like to also take a spacewalk. Anderson said, "One of the consequences of the spacewalk is that you get a little bit more time up there. Instead of a week to 10 days, you'll probably get close to three weeks." The private spacewalker would exit the ISS out of a Russian airlock, outfitted in an Orlan space suit. [4]

So what's next? A Space Adventures team has blueprinted a circumlunar mission using a unique blend of existing and flight-tested Russian technology. At the heart of the lunar leap is Russia's Soyuz spacecraft. A pilot and two passengers would depart earth in the Soyuz, linking up in orbit with an un-piloted kick stage for a boost outward to the moon. The project is called Deep Space Exploration (DSE)-Alpha and will be the first in a series of deep space missions being planned by Space Adventures at the cost of $100 million per person. The mission as now planned would be conducted in cooperation with Roskosmos and the Russian space design bureau, Energia. Anderson explained, "The Soyuz was originally designed as a circumlunar spacecraft. It hasn't flown with people around the moon, of course. But the Soyuz would fly a free-return trajectory—a boomerang course—around the moon. So there's not a lot that needs to be done to the Soyuz to accommodate for that ... it could probably fly around the moon right now." [5]

How Tito Led the Way

Tito, the first fare-paying astronaut, started his career in the industry as an aerospace engineer at NASA's Jet Propulsion Laboratory in 1963. There, he calculated trajectories for interplanetary probes of the Mariner series.

He was therefore not only a long-term space junkie but also a very well-informed one. As a space scientist, he had participated in the robotic exploration missions to Mars and Venus. Then, in 1972, he founded his own company, Wilshire Associates Inc., in Santa Monica, California. The company went on to become a leading provider of services in the areas of management consulting and investment technologies. Tito had become a highly successful businessman, and his enthusiasm for space travel led him to pay over $20 million for the privilege of becoming the first paying customer in space.

Tito's epic trip was arranged when Anderson of Space Adventures managed to broker a deal with the cash-strapped Russian Federation's space agency. It started to become reality in February 2000, when Anderson had a meeting with Tito at the offices of Tito's company in California. The previous year, Anderson had paid the Russians $100,000 to study whether the Soyuz could be used to transport tourists to the ISS. The Russians said it could be done, and Anderson happened to have the study's results on him when he stopped by to ask Tito to invest in Space Adventures. "We talked a long time," recalled Anderson. "Then Tito said, 'I think what you're doing is great—it's the future—but I'm completely uninterested in investing in your company. But I do want to orbit the earth now. Can you help?'"

So began 10 months of intensive negotiations with the Russians, culminating in seven signatures on the final agreement. Anderson was at one of Tito's first meetings with the Russians, a typical 11 a.m. brunch at Star City, with bottle after bottle of vodka being consumed. After three hours, the general in charge, now tipsy, suddenly announced it was time for Tito's medical exam. They went out to the vestibular chair, basically a barbershop seat that spins at 40 revolutions per minute for 15 minutes, while an engineer barked out commands like "Move your hand forward, touch your chest," at the poor soul being spun. After Tito got off the chair without losing his brunch, the general said, "Mr. Tito, he's a real man. No problem for spaceflight." [6]

First, Training at Star City

But this was not like booking an airline ticket to Moscow. Tito actually began his preparations to be the first "Citizen Astronaut" at the Gagarin Cosmonaut Training Center in October 2000. After rigorous training at the Star City facilities, just outside Moscow, he was judged fully trained and competent to make the trip to LEO, and on April 20, 2001, he blasted off for his journey into the annals of history. Despite NASA's objection that only fully trained astronauts and cosmonauts could fly into orbit, he discharged his duties competently and returned safely from the eight-day mission. After his return, the American businessman appeared before a U.S. Congressional Committee to tell his story, in which he recalled the following:

> "There was one thing not even the most extensive training could prepare me for: the awe and wonder I felt at seeing our beautiful earth, the fragile atmosphere at its horizon and the vast blackness of space against which it was set. Just imagine being able to watch 16 sunrises and sunsets each day. And, thanks to a team of generous ham radio operators and the crew on the ISS, I was able to connect more clearly with my sons down on earth than I had previously when we were face-to-face. As any one of the 400-plus people who have traveled to space will tell you, no amount of training can prepare one for the experience of weightlessness and the freedom of effortless movement. It remains something that's still hard to describe to others. I can say that you get a sense of total relaxation. The nights I slept in space were the best nights' sleep I've had since I was a baby." [7]

Since his history-making flight, Tito has continued to promote the virtues of human spaceflight in various ways, and in February 2013, he announced an ambitious new plan for a mission to Mars. He said, "We have not sent humans beyond the moon in more than 40 years—it is time to put an end to that lapse." His new operation, "Inspiration Mars," plans to select a married couple for a 501-day "fly by" in a 14x12-foot Dragon space capsule, with supplies including a ton of dehydrated food and 60 pounds of toilet paper! According to Tito, the historic voyage to the Red Planet is scheduled for January 5, 2018, when the alignment of the planets is favorable. The cost of the voyage is expected to be between $1 and $2 billion. [8]

Now Follow That ...

Following Tito's spaceflight, the next client for Space Adventures and Soyuz was a South African computer multi-millionaire, Mark Shuttleworth. He was born in Welkom, Free State, South Africa, and holds dual citizenship for South Africa and the United Kingdom. He spent more than a week on the ISS in April 2002 participating in experiments related to AIDS and genome research. To make the flight, Shuttleworth had to undergo one year of training and preparation, including seven months at Star City. Just as Tito before him, he recognized he was purchasing not only the chance to undergo a unique experience afforded few humans but also a place in history.

Then, over three years later, in October 2005, Gregory Olsen became the third fare-paying space traveler. Olsen, like Tito, was a trained scientist. His company in California produces specialized high-sensitivity cameras. Olsen cleverly conceived his trip to the ISS in the context of promoting his cameras and his company. Olsen thus used his time in orbit to conduct a number of experiments and in part to test his company's products.

Figs. 3.3 and 3.4 Mark Shuttleworth and Gregory Olsen on their Soyuz missions to the ISS. *(Courtesy of NASA.)*

By this time, the story of fare-paying astronauts had started to appeal to the journalists who report on outer space as an ongoing story.

To some, however, these aspiring "spacemen" were no more than a bunch of extremely rich guys willing to pay a fortune to realize a childhood dream. In the early days of NASA's space exploration, the special quality required for an astronaut was defined as "The Right Stuff," the title of Tom Wolfe's 1979 book and the movie made in 1983. Now, it seemed that "the right stuff" was a large bagful of dollars!

Anousheh's Dream

Then, in 2006, "Citizen Astronaut" number four definitely created a new series of headlines. This time, the wealthy high-tech guy in space was a very attractive, high-tech young woman who did not even look her 39 years of age. It was Anousheh Ansari, chairman and co-founder of the Plano, Texas-based Prodea Systems, Inc., who realized her dream of an ascent to earth orbit.

Anousheh, who not long before had graduated with flying colors as an engineer from George Washington University, has done a lot in her short life, and in September 2006, she took her turn to fly aboard a Soyuz TMA spacecraft from the Baikonur Cosmodrome in Kazakhstan, en route to the ISS. [9]

Fig. 3.5 Anousheh Ansari, the first female paying passenger on Soyuz.
(Courtesy of NASA.)

On her 10-day mission, NASA astronaut Michael Lopez-Alegria and Russian cosmonaut Mikhail Tyurin were her companions in conducting a series of experiments on behalf of ESA. Seven years later, in 2013, and under the banner headed "there are no coincidences," Lopez-Alegria went on to become the new president of the CSF—although he served in that post only until the summer of 2014, when he was succeeded by Eric Stallmer, the former president of the Space Transportation Association and AGI executive. [10]

After Anousheh's return, she told *Space Future*:

> "Fly[ing] to space and see[ing] earth as a planet, instead of a city or a country, changes the way you look at things…. Instead, you start looking at everything differently and from a bigger perspective. You realize that you're part of a bigger universe, but at the same time you realize how vulnerable and fragile you are…. You just think how silly it is that people on earth fight for small pieces of land or things that seem unimportant….I would like to thank Space Adventures for providing the flight opportunity, the crews of Expedition 13 and 14 who made me feel very welcomed during my time spent aboard the ISS and all those who helped me prepare for this adventure. I hope those around the world who followed my mission consider what their own dreams are and pursue them, as I have done with mine."

Creators of the XPRIZE

Anousheh, who is a member of the Ansari family, had already played a key role in the development of new spaceplane technology before she herself went aloft. As described earlier, it was the family acting on the request of Peter Diamandis to put up the $10 million for the original XPRIZE money who then found a very innovative way to finance what became "The Ansari XPRIZE." For $1 million family bought an insurance policy that ultimately had a chagrined insurance firm paying out the $10 million in XPRIZE money when Rutan and

Allen, against the odds, won the competition for the first successful manned spaceflight in 2004.

This XPRIZE competition did as much as any other single factor to fuel the furious rush to create viable, safe, and much lower-cost access to space. Dozens of "can do" start-up firms were inspired to compete for the prize, and even those who did not succeed helped propel the new technology of designing and building spaceplanes forward. Diamandis, now head of ZERO-G, is also on to his next space adventure. He has announced plans to seek to mine asteroids by teaming up with Hollywood producer and director James Cameron. This effort is known as Planetary Resources. [11]

As Diamandis explained when he received the Arthur C. Clarke Innovators Award in 2008, his prime objective is to bring an asteroid that is largely comprised of platinum back close enough to earth to realize its retail market value of billions of dollars. Ansari, Anderson, and Diamandis are the "anything is possible" type of entrepreneur, hell-bent on making space tourism a reality—and sooner rather than later. The three of them have all become a new kind of space rock star.

Ansari's Prodea company has invested in other space-related activities including a planned spaceport in the United Arab Emirates and another in Singapore. Prodea was also involved with Space Adventures and Roskosmos in developing the Explorer spacecraft, based on the C-21 concept produced by Russia's Myasishchev Design Bureau for Space. This rocketplane would also be flown up by a carrier aircraft and then launched to the edge of space—in much the same way as SpaceShipOne and SS2. However, in 2006 it was announced that this project has gone quiet, and so it is not certain what progress has been made on this front.

At the Farnborough Air Show in 2012, Roskosmos also announced its plans for a new manned spacecraft to replace the Soyuz series. Further details on the Explorer spacecraft and the Roskosmos plans for replacing the Soyuz will follow at a later date. [12]

The Space Nerd

In April 2007, Charles Simonyi, Ph.D., became the fifth paying passenger to fly on board a Soyuz mission to the ISS. The total price for his 13-day mission to visit the orbiting ISS was not revealed, but it was reliably reported by Space.com to be in the $25 million range. Then in March 2009, he repeated the trip—and to date he is the only "Citizen Astronaut" to have made two flights to the ISS. As a software pioneer who developed some of the early Microsoft systems, Dr. Simonyi laid claim to being the "First Nerd in Space." He made his name as the chief architect of Microsoft Word, Excel, and several other widely used application programs. Later, he left Microsoft and founded a new company, International Software, which develops and markets computer software for knowledge processing.

His first mission grabbed quite a few headlines, but the press focused mostly on the romantic angle. This was because "astronaut nerd" Simonyi was also the boyfriend of Martha Stewart, the highly successful American media chef and home designer. It was reported that Martha was busy preparing duck pâtés and other exotic foods for his return to earth.

However, there was also a serious element in his ascent to orbit. He was actually trained to assist several international space agencies by conducting experiments during his nearly two weeks in space. Like his predecessors, he carried out his final preparations for the mission at the Yuri Gagarin Cosmonaut Training Center in Star City, Russia, undertaking a comprehensive overview of the mission in advance. It is not too surprising that Simonyi has also joined up with Cameron, Diamandis, and other well-known space enthusiasts in the Planetary Resources undertaking.

Fig. 3.6 Charles Simonyi, the space "nerd." *(Courtesy of NASA.)*

Like Father Like Son

Number six on the list of fare-paying passengers was Richard Garriott, a 46-year-old British-born American citizen. A successful businessman, he has worked in the design and development of computer games. His flight to the ISS, on October 12, 2008, was inspired by his ambition to follow his father, Owen Garriott, who made two spaceflights as a NASA astronaut, in 1973 and 1982. Richard indeed lived up to his father's example and performed his space duties on the ISS very well. By this time, NASA's insistence that paying customers could not be trained to perform competently seemed to be a rather weak argument in light of six examples to the contrary.

Fig. 3.7 Richard Garriott, son of a NASA astronaut. *(Courtesy of NASA.)*

Then, in September 2009, Guy Laliberte became the first Canadian space tourist when he flew into orbit aboard Soyuz TMA16. During his 12 days on the space station, he proclaimed his flight a "Poetic Social Mission" and conducted the first-ever artistic event in space, called "Moving Stars and Earth for Water." This was a two-hour event that featured celebrities including Salma Hayak, Shakira, and Bono.

Fig. 3.8 Guy Laliberte became "Citizen Astronaut" number seven. *(Courtesy of NASA.)*

A Soprano in Space?

These first seven "Citizen Astronauts" were variously reported to have paid anything between $20 and $35 million, but Space Adventures, the driving force behind this new business, reported that despite the cost, there was still a list of adventurers just waiting for the opportunity to experience the flight. Laliberte appeared likely to be the last in the series of paying passengers on Soyuz, because with the end of the U.S. Space Shuttle program in 2012, the Russian spacecraft became the only link with the ISS, and it was committed to focusing on the essential task of ferrying astronauts and supplies.

However, in October 2012, it was announced that at least one more would be added to the list—and this time it is a name with celebrity status. At a news conference in Moscow, Sarah Brightman, the world-renowned soprano, announced that she was starting preparations to make her spaceflight during 2013. However, the Russian agency said later that it would not take place before 2015 (amid comments in the U.K. that it was all a publicity stunt for her next album). She became famous for her role in "Phantom of the Opera" and is the former wife of impresario and composer Andrew Lloyd Webber. She is now a UNESCO Artist for Peace Ambassador, and in her announcement, she said:

> "I don't think of myself as a dreamer. Rather, I am a dream chaser. I hope that I can encourage others to take

inspiration from my journey both to chase down their own dreams and to help fulfill the important UNESCO mandate to promote peace and sustainable development on earth and from space. I am determined that this journey can reach out to be a force for good, a catalyst for some of the dreams and aims of others that resonate with me." [13]

But Was Tito *Really* the First?

But who really was the first paying passenger to travel into space? Ten years before Tito's groundbreaking spaceflight, the Russian space agency decided to allow Toyohiro Akiyama, a reporter for the Japanese television company Tokyo Broadcasting System (TBS), to fly to the Mir space station and return a week later, for a price of $28 million. Akiyama gave a daily TV broadcast from orbit and performed scientific experiments for Russian and Japanese companies. However, his company paid for the flight, making Akiyama a sort of "business traveler" rather than a space tourist.

And a Japanese Advertising Executive Is Next

In early January 2015, Space Adventures announced that Satoshi Takamatsu was in his final stages of training to be the next to fly after Ms. Brightman. Takamatsu is, like so many in the commercial space travel business, yet another serial entrepreneur. He started out with the largest advertising firm in Japan, Dentsu Inc., but a decade ago he branched out and started his own firm. Takamatsu has long been linked to space travel. He indeed conceived of and directed the first television commercial filmed in space. He is thus following in the intellectual and physical footsteps of astronaut Akiyama, who went up in 1991. Today Takamatsu heads his mainline advertising production company as well as Space Films and Space Travel. [14]

The First 500 Seats on SS2 Sold Out and a Total of 700 Booked

The next group of "Citizen Astronauts" is expected to be on Virgin Galactic's maiden flight, which will reach the edge of space, but they will not be flying into orbit like the intrepid group that paid millions to experience the journey on board Soyuz. It is reported that the first 500 seats for the first flights of SS2 have been fully sold out at $200,000 each. The current booking price is $250,000, with another 200 booked, and so at least 700 people are committed to flying into space on Virgin Galactic.

Even after the fatal accident of the SS2 spaceplane on October 31, 2014, there were only a few requests for refunds of the flight deposits for those booked to fly. [15]

Some of the intended passengers, including Richard Branson himself and his son Sam, started their training as long ago as 2008 at the U.S. National Aerospace Training and Research Center. To illustrate his confidence, Branson had also once said that the first passengers would also include his 92-year-old father and 89-year-old mother, but because of their advanced age, this has now been rethought. Nevertheless, he certainly knows how to hype the safety of his newest enterprise. Risking his own life and the lives of his family members seems to be his ultimate "safety guarantee."

Eliminating Hassle, Inconvenience, and Fear

Many now see the prime way forward as a "tourist experience and entertainment enterprise." This vision will stress that all offerings to the public market must be safe, secure, convenient, and affordable—and yet still be exciting and awe inspiring. Such a business model requires the elimination of hassle, inconvenience, and fear from the space tourism experience while making passengers feel they have undertaken a once-in-a-lifetime journey that sets them apart from mere earth-bound dwellers. Of course, after the training and the briefings, and after the corporate and FAA waivers are signed to acknowledge they are undertaking a death-defying feat, space tourists may still feel a bit less than safe—unless they are entirely clueless.

In the United States, the FAA/AST undertook an elaborate rule-making process. It adopted regulations to govern space tourism flights, pilots, crew, and launch range safety officers. During this process, it tried very hard to accommodate these new business models and to respond favorably to the comments received from CSF members.

And so, following on from those first pioneering and fare-paying flights from Russia to the ISS and back—all of which were mishap free—now is the time to see if these commercial systems can actually work safely on an ongoing basis; if so, there will be a dramatic increase in the number of "Citizen Astronauts."

REFERENCES
[1] "Virgin Galactic Acquires Full Ownership of The Spaceship Company," Space Travel.com, October 8, 2012, http://www.space-travel.com/reports/Virgin_Galactic_Acquires_Full_Ownership_of_The_Spaceship_Company_999.html.
[2] ZERO-G Corporation, The Weightless Experience, http://www.gozerog.com.
[3] Isachenkova, V., Associated Press, April 11, 2008.
[4] *Space News*, July 15, 2007.
[5] Space.com, August 10, 2005, http://www.space.com/news/050810_dse_alpha.html.
[6] Dennis Tito testimony, http://www.spacefuture.com/archive/hearing_on_space_tourism_testimony_by_dennis_tito.shtml.
[7] Ibid.
[8] Inspiration Mars, http://www.inspirationmars.org/.See also Inspiration Mars stories in *Daily Telegraph* (London) February 26, 2013, and *Washington Post*, February 26, 2013.
[9] Biography of Astronaut Anousheh Ansari, Space Tourist 4, http://www.spacefacts.de/bios/astronauts/english/ansari_anousheh.htm.
[10] Bandla, S., "Eric Stallmer Named President of the Commercial Spaceflight Federation," http://www.commercialspaceflight.org/2014/07/eric-stallmer-named-president-commercial-spaceflight-federation/.
[11] Planetary Resources, The Asteroid Mining Company, http://www.planetaryresources.com/.
[12] *Sunday Telegraph* (London), March 24, 2013.
[13] "British singer Sarah Brightman to be Russia's next space tourist," Space Adventures, http://www.spaceadventures.com/index.cfm?fuseaction=news.viewnews&newsid=868.
[14] "Space Adventures announces that Satoshi Takamatsu will begin orbital spaceflight training in Star City, Russia," January 7, 2015, http://www.spaceadventures.com/press-releases/.

[15] Kramer, M., "Virgin Galactic's SpaceShipTwo Crashes in Test Flight: 1 Dead, 1 Injured," October 31, 2014, http://www.space.com/27618-virgin-galactic-spaceshiptwo-crash-kills-pilot.html.

CHAPTER 4

The Gigabuck Space Entrepreneurs

If you want to start a new industry, there are many good ways to begin. A totally new idea perhaps, such as the Internet, or a new consumer mass-market product such as plastics—these are all good beginnings. The space tourism industry, however, has attracted an extra-special ingredient. Many of the new "space tourism entrepreneurs" are self-made multi-billionaires. The mainstays of this new business are very successful business people, all with a touch of exotic glamour and PR élan, plus gigabucks at their command.

Most billionaires become wealthy by innovative thought and shrewd investment. The space tourism leaders are certainly no exception to the rule. As we've already described, this amazing line-up includes Virgin Atlantic head Sir Richard Branson, Amazon.com founder Jeff Bezos, PayPal and Tesla Motors wizard Elon Musk, Microsoft co-founder Paul Allen, video game designer John Carmack, hotel magnate Robert Bigelow, and the Iranian-Dubai-American Ansari family, whose wealth springs from a number of sources. Just to add some additional spice, there are also some old-fashioned mere multi-millionaires such as the late Jim Benson, who made his first fortune in the computer industry. His SpaceDev company was one of the driving forces in the early days of the commercial space sector, and it is now part of the equally dynamic Sierra Nevada Corporation. Then there is the California-based technology entrepreneur Naveen Jain (who briefly became a billionaire during the dot-com boom) and Eric Anderson, who has already made a handsome sum in the space tourism business with his Space Adventures company.

Another highly successful and innovative entrepreneur is David Thompson, whose OSC won a re-bid contract from NASA for the COTS development. Thompson's company was selected to go toe-to-toe with Musk's SpaceX company to develop a truly commercial vehicle to take cargo to the ISS. However, SpaceX's Falcon launcher and Dragon capsule got there first. OSC's Antares rocket and Cygnus spaceship ran into delays and then experienced a catastrophic loss on October 27,

2014. This has ironically been traced back to faulty Russian rocket motors that had been scheduled to be replaced. [1]

OSC and Thompson, in looking for a new strategic thrust, has opted to join forces with Alliant-ATK, with both companies needing a rebound after being passed over by NASA in the later rounds of new commercial spacecraft development to provide astronaut launching capabilities to the ISS. The new company, Orbital ATK, brings new strengths to each of them, but the resulting $5 billion corporation no longer resembles the agile and entrepreneurial company that Thompson started some three decades ago. [2]

In the final NASA competition for the human launch system between the two entrepreneurial firms of Sierra Nevada and SpaceX plus Boeing, SpaceX and Boeing came out the winners. [3] Thus the despite the surge from entrepreneurial companies in the new space race, the establishment firms of Boeing, Lockheed Martin, and Northrop Grumman are still playing hard to stay in the game. We will go into detail below and in subsequent chapters.

Discarding Conventional Ways

Virtually all of these entrepreneurial drivers of the new space tourism industry share a common trait. This is the dogged determination to find new and innovative ways to access space—plus an inclination to discard outmoded concepts and conventional ways of thinking. These are the people who, like Alexander the Great, tend to pull out a sword and hack through the Gordian knot rather than taking the slow and conventional approach to problem solving. They are not interested in bureaucratic red tape, and, as a rule, they throw out of the window the cumbersome processes found within traditional governmental space agencies.

However, this straightforward but sometimes cavalier attitude leads to some reasonable jitters when it comes to space safety concerns. The explosion in the Mojave Desert in 2007, when a nitrous oxide tank blew up, resulted in multiple deaths and injuries plus fines from OSHA. Clearly, this tragedy ignited new concerns, especially since the reason for the explosion was never completely understood. Then came the fatal crash during a test flight of Virgin Galactic's SS2 on October 31, 2014.

Those who are part of this new breed of spaceplane developers believe they can accomplish their mission by hiring dozens of the most talented and dedicated employees and by focusing on a very clear and precise mission—to enable paying passengers to fly into space. It might be near-space or protospace, and the parabolic weightless flight above 100 kilometers might last only about four minutes, but it is still high enough to soar above the atmosphere. These high-flying space tourists will be able to see the brilliant blue ball of earth in the sky against the darkness of space and witness the world as a whole rather than as a crazy quilt of countries with artificially drawn borders. For many starry-eyed "Citizen Astronauts," this is more than enough reason to fork over $200,000.

This new "commercial space" approach stands in contrast to creating and staffing huge national space agencies and large aerospace corporations that command vast complexes and labs with thousands of employees. These old-guard space enterprises look like vast, high-tech armies from the military-industrial complex. On the other hand, the approach of the new entrepreneurial organizations, with dozens or perhaps hundreds of employees, looks, acts, and feels very different from that of the space agencies. These new and sometimes brash organizations put their "can do" bravado out front.

But the most ambitious of the new billionaire-backed space startups, namely SpaceX and Virgin Galactic, are now aspiring to do much more. SpaceX is now seeking to develop spacecraft that could travel to Mars, and Virgin Galactic is moving to develop Space One low-cost orbital launchers. And another of them, Allen, is backing the amazing Stratolauncher system.

New Commercial Space Industries—The "Anti-NASA"

The "lean and mean" entrepreneurial space companies that these billionaire business people are creating might well be called the "anti-NASA" because of their mood and temperament. This is to say their approach is antithetical to the typical NASA approach to space-related activities. They are thinking differently, and their approach is innovative and unconventional in terms of organization, staffing, and goals. The starting point for most projects is "outside-the-box" thinking. These new organizations are seeking to provide mass public access to space that is low in cost, safe, hassle free, *and* achievable in the near term. They are also producing new low-cost access to space for applications and exploration. Innovations, interventions, and ingenuity permeate this billionaire-driven new space economy.

In this quest, they are looking for ways to extend the technology and reliability of the aerospace industry. They also seek to upgrade it to achieve sub-orbital flight and to extend beyond to orbital flight and even to true space exploration. As shown in Table 1.1 in Chapter 1, their initial stage mission is to evolve new systems in the tradition of Burt Rutan's innovative design of SpaceShipOne rather than the gigabuck expenditures that NASA devoted to the Apollo program. Missions like Apollo or a Mars landing are seen as being better left to governmental agencies. For the most part the new space entrepreneurs only ask the government to stay out of their way. A few years back, when we talked to the former president of the CSF, Bretton Alexander, we asked him if his members were closely studying safety standards for structural materials, fuels, and life support systems that NASA and ESA had developed over the years. He quickly replied, "We are looking for new solutions. Why would we want to follow the example of the space agencies, where 4 percent of their astronauts have been killed? We are looking for new approaches and new solutions to space transportation." [4]

Confident? Yes! Brazen? You had better believe it! Irresponsible and hopelessly cocksure? Time will tell. One thing for sure is that they are looking for new and innovative answers.

High-Flying, High-Testosterone Guys

The space billionaires' "business models" are based on the idea that they want to reduce costs and develop affordable transportation systems that are safe and consumer friendly. They are quite willing to leave "research" and bleeding-edge technologies to NASA and the other space agencies. The high-flying, high-testosterone, and high-pressure guys with deep pockets want to convince potential space tourists that not only will their new rocket planes fly into space, but also that they will come down—safely. As the billionaire businessman and adventurer

Branson has said, "I'm absolutely sure that millions of people want to go into space and it's up to us to make it affordable." [5] Allen took a more matter-of-fact approach when he helped launch the Stratolauncher enterprise, saying, "There is a new age in space occurring and we are making it happen." There are few more "can do" guys around than Allen. He seems to have the truly magical entrepreneurial talent that only a Steve Jobs or a Bill Gates can match. His presence in the space tourism industry as the "grownup in the room" certainly inspires confidence.

The Safety Factor in Space Tourism

When Alexander, then head of the CSF, was pressed about following the safety advice of NASA and the other space agencies, he summed it up this way: "When we look at their safety record, we realize we need a new model and a new approach." [6]

There is clearly a long way to go to achieve the "still to be met" space safety goal, set over two decades ago. This goal was to experience only one space fatality per 1000 spaceflights. The reality is that the national space agencies worldwide have had fatalities resulting from 1 percent of all flights with humans aboard. In fact, of the 500 or so astronauts and cosmonauts that have flown to date, 22 have died. This is 4 percent of those who have made it to space and back. (The difference between the 4 percent fatalities for those that have flown and the 1 percent mortality per flight is that many have flown on multiple missions, and most flights have had several people on board.)

The new billionaire space entrepreneurs see the prospect of developing not only new approaches to spaceflight but also innovative ways to embark on profitable space enterprises based on mass-market volumes. These initiatives require an entirely different business model from that pursued by NASA or the other space agencies around the world. The starting point has been to find economical ways to exploit the safest jet and rocket propulsion, using conventional vehicles to reach "spy plane" altitudes, proven rocket technology, and modern and highly cost-effective manufacturing and testing facilities. In short, the space agencies think in terms of multi-billion-dollar "research" programs lasting decades, whereas the new commercial space people think in terms of million-dollar "development" projects with turnaround schedules of three to four years at most, and perhaps months in some cases. The sobering reality is that the development of SS2 started in 2005 and has now taken more than a decade. Virgin Galactic and the SpaceShip Corporation have found that human spaceflight—even sub-orbital flight—is hard.

A famous concept attributed to Englishman Thomas Occam is often called "Occam's Razor" (or sometimes "Ockham's Razor," since the British have peculiar ideas about spelling). Anyway, Occam wisely advised, "If there are two solutions, pick the simplest." This advice seems to be the constant guide to the emerging space tourism business.

The Space Entrepreneurs Can Make a Difference

The new breed of space entrepreneurs, with their business acumen and marketing expertise, may not only identify many more commercial applications but will also exploit them sooner than governmental agencies would. We can at least have high hopes for the space exploits of successful men like Allen (who financed SpaceShipOne), or Musk, Bezos, and Branson. Such innovative and fearless thinkers can reasonably be expected to be more adept at bringing these new products or services to market sooner and with greater flair.

The key to focus on here is that these billionaires can make a difference. At the same time, there are quite a few other spaceplane pioneers. They may not have the big bucks and fame of their super-wealthy colleagues, but they still have lots of determination and a good deal of expertise. These dozens of smaller-scale developers that are working on spaceplane systems have also dared to envision totally different business models for their enterprises—often on shoestring budgets. Their approach, too, has been much different from that of NASA or other space agencies.

This "second tier" of developers has also looked at market demand, service needs, and product development from the perspective of business innovators, not from that of governmental research scientists. They see space tourism as not just a scientific and technical mission, but rather as a business that will first be based on providing entertainment or a "vacation experience." This is admittedly a big step beyond hot air balloons or bungee jumping. Nevertheless, this innovative thinking has led to creating innovative designs involving lighter-than-air craft, ion engines, and hybrid fuels; to combining high-altitude spy plane technologies with rocket systems; and to using other unconventional approaches.

The CSF and Its Members

The CSF, with its membership of spaceflight billionaires, is one of the most exclusive "clubs" in the world. As of February 2015, it lists 19 Executive Members, including key players such as Bigelow Aerospace, Blue Origin, Mojave Spaceport, Sierra Nevada Corporation, Space Adventures, SpaceX, XCOR Aerospace, and Virgin Galactic. In addition, the current breadth of the industry is illustrated by the list of a further 32 companies as Associate Members. The full list of current members appears in Appendix B.

The founding chairman of CSF was Anderson of Space Adventures, and in September 2012, he was succeeded in this role by Stuart O. Witt, general manager of the Mojave Air and Space Port. Mojave Space Port was designated the nation's first inland spaceport and played host to the world as Scaled Composites qualified for and won the $10 million Ansari XPRIZE and furthermore gave birth to the first man-rated commercial space program in the world. [6] Frank Dibello, the president and CEO of Spaceport Florida, succeeded Witt, and Mark Sirangelo, the head of Sierra Nevada Corporation, succeeded him.

In March 2012, the CSF announced the appointment of a former NASA astronaut, Michael Lopez-Alegria, as its president, but he served only a short while. In July 2014, he was succeeded by Eric Stallmer, a former government relations executive with Analytical Graphics Inc. and former president of the Space Transportation Association. Stallmer now leads the federation's staff and works with the 50-plus CSF members, which include providers of commercial orbital and sub-orbital spaceflight, spaceports and launch facilities, suppliers, and educational and research institutions.

The International Reach of New Developments

The commercial spaceflight industry is not just a U.S. enterprise, as shown by the membership of the CSF. Although three-quarters of the space tourism and spaceplane businesses are based in the United States, quite serious efforts are also underway in Canada, China, Israel, Russia, the United Kingdom, France, Germany, and Switzerland, among others (as described in more detail in Chapter 7). In some countries, such as China, France, Germany, India, Japan, and Switzerland, the emphasis on development in space tourism and spaceplanes has often been in the public sector as opposed to truly commercial enterprises. In the case of Russia, Roskosmos has indicated a willingness to work in parallel with commercial aerospace concerns while also maintaining a governmental program.

Although most of the first commercial space tourism businesses will be U.S. owned, some, like Virgin Galactic, will be U.K. owned but operated from New Mexico. Others, such as Space Adventures, may operate from spaceports in the United Arab Emirates, Singapore, or Malaysia but will still be U.S.–owned enterprises. Even so, the rest of the world, in terms of trade considerations, national security concerns, and international aviation regulation, will necessarily be involved. At this point, the rest of the world seems willing to follow U.S. regulations until they are proven wrong.

Branson—The British Knight

So let us take a closer look at the remarkable group of entrepreneurial billionaires—up close and personal. So far, the U.K.'s Branson has grabbed the most headlines. As in all his ventures, Branson has set his goal to be number one, and despite years of slippage he is now set to be first with his Virgin Galactic flights on the SS2 craft. Branson began his involvement as an investor in the SpaceShip Corporation, following the XPRIZE success of Rutan and Allen. He thus managed to become both a spaceplane manufacturer and a thrill marketer. And by exercising his options, Virgin Galactic now owns the SpaceShip Corporation 100 percent.

Branson invested in the second generation of a proven concept by placing an order for the SS2 spacecraft. Not one to do things by half measures, he ordered not one but a fleet of five SS2 vehicles at a cost exceeding a quarter billion dollars. On the basis of this order, he launched his Virgin Galactic business.

Today there is an impressive list of prospective clients—now over 700 in number. These prospective passengers include movie stars, sports figures, and a British royal, and two bookings involve marriage:

One couple plans to get married in space, and another is planning to have their honeymoon in space. In addition, as described in Chapter 1, Professor Stephen Hawking, the world's leading astrophysicist and cosmologist, completed a challenging ZERO-G flight in 2009, despite his afflictions, and he has said he plans to take a spaceflight on Virgin Galactic as well.

A "Transformational Leader"

The multi-faceted Branson has had a dynamic and widely publicized business career. He was born in 1950, in Surrey, U.K., and is now Britain's most famous entrepreneur. He is best known for his Virgin brand name, which now encompasses a wide variety of businesses. According to the *Forbes* 2013 list of billionaires, he is the fourth richest U.K. citizen. He is described as a "transformational leader," with his maverick strategies for the Virgin Group as an organization driven by informality and information, not strangled by top-level management.

An entrepreneur from an early age, he is reputed to have started two failed ventures by the age of 15: a Christmas tree–growing business and a pigeon-raising farm. Surprisingly, he was not a very good student, suffering from dyslexia. He was a good sportsman at school, but a serious athletic injury helped to launch his illustrious career. At 16, he decided to quit school and moved to London, where he began his first successful entrepreneurial activity. He identified with the energy of student activism in the late '60s and started his own newspaper, called "Student Magazine." The headmaster of his school wrote, "Congratulations, Branson. I predict that you will either go to prison or become a millionaire." [7]

In 1970, he set up a record mail-order business and two years later opened a record shop in Oxford Street, London. Next, he launched the record label Virgin Records and opened a recording studio. Apparently, the name "Virgin" came about when one member of his group said, "We're complete virgins at business."

From Pop Music to Banking, Transport, and Space

The company's first disc was multi-instrumentalist Mike Oldfield's *Tubular Bells*, which was to become a best seller. In fact, Virgin secured the album because no other company was prepared to produce such an unconventional record. Branson's company also courted controversy by signing bands like the Sex Pistols and won praise with obscure avant-garde music such as the so-called "krautrock" bands Faust and Can. Later, in 1992, to keep his airline company afloat, Branson sold the Virgin label to EMI. [8]

Fig. 4.1 Sir Richard Branson proudly displayed his Virgin Galactic project in 2009. *(Courtesy of Virgin Galactic.)*

He created Virgin Atlantic in 1984, offering competitive fares on scheduled transatlantic flights. It grew to become Britain's second-largest carrier, and once again he had spotted a gap in the market and filled it spectacularly. In 1997, he took what many saw as one of his riskier business exploits by entering the railway business. Virgin Trains won the franchises for two sectors of the former British Rail network. This venture promised new high-tech trains and enhanced levels of service, but in practice, this has proved hard to deliver.

In 1999, he moved into telecommunications by launching Virgin Mobile. Then, in 2006, he combined Virgin Mobile with a leading U.K. cable TV company, NTL, to create Virgin Media, which became a major player in the U.K. broadband market and was then acquired in early 2013 by the giant U.S. company Liberty Global.

Among many other enterprises, Branson entered the financial services market in the U.K. with Virgin Money, which went on to purchase a troubled British bank in the wake of the 2008 financial meltdown. He has also been involved with enterprises ranging from health services to motor racing. In short, Branson, who is also an avid explorer, balloonist, and risk taker *par excellence*, seems to have no limits to his diverse interests and commercial ventures. In 1999, he received a knighthood from the Queen for "services to British entrepreneurship." He is married and lives with his wife, Joan, and their two children, Holly and Sam, in London and Oxfordshire, U.K.—and on his own Caribbean island.

Branson's views on commercial spaceflight have ranged widely. He talks about his spaceflight venture as part business, part amusement park ride, and part visionary mission to save the earth. He has speculated about developing rockets for supersonic transport between New York and London or perhaps colonizing the "Virgin Moon." He has often returned to the theme of global ecology, saying, "I hope that people going into space will come back and appreciate this beautiful world more." He has not revealed, however, that the hybrid solid rocket fuel mixture of nitrous oxide and aluminum polyimides that his rocket plane uses for its fuel spews out "dirty particulates" and will create more pollution to the upper stratosphere than that produced by liquid-fueled rocket systems.

Under the initial deal with Rutan and Allen, Branson agreed to pay for a license and five SS2 craft, but now that Virgin Galactic owns the company outright the original financial terms presumably no longer apply. The current business plan envisions 50 passengers a month, paying $200,000 to $250,000 each (depending on when they signed up for their spaceplane ride). The payment includes "astronaut training" and a two-hour flight to an apex beyond earth's atmosphere. The "ride" is actually wrapped up in a three-day astronaut experience. If this program holds true, then $10 million a month seems like good income, but against considerable expenses and risk management fees, it may be a marginal business plan, especially if a mishap closes down operations for a significant period.

Four Minutes of Weightlessness

Ironically, one of Branson's largest problems has been the fact that the U.S. government's International Trade in Arms Regulations (ITAR) processes slowed the transfer of technology to the U.K.–owned company. However, work eventually proceeded and licenses were granted. Virgin Galactic will fly "space tourists" on sub-orbital flights to an altitude of over 100 kilometers so they can experience weightlessness, see the dark sky, and see the earth as a large blue marble in space. (Actually, it should be pointed out, in terms of truth in marketing, the "payoff" period of the flight with weightlessness is really very short—only about 4 to 5 minutes.)

After it separates from the White Knight mother aircraft, the SS2 vehicle will fly almost vertically until it reaches maximum velocity. The passengers and crew will then coast upwards into a parabolic arc, where weightlessness will be achieved. Then, on the descent, rocket engines will be fired to pull out of the nosedive, and this will be the most critical part of the mission as the crew and passengers will be subjected to very high g forces.

If all goes well, the craft will ultimately land at the spaceport from which it took off. The entire experience will take about the same amount of time as an airline flight from Washington D.C. to Atlanta, Georgia. It will be more like an experimental test pilot trying out a very high-altitude jet than an astronaut going to the moon or even the ISS. A whole lot of sensations will be packed into a very condensed period. If it were taking place in an amusement park, a teenager would be inclined to say, "Let's do that again." The problem is that the price tag will prevent many people from signing on for the next available flight.

The high price means that a lot of "value-added" will be provided on the ground in terms of training and preparation so that passengers will better appreciate the worth of their investment. If anyone can pull this off as a successful business, Branson now seems the one most likely to do it.

Allen and Rutan—The SpaceShipOne Story

Branson is not alone in a club of billionaires with the itch to soar into the stratosphere via a space tourism business. Allen has already made his mark in the world of space by claiming the Ansari XPRIZE together with Rutan. He was the money and part of the brain trust behind the spectacular performance of SpaceShipOne.

Allen was born in Seattle, Washington, in January 1953 and went on to co-found (with the equally legendary Bill Gates) one of the world's most lucrative and influential companies, the Microsoft Corporation, in 1974. The company's products revolutionized personal computing and made billionaires of both men. Allen actually left the company in 1983 due to illness and has since invested in a wide variety of projects covering the technology, entertainment, sports, and aerospace fields. His funding of the Science Fiction Museum in Seattle is just one way Allen has enriched the world culturally and intellectually. His extravagant purchase of the L.A. Clippers is how he has commanded the most headlines.

He reportedly invested many times more than the $10 million prize money in backing Rutan's team in its bid to win the XPRIZE. Although this may be merely chump change for NASA, it was serious money for a private entrepreneur. However, this imaginative team prevailed. Rutan and Allen became the winning combination that not only won the XPRIZE but also went on to join with Branson to create the new SpaceShip Corporation. Then, in 2012, Allen joined Rutan and another wealthy space pioneers, Musk, to launch the innovative Stratolauncher enterprise. However, during 2012, Musk and SpaceX announced that he was leaving this venture, and David Thompson of OSC was recruited to fill the void.

Allen's Partner—The "James Bond" Innovator—Rutan

Allen's partner, Rutan, has made a habit out of doing the impossible. He designed the Voyager aircraft. This unconventional airplane had long and elegant wings that were also high-capacity fuel systems. Back in 1986, this flying fuel tank was able to circle the world non-stop without refueling. Rutan has also designed various Unmanned Autonomous Vehicles (UAVs), military aircraft capable of flying long-duration missions. It is great fun to tour his menagerie of space toys out in the Mojave Desert; these include one of the personal rocket craft flown by 007 in the James Bond movies. Supposedly, these were designed by "Q" and the zany band of inventors in the Bond movies, but in reality the ultra-light rocketplanes for these spy films came out of Rutan's shop.

Quite a few strange experimental craft, a number of them officially "classified," have emerged from his Scaled Composites facility over the years. It was one of Rutan's experimental planes that the singer John Denver was flying when he tragically crashed and died in October 1997.

Of the dozen and a half entries for the Ansari XPRIZE contest, Scaled Composites was considered one of the few teams that might succeed. Rutan had already achieved almost legendary status as a designer of new types of aircraft. His reputation was firmly established even before he successfully met the challenge of flying crew and passengers above 100 kilometers and repeating the flight within eight days to claim the XPRIZE. His innovative design involved a two-tiered approach, by flying the so-called White Knight "carrier vehicle" up into the stratosphere and then releasing SpaceShipOne to fly into "outer space." One of the key features to the safe re-entry of the SpaceShipOne vehicle was to achieve a relatively slow descent that did not involve high temperature gradients. Rutan diagnosed high temperatures as a key safety problem and sought an alternative instead of developing a thermal protection system to protect against dangerous high and ablative heat (as NASA and other space agencies have always done). Instead, in his pursuit of safety, he concentrated on getting the heat levels down.

Fig. 4.2 An unstoppable partnership, Burt Rutan and Paul Allen.
(Courtesy of Space.com.)

SpaceShipOne also used an innovative engine that employed a hybrid fuel system, with neoprene rubber as the fuel and nitrous oxide (that is, laughing gas) as the oxidizer. This system was developed by Jim Benson's SpaceDev Corporation, and it allowed the solid-fuel system to be "throttled" so that if a problem developed, the laughing gas would shut down and so would the engine. The down side is that this fuel is actually a pollutant. The next step is to find a safe-performance, throttleable fuel that is also clean.

Most recently, SS2 has shifted to using an aluminum polyimide fuel in place of the special version of neoprene rubber as a fuel. This is because there is a higher thrust output, but unfortunately it still produced polluting particulates.

Allen's Place in Aviation and Spaceflight History

Allen spent a great deal of money to claim the XPRIZE, but what he actually bought was a place in aviation and spaceflight history. Apparently he considered the partnership with Rutan a bargain.

In an interview back in 2006, Rutan said, "We have shown that you can construct a vehicle like this with a modest budget. The big question is, how many people will sign up, and will they pay $50,000 to $200,000 to go on one of those flights? It's not something I would contemplate unless I had partners willing to share the risk … right now we are doing all these things as experimental flights. We have permission from the FAA to go Mach 2.5 straight up. There aren't many vehicles that do that. I think it will be good for the government to encourage something like space tourism. Having a space-tourism experience, whether sub-orbital or orbital, within the reach of people would be an exciting prospect." [9]

Let's put Rutan and Allen's accomplishment into perspective. From the 1970s through the early 2000s, NASA spent many hundreds of millions—if not billions—on spaceplane development, with excruciatingly poor results. NASA managed to cancel contracts for the development of a half dozen spaceplanes or escape vehicles over three decades of failure. If you want to keep count, these were the HL-20, X-33, X-34, X-35, X-37, X-38, and X-43—actually a baker's half dozen of seven.

SpaceShip One—NASA Zero

When the Rutan/Allen team claimed the XPRIZE, NASA emerged with a red face to match the red ink of its own attempts. As noted earlier, an impromptu sign at the landing site at the Edwards Air Force Base in California summed up NASA's embarrassment with this succinct assessment: "SpaceShip One—NASA Zero." There are several reasons why so many feel that the time has arrived to allow entrepreneurs to come to the fore and let the space agencies play a supporting role when it comes to LEO and spaceplane development. These reasons include at least the following three: NASA's inability to truly fix the Space Shuttle "foam insulation" problem, the cost overruns on the ISS, and the incredible string of failures with spaceplane development.

Although this experimental sub-orbital spaceflight system performed well enough to claim the prize, it was certainly not a problem-free program, and the stability of the SpaceShipOne vehicle was sufficient to delay several of the test flights. The greatest challenge has been to convert the experimental test vehicle into a craft that is reliable enough to fly space tourists safely into space on a routine basis.

And now Rutan and Allen are also involved in the Stratolauncher project, which takes this same strategy to much higher economies of scale. This space project involves the design and building of the world's largest aircraft. Powered by six 747 engines, it is to be used as a launcher system for larger rockets or spaceplanes. This is the White Knight carrier vehicle cubed!

Northrop Grumman Swallows Scaled Composites

There was a key development in July 2007, when Los Angeles–based aerospace giant Northrop Grumman agreed to buy Rutan's company, Scaled Composites LLC. According to the *Los Angeles Business Journal*, the deal pairs an aerospace powerhouse with an emerging player in the nascent space tourism industry whose highly respected technology is advancing at a steady clip.

Northrop had first invested in Rutan's company in 2000, but through this transaction, it agreed to increase its stake from 40 percent to 100 percent. Rutan said, "My company has been owned by other corporations in the past and we have maintained our research and development culture that does the most efficient work in prototypes, and we expect no changes under the most recent equity revisions." [10]

Northrop is a $30 billion industry giant, with a major part of its revenues coming from its space technology division. Over the years, about 90 percent of that came from government contracts, but after the company lost a bid to build the successor to NASA's Space Shuttle, the company has not announced plans to build manned rockets. It is, however, developing unmanned vehicles as well as satellite and space radar technologies. At the time, Northrop spokesman Dan McClain declined to characterize Northrop's interest in space tourism. He said the company was not seeking any specific technology by buying Scaled Composites. He also said the deal would not affect Scaled Composites' operations and its entire management team would remain intact.

"We really value their current mode of operation," he said. "Northrop has always valued the innovative and entrepreneurial qualities of Scaled Composites and we think it's a good fit with our company's ongoing efforts with aeronautics and spaceflight." [11]

What went unsaid is that Northrop is clearly interested in designing and building hypersonic jets that could fly executives and passengers around the globe. Currently Boeing, Lockheed Martin, and Northrop Grumman have design contracts from NASA to develop conceptual designs for such vehicles. The ability to draw on Rutan's brainpower for this activity is clearly a plus.

Allen and the Birth of Microsoft

As the financial strength behind Rutan's enterprise, Allen has played a critical role in moving the space tourism business from science fiction to science fact. But the story of his success and wealth began more than 40 years ago. It was in 1968, at Lakeside School, a prestigious private school in Seattle, that Allen met eighth-grader Bill Gates, who, like Allen, spent most of his free time figuring out the inner workings of their school's new computer. "Our friendship started after the mothers' club paid to put a computer terminal in the school in 1968," Gates told *Fortune* in 1995. "The notion was that, of course, the teachers would figure out this computer thing and then teach it to the students. But that didn't happen. It was the other way around." The pair became so adept with computer technology that, while still in school, they were both invited to serve as amateur technicians at a local computer center in exchange for free computer time.

Allen graduated from high school in 1971 and entered Washington State University. That same year, he read about the Intel Corporation's 4004 chip, the first computer microprocessor. In 1972, he and Gates purchased the next generation of the chip, the 8008, for $360. The pair used the chip to develop a special computer that conducted traffic-volume-count analysis and started a company called Traf-O-Data, planning to sell the computers to traffic departments. They eventually abandoned the company, and in 1974, Gates left Washington for Harvard University in Cambridge, Massachusetts. Allen followed, dropping out of Washington State and accepting a job as a computer programmer at the Honeywell Corporation in Boston. Allen hit upon the seed for their next business move in a *Popular Electronics* magazine cover story describing MIT's Altair 8800 minicomputer. Recognizing that the computer would need a programming language, Allen and Gates set out to write a version of BASIC, a widely used computer language, specifically geared toward the Altair. They convinced MIT to buy their programming language. According to the *Fortune* interview, the credit line in the source code of their first product read: "Micro-Soft BASIC. Bill Gates wrote a lot of stuff; Paul Allen wrote the rest." [12]

The Young Entrepreneurs in New Mexico

Allen and Gates soon changed their company's name to Microsoft and moved their business to Albuquerque, New Mexico. The young entrepreneurs quickly built up an impressive client list that included Ricoh, Texas Instruments, Radio Shack, and another new start-up, Apple Computers. In 1978, with sales already over $1 million, they relocated their company to Seattle, where, by 1979, they had hired more than 35 employees and a professional manager.

Microsoft entered into one of the most significant business deals in its history when International Business Machines (IBM) approached the company seeking a programming language for its new personal computer, which was secretly under development. That same year, Allen negotiated the purchase of Q-DOS, a little-used operating system produced by Seattle Computer. Microsoft paid $50,000 for Q-DOS and, in turn, licensed the product to IBM for use with its new PC. In addition, Gates and Allen convinced IBM to allow other companies to copy the specifications of their PC, spurring the ensuing flood of PC "clones." The widespread availability of PCs necessitated compatible software programs, which, in turn, required a universal operating system. The rest is largely history.

Hit by Cancer

In 1982, Allen was diagnosed with Hodgkin's disease, a form of cancer. He continued to work part-time at Microsoft during 22 months of radiation treatments, but in March 1983, he retired from the company and spent the next two years traveling, scuba diving, yachting, skiing, and spending time with his family. He retained a 13 percent share of the company and continued to serve on its board. Finances were not a concern—as of 2003, he was estimated to be the world's fourth wealthiest citizen, worth $21 billion—and Allen sought out new business and investment opportunities. These included Asymetrix, which produced applications that allowed both programmers and non-programmers to develop their own software, and Vulcan Ventures, an investment firm focused on technology. Allen also invested in numerous companies, including Ticketmaster, America Online, Egghead Software, and the pharmaceutical company Darwin Molecular Corporation. Later investments focused on cable television, wireless modems, and Web portals.

In 1992, Allen founded Interval Research, a think tank focused on the Internet and compatible technologies. He also began to channel funds into entertainment and sporting ventures. He purchased the National Basketball Association's Portland Trailblazers in 1988 and the National Football League's Seattle Seahawks in 1996. He also purchased a reported 24 percent share of the film and television studio DreamWorks SKG. And internationally, he is following in the footsteps of Russian, Asian, and other American billionaires by negotiating for the purchase of a major U.K. soccer team.

Allen celebrated an even earlier passion with his support of the Science Fiction Museum and Hall of Fame, which opened in Seattle in

2004. He further indulged his interest in otherworldly phenomena with a $13.5 million donation to the Search for Extra-Terrestrial Intelligence. Then he funded the development of SpaceShipOne, the world's first private spacecraft, for an amount estimated to be in the range of $20 to $30 million. SS2, which has had its ups and downs, literally and figuratively, was essentially built based on a scaled up and improved version of the SpaceShipOne design. This technology is being licensed to the SpaceShip Corporation from Allen's Mojave Aerospace Ventures Company. Thus, Allen recouped not only the $10 million XPRIZE money but the licensing fees associated with the SS2's design and production.

The Amazon Man Creates "Blue Origin"

Another space industry billionaire from the Internet era is Jeffrey Bezos, the founder and CEO of Amazon.com. His groundbreaking dot-com company is consistently ranked as one of the top retail sites on the Internet and offers over one million book titles, and now a dizzying array of other consumer products. His new company, Blue Origin, was created in 2000; its headquarters are in Kent, Washington, with its launch and test facilities in Culbertson County, a remote part of North Texas, very near the New Mexico and Texas border. The objective of Blue Origin was to start offering sub-orbital space tourism flights, but the company's slogan suggests a more ambitious goal: "Creating an enduring human presence in space."

Blue Origin began by developing a launch system designed to take off and land vertically. The first test flight of a development vehicle for the company's New Shepard spacecraft occurred in November 2006. Few details were provided, as has been the case with this secrecy-shrouded project to date. But apparently it did achieved neither the desired test flight results nor the needed speeds for even a sub-orbital flight. It was reported that this very short test firing rose to the height of the Washington Monument (about 87 meters)—and then landed. [13]

According to documents Blue Origin submitted to the FAA during 2006, the New Shepard Reusable Launch Vehicle would stand about 15.25 meters (50 feet) tall and about 7 meters (22 feet) wide at the base. The fuel system is thought to be kerosene and hydrogen peroxide. The vehicle, for safety and escape purposes, consists of two stacked modules. One is to provide system propulsion, and the other provides escape capability for the flight crew.

Bezos Wins Recognition and Funding from NASA

Bezos continued to press ahead, and in 2010, Blue Origin received an award under NASA's Commercial Crew Development (CCD) program to develop an escape system for a biconic capsule to be launched on an Atlas V rocket. Then in April 2011, the company received a commitment from NASA for $22 million under phase 2 of the CCD program to accelerate development of escape technology. This involved testing and developing the liquid hydrogen and liquid oxygen engine for the reusable booster of the Atlas V rocket. [14]

Fig. 4.3 Jeff Bezos in a celebratory mood. *(Courtesy of Blue Origin.)*

In 2011, Blue Origin suffered a setback when an unmanned test rocket exploded over the West Texas desert. The *Wall Street Journal* said the mishap "dealt a potentially major blow" to Bezos's space plans. But on the Blue Origin website, Bezos was more phlegmatic and said that although it was "not the outcome any of us wanted," he had "signed up for this to be hard."

The work continued, and in October 2012, the company announced that it had completed a successful test of its pad escape system, a "pusher escape motor" that safely launched the empty crew capsule to an altitude of 703 meters and then landed by parachute. This has been designed for use at any time in the ascent phase of the New Shepard sub-orbital spaceplane, and Bob Meyerson, the president and program manager for Blue Origin, said, "Providing crew escape without the need to jettison the unused escape system gets us closer to our goal of safe and affordable human spaceflight." [15]

With such a short heritage the Blue Origins enterprise has now established itself as a key player in the U.S. space game. In September 2014 United Launch Alliance, which provides both the Delta and Atlas launch vehicles, announced that it had reached an agreement with Blue Origin to develop a new hydrogen and oxygen rocket motor for the Atlas vehicle [16]. Thus Blue Origin is to supply the American-made motor for what many consider to be the most reliable launch vehicle in the U.S. arsenal, and this agreement weds Blue Origin to both Boeing and Lockheed Martin, the partners in the United Launch Alliance.

Fig. 4.4 A test firing of the Blue Origin sub-orbital spacecraft.
(Courtesy of Blue Origin.)

The Early Days of Bezos

Bezos was born in Albuquerque, New Mexico, and at an early age, he displayed a striking mechanical aptitude; even as a toddler, he showed a remarkable talent by dismantling his crib with a screwdriver. [17] The family moved to Miami, Florida, and in high school, he fell in love with computers. He went on to Princeton University, where he graduated with a degree in computer science and electrical engineering.

In 1986, Bezos joined FITEL, a high-tech start-up company in New York. Then, in 1988, he joined Bankers Trust Company in New York to lead the development of computer systems that helped manage $250 plus billion in assets. In the process he became the company's youngest vice president as of February 1990. From 1990 to 1994, he helped build one of the most technically sophisticated and successful quantitative hedge funds on Wall Street for D. E. Shaw & Co., New York. In 1992, he was elevated to a senior vice president for D. E. Shaw. Then, he found the Internet.

This remarkable new computer network was first envisioned by the U.S. Defense Department's Advanced Research Projects Agency (ARPA) and was first known, not too surprisingly, as ARPANET. The concept, as envisioned by Vint Cert and others, was to create a maze of interconnected networks that would stay connected during an emergency or surprise nuclear attack even if many of the other connections should fail. In a surprisingly short period of time, it was adopted by government scientists and academics at research universities for the exchange of data and messages. In 1994, however, there was still no Internet commerce to speak of, and Bezos observed that Internet usage was increasing by 2300 percent a year and growing, in the United States and around the world.

He saw an opportunity for a new sphere of commerce, and he immediately began considering the possibilities. In typically methodical fashion, he reviewed the top 20 mail order businesses, and identified books as a commodity where no comprehensive mail order catalog existed, because any such catalog would be too big to mail. He concluded that books, as well as music, were perfect for the Internet since he could create a sales company that would be able to share a vast database with a virtually limitless number of people, and the customer base would only expand as the Internet grew across the world.

At the American Booksellers Convention, he found that the major book wholesalers had already compiled electronic lists of their inventories. All that was needed was a single Internet location where the book-buying public could search and place orders directly. Bezos knew the only way to seize the opportunity was to go into business for himself. It would mean sacrificing a secure position in New York, where a life of luxury was already his, but he decided to make the leap.

Amazon—The Electronic River That Spans the World

Bezos set up shop in his two-bedroom house, with extension cords running to the garage and three Sun MicroStation processors. When he had the test site up and running, he asked 300 friends and acquaintances to test it. The code worked seamlessly across different computer platforms. On July 16, 1995, Bezos opened his site to the world, and he told his 300 beta testers to spread the word. In 30 days, with no press, Amazon had sold books in all 50 states and 45 foreign countries (he called the company Amazon after the seemingly endless South American river, with its almost infinite number of branches). The business grew faster than Bezos or anyone else had ever imagined. The company went public in 1997, and two years later, the market value of shares in Amazon was greater than that of its two biggest retail competitors combined. Amazon moved into music CDs, videos, toys, electronics, and more. When the Internet stock market bubble burst, Amazon re-structured, and while other dot-com start-ups evaporated, Amazon even began to post profits.

Today, Bezos and his wife, Mackenzie, live north of Seattle and are increasingly concerned with philanthropic activities. Bezos has consistently been a person of vision, and his new dream of moving people into space is fully consistent with his entrepreneurial view of the world. He consistently portrays the type of person who looks at the world and instead of asking, "Why," asks, "Why not?"

In the Beginning—Genesis

Like Bezos, Robert T. Bigelow is a highly successful entrepreneur who has spotted a new opportunity. This time, however, the owner and developer of the Budget Suites hotel chain has his experienced eye on the long-term prospect of hotels in space. In 1999, he founded Bigelow Aerospace, described as a general contracting, investment, and research and development company that seeks to achieve economic breakthroughs in space. His objective is to design, develop, fabricate, and deploy habitable commercial space complexes at viable costs.

Bigelow has been granted eight patents, and more than a dozen others have been submitted and are awaiting approval. He has already invested many tens of millions toward R&D and says he is prepared to invest a further $500 million to realize his goals.

In July 2006, Bigelow Aerospace launched a 33-percent-scale model prototype space station named Genesis I. It was some 2 meters in diameter, but the final version will be nearly 6 meters in diameter and thus will be some 27 times larger in total volume (3 times longer x 3 times wider x 3 times taller). Next came the launch of Genesis II in June 2007 and the deployment of the first habitable commercial space structure. Like its predecessor a year earlier, Genesis II flew into space on a Dnepr rocket, a converted ballistic missile from Russia's military arsenal, and it went into a near-circular orbit about 350 miles high. Compressed air from several on-board gas tanks began inflating the module shortly after arriving in space. Genesis II is more than 14 feet (4.5 meters) long and eight feet (2.5 meters) in diameter when fully expanded, according to Bigelow Aerospace. It carries an upgraded suite of internal avionics, a revamped inflation system, and the various objects contributed by paying customers. Twenty-two cameras mounted both inside and outside the module are beaming back imagery to Bigelow ground stations in Nevada, Virginia, Alaska, and Hawaii. Genesis II also carries Biobox, an animal habitat housing colonies of ants, cockroaches, and scorpions.

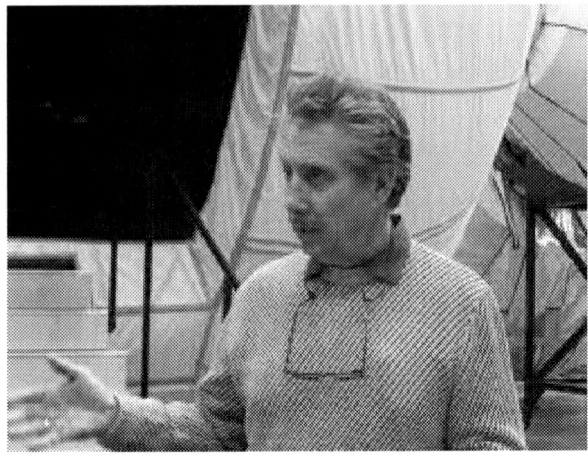

Fig. 4.5 Robert Bigelow explains his orbital hotel plans.
(Courtesy of Bigelow Aerospace.)

Today, Genesis I and II are still flying in space, and the Bigelow Aerospace website shows a wide range of space images from on-board cameras. The ultimate goal is to deploy in LEO a three-stories-tall inflatable spacehab that will have greater volume than the ISS. These inflatable bio-systems will represent a small fraction of the cost and mass of the dense, metal-based ISS platform.

Making a Dream Come True

The launch of Genesis was the culmination of a long dream. The whole time Bigelow was building hotels in Arizona, Nevada, and Texas, he was secretly hoping to build in outer space. When he told his wife that he was starting a new company, Bigelow Aerospace, she thought: "Well, okay, we'll see. You know, it might be a passing fancy."

It wasn't. Bigelow now has a giant facility in Las Vegas and more than 100 employees. His goal is to build an orbiting complex of rooms that can be used by private companies, foreign countries, or tourists—a type of public space station. He licensed the inflatable technology from NASA for his space habitats. "We stand on the shoulders of a NASA program that was canceled called TransHab," he says. TransHab was originally proposed in 1997 as possible crew quarters for NASA's space station. Congress eventually told NASA to stick with traditional aluminum cylinders, rather than spending money to develop inflatable technology. So once again, we see the willingness of entrepreneurs to pursue new, innovative, and more cost-effective technology that NASA or other traditional space agencies have abandoned. Once again, we also see that political pressures on NASA by Congress are one of the reasons that NASA cannot be more innovative and why NASA has to settle for a political solution rather than what is most technically and operationally most efficient. Political pressures on the design of the Space Shuttle very likely resulted in a system that was more dangerous. [18]

Bigelow has plans for more experimental versions of these inflatable spacecraft in preparation for the launch of the full-scale, three-story-tall capsule. George Whitesides of the National Space Society, a group that promotes space exploration, says the full-scale spacehabs are impressive. "I've been inside the full-scale versions of Bigelow's space hotel, and it is huge," Whitesides said. "Other entrepreneurs are building small rockets and spaceships that will give people a way of getting into space. But only Bigelow is working on building somewhere that people can stay and enjoy it." [19]

This, however, is not completely true. Although Bigelow is the current leader in the field, he is not the only pioneer pursuing the development of spacehabs. Inter Orbital Systems Inc. (IOS), with many fewer resources at its command, has sought to develop a large-scale, one-and-a-half stage-to-orbit vehicle that would use the evacuated oxygen tanks as a habitat for space tourists once the rocket is in LEO.

In addition to space tourism and orbital hotels, other expected uses for Bigelow's expandable modules include microgravity research and development, and space manufacturing. The company plans to sell its BA 330 modules for $100 million apiece. He told *Space News Business Report* that he has invested more than $90 million in Bigelow Aerospace. "As a general contractor for 35 years," he said, "we're not strangers to contracting, to banking, to the financing of major projects. That's crucial if you really want to get the financial horsepower involved. Number one, the business model has to serve a customer. Number two, it has to be very cost-effective, and number three, it has got to do what it says it's going to do. The banking world appreciates that and they respond … Wall Street responds in predictable ways." [20]

A $50-Million Challenge—And Nobody Came?

Bigelow also announced a bold move into the space competition business with "America's Space Prize," which sought to stimulate the development of new technology for going into space for the longer term. The award amount was set at the very compelling level of $50 million. [21] However, the winner had to jump through a daunting number of hoops to succeed and win the large prize.

First of all, the entering corporations had to be U.S.-based entities, and they had to develop the winning launcher and spacecraft without government funding—although use of governmental test facilities was allowed. The extremely difficult challenge involved building and launching a new crew-capable spacecraft that could carry a crew to an orbit of at least 400 kilometers' altitude. This craft had to demonstrate the capability of carrying a crew of five and docking with a space station. The craft had to then return safely to earth and repeat the accomplishment again within 60 days but with a crew of five people aboard the second time.

Finally, the deadline for claiming the prize was essentially five years in the future from the time the competition was announced in November 2004, or January 2010. In light of all the conditions—including having a crew of five to make the second journey into space—it is not surprising there were no entries or even test flights by the target date. It was thought that the cost of developing the systems and carrying out the tests would have likely exceeded the prize money by a large margin. Perhaps Bigelow learned from the XPRIZE competition to make the conditions to win as demanding as possible.

Bigelow, ever the entrepreneur who invests and risks his money cleverly, did offer an additional incentive. He said his company was prepared to offer $200 million in conditional purchase agreements for six flights of the prize-winning vehicle. He explained this new launch capability was a part of his ultimate plan to deploy a private space station for space tourist flights. Clearly he needs a reliable private and cost-effective vehicle for his space habitat venture to succeed. He explained,

"It could be somebody who doesn't win the competition, but who comes in late, but we like their architecture better than the winner's architecture."

Bigelow has received several honors for his spaceflight efforts. In 2006, he received the "Innovator Award" from the Arthur C. Clarke Foundation. [22] The award recognizes "initiatives or new inventions that have had recent impact on or hold particular promise for satellite communications and society, and stand as distinguished examples of innovative thinking." Bigelow was presented with the award at the Arthur C. Clarke Awards in Washington D.C. alongside Walter Cronkite, who was honored on the same night with the Arthur C. Clarke Lifetime Achievement Award. That night at the Cosmos Club, we had a true American icon appearing together with a new space icon in the making.

The Young Billionaire and SpaceX

Elon Musk, another Internet entrepreneur, took a rather different approach to developing a new space capability. Born in South Africa, he taught himself how to write computer code, and at the age of only 12, he created a game he called "Blast Star"—which he sold to a computer magazine for $500. In South Africa, he felt isolated from the burgeoning software industry, and at 17, he first moved to Canada and then entered the University of Pennsylvania to earn degrees in physics and business. In 1995, he was ready to begin his doctorate at Stanford when he was attracted by the Internet madness going on around him in Silicon Valley. He looked for a business that would generate cash quickly and founded Zip2, which helped newspaper companies put classified ads and other local information on websites.

After selling Zip2 for $307 million, he started his next business, X.com, which offered a variety of banking services to consumers, but the most popular feature was the ability to e-mail money. Later, X.com acquired the rights to the name "PayPal" and focused on improving its e-mail money feature. PayPal soon became a favorite among users of eBay to make payments, and just months after PayPal went public, eBay bought it, for $1.5 billion.

Musk was suddenly a billionaire in search of his next challenge—and space became his newest frontier.

He had no prior space systems experience but with a compelling amount of self-confidence, he started SpaceX, or Space Exploration Technologies, in 2002. He started out with some lofty objectives. He aimed to build a rocket that he said would cost only a third as much as current models. He envisioned a possible world in which governments, universities, and businesses in need of an inexpensive way to get satellites into space would turn to SpaceX over more established rocket suppliers.

Fig. 4.6 Elon Musk made history when the Dragon 9 spacecraft docked with the ISS. *(Courtesy of SpaceX.)*

From the outset, Musk self-funded SpaceX. Even from the start he expressed no small thoughts. When one of the authors first met him at the Japanese U.S. Science, Technology and Space Applications Program (JU.S.TSAP) in Hawaii, he explained that he aspired to much more than to build rocket launchers: spaceships that could travel to Mars. His first initiative was a 68-foot-tall rocket called the Falcon Explorer. His first-stage miracle was to come up with a rocket that could be launched for just $6 million—a remarkable price that was less than a third of the going rate. His succinct vision was also quite simple: "...the Falcon is designed to be a truck, not a Ferrari." This very low cost first attracted the U.S. Air Force as a backer. To get the job done, Musk lured top talent from large defense companies who had experience in building rockets. One of these was Tom Mueller, the company's vice president of propulsion, who previously headed liquid rocket propulsion development at TRW Space & Electronics.

Although Musk's plans sometimes seemed crazy, he had some early successes. The first stage of the two-stage rocket was successfully fired at a 300-acre testing facility in McGregor, Texas, and in March 2006 he launched a Falcon that made its way almost to orbit before it experienced a failure. A second test flight in March 2007 also failed to reach orbit, but Musk apparently put a good spin on the failure, saying that the problem would be "pretty straightforward to address—we feel like there's really no need for an extra test flight."

Musk's company's big breakthrough was when he became a winner in 2007, along with RpK, of the NASA COTS competition. The company was thus committed to providing a launch capability to take cargo to the ISS, to replace the Space Shuttle after 2010. When RpK failed to meet the NASA-set targets to raise sufficient financial resources in late 2007, OSC was selected as the new competitor to develop a commercial launch system and "dockable spacecraft" to go to the space station. OSC's late entry proved a major liability. Its Antares rocket, despite being based on the Taurus vehicle, has constantly been behind schedule, and its failure on October 27, 2014, has perhaps put its program permanently in the shadow of SpaceX's Falcon 9.

A First-Round Victory in 2012

SpaceX, which operates out of Hawthorne, California, has focused on winning out on the CCD competition. The company expanded from 160 employees in 2005 to over 1100 by 2010 and over 3000 by February 2015. SpaceX became the first privately funded company to successfully launch, orbit, and recover a Dragon spacecraft, thereby winning the 2011 Space Achievement Award from the Space Foundation. And when the company's Falcon 9 launch vehicle carrying the Dragon spacecraft docked with the ISS successfully in May 2012, it could justly claim a first-round victory.

Musk's spacecraft successfully docked within reach of the station's robotic arm and thus was able to supply food, water, and other provisions for the astronauts on board. And this was just the first of many planned resupply missions. This flight marked the entry of commercial space companies into the big leagues and placed SpaceX at the heart of ISS operations. [23]

However, docking with a robotic arm will not be the preferred system if SpaceX (and other commercial operators) move on to the next stage of ferrying crew to and from the ISS. Commercial operators will need to adapt their Dragon spacecraft to use the NASA Docking System (NDS), so far the only docking system that is compliant with the International Docking System Standard. As Skip Hatfield, NASA's manager, has said, "In the event that the crew needs to leave for some reason, you don't want to be dependent on a system on the ISS like the arm. You want to be able to jump in the thing and just depart, in case you're having a bad day, so to speak." [24] As of March 2013, the Falcon 9 and Dragon spacecraft had completed three docking operations using the station's robotic arm.

And Musk, true to form, is still pushing the pedal to the metal. His latest space coup has been to launch the amazing Deep Space Climate Observatory (DSCOVR) satellite to the L-1 Lagrange Point, a million miles (1.6 million kilometers) out in space. This remarkable satellite, the renamed Triana spacecraft (which began as a project advocated by Vice President Al Gore), has a high-definition space telescope that constantly monitors the earth's atmosphere. It also has two solar sensors that allow constant monitoring of the sun for coronal mass ejections. These could bring violent and even devastating solar storms to earth, threatening our satellites, power systems, and computer and communications networks. The heavy-lift Falcon 9 rocket, despite some delays, flawlessly delivered this vital new earth and solar observatory on February 11, 2015. And that wasn't all; the recoverable first-stage booster landed within 10 meters of the targeted ocean platform. This showed that recovering and recycling the expensive first-stage booster has great promise. [25]

This is a remarkable set of accomplishments for someone just entering his forties, but this hardly completes his profile. In case no one has noticed, Musk has launched and then conducted a highly successful IPO for the Tesla Motor Company. This company has developed both a high-performance sports car and a sedan, and its stock price has soared. Tesla, under Musk's innovative guidance, has moved to create in Nevada a new billion-dollar state-of-the-art battery "gigafactory." The word "gigafactory" may seem a bit of hyperbole, but Musk intends by 2020 to dominate the world battery market. His vision is to annually produce, within only a five-year window, the equivalent of all battery capacity that was produced in 2013. [26]

Anderson's Early Days

As described in Chapter 2, Eric Anderson of Space Adventures is also among the new breed of space entrepreneurs. He began thinking hard about space tourism while interning at NASA in the summer of 1995. He was a junior in the aerospace engineering program at the University of Virginia, and these internships taught him two tough lessons: one, his 20/40 eyesight would always keep him from taking the test to gain astronaut status, and two, you'll never be able to trust NASA-crats to figure out how to put civilians into space for a reasonable cost. "Things there are massively over-inflated," says Anderson. "I'd see a million-dollar study produce a 100-page report I could have written in college. The government has no incentive to make things cheaper. The bigger the budget, the more power they wield." [27]

After graduating at the top of his class from the University of Virginia's engineering school in 1996, Anderson raised $250,000 from investors—including Peter Diamandis, founder of the XPRIZE Foundation, and Michael McDowell, who started the Arctic cruise company Quark Expeditions—to launch Space Adventures from a room in his Arlington, Virginia, townhouse. Inspired by McDowell's idea of selling adventure vacations aboard Russian icebreakers and submersibles, Anderson figured he could connect rich thrill-seekers with a Russian government barely able to afford its space program. McDowell introduced him to military types in Moscow, but the skies were new territory, and the Russian bureaucracy proved a nightmare. "We found people who said they could arrange the flights," says Anderson, "only to learn later they had no authority. Half the battle was getting to the right person." [28]

It Began with MiG Flights

Once Anderson had secured permission, he set about finding clients for flights in a Russian MiG fighter plane. At an Explorers Club gala in New York, he got his first three: Lotsie Holton, an heiress to the Anheuser-Busch fortune, and her son and husband. The trio had such a good time on their MiG flights that they spread the word among the wealthy. Soon Anderson had full flights. He had no idea he'd be able to sell tickets to the ISS until that February 2000 meeting with Dennis Tito at the offices of Tito's company, Wilshire Associates, in Santa Monica, California. The previous year, Anderson had paid the Russians $100,000 to study whether the Soyuz could be used to transport tourists to the ISS.

Once Tito had flown and then successfully touched down in the Kazakhstan desert, Anderson felt euphoria and intense relief, but also a little envy. A big part of him wished he had been in Tito's shoes. "I'll go on a sub-orbital test flight in two or three years," he said. "I would never sell anything I wouldn't do myself." [29]

From Video Games to Rocketry Enthusiast

Next, we look at the remarkable career of John D. Carmack II, who is a widely recognized figure in the video game industry. Though he is best known for his innovations in 3D graphics, Carmack is now also a rocketry enthusiast and the founder and lead engineer of Armadillo Aerospace. He had aspirations of sub-orbital space tourism in the short term, eventually leading to orbital spaceflights, and he began by developing the Black Armadillo project to compete for the Ansari XPRIZE.

Born in August 1970, he became a prolific programmer and co-founded id Software, a computer game development company, in 1991. He was the lead programmer of the highly successful id computer games *Wolfenstein 3D*, *Doom*, *Quake*, and subsequent sequels to *Doom* and *Quake*. His revolutionary programming techniques, combined with the unique game designs of John Romero, led to a mass-popularization of the first-person shooter(FPS) genre in the 1990s. This has allowed him to achieve significant wealth. He is not the type of person to rest on his laurels, and he is clearly someone who rises to a new challenge.

"It was exciting to move to a new field I didn't know anything about," Carmack said. "I am drawn to the engineering. I enjoy solving problems and finding novel solutions to things, and I've been at the top of my field in software for so long. The challenges, while they evolve, are not so novel anymore."

Fig. 4.7 Not a computer game, but the real thing. John D. Carmack experiences a weightless flight. *(Courtesy of ZERO-G Corp.)*

The Aim—To Simplify Controls and Costs

Armadillo Aerospace was founded in 2000 by Carmack in Mesquite, Texas. Its initial goal was to build a manned sub-orbital XPRIZE-class spacecraft, but with the long-term ambition of orbital spaceflight. The company placed a strong emphasis on a rapid build and test cycle. Armadillo has designed and built a number of different vehicles using a variety of propellants. Each design had several features in common. One was the use of modern computer technologies and electronics to simplify rocket control and to reduce development costs. Another was the use of liquid propellants and vertical take-off and landing (VTOL) to facilitate short launch-to-launch times.

Armadillo's XPRIZE vehicle was unorthodox among modern rockets in that instead of using stabilization fins, which complicate the design and increase drag, it used an aerodynamically unstable design, where the computer-controlled jet vanes are based on feedback from fiber optic linked gyroscopes. A preference for simplicity and reliability over performance was also evident in the company's choice of hydrogen peroxide (50 percent concentration in water) and methanol as a mixed monopropellant for the vehicle.

When the Armadillo designers learned that they were not going to win the XPRIZE, they changed gears. They opted to switch to liquid oxygen because of difficulties with peroxide catalysts and the lack of availability of high-concentration peroxide in the United States for small companies. In June 2004, they successfully demonstrated a computer-controlled VTOL flight of its prototype vehicle, becoming the third unmanned rocket in history to have done so, after the McDonnell Douglas DC-X and Japan's Institute of Space and Astronautical Science (ISAS) Reusable Vehicle Test (RVT).

Failures and Then "Hibernation"
Armadillo Aerospace competed in the NASA Lunar Lander prize challenge, taking two similar vehicles, Pixel and Texel, to the event. The vehicles narrowly failed to win the Level 1 prize, after making three dramatic attempts totaling over 5 minutes in the air, but finally crashing out on the final attempt. Persistent landing problems were the main cause of failure, with the undercarriage breaking several times, and landing slightly off the pad on one occasion due to guidance issues. Later, Carmack reported progress in his group's modular rocket work and expressed confidence that his lander would be able to make precision landings in the future. [30]

The company's Stiga launcher project had mixed results in January 2012. After a successful launch from Spaceport America, it completed a 169.5-second ascent and reached an apogee of between 90 and 95 kilometers above sea level, returning views stretching into Colorado. However, on the way back down, the recovery system failed and the rocket was destroyed by impact with the ground. [31]

Following this failure, Carmack announced in 2013 that he was putting Armadillo Aerospace into "hibernation mode." But in 2014, a group of his former employees—and several financial backers—announced the creation of Exos Aerospace. They set up a new operation in a former Armadillo facility at Caddo Mills Municipal Airport in Texas and set out to build on their years of experience. The founders of Exos, Russell Blink and Phil Eaton, announced that their aim is "to develop an affordable human-rated rocket." [32]

"Moon Express" and Beyond?

And finally we have the story of Naveen Jain, introduced earlier as the California-based native from India. His "Moon Express" project is targeted at mining precious metals and minerals from the moon, but he told NBC News that in 15 to 20 years he envisions a day when the moon is used as a way station enabling easier travel for exploration of other planets. [33] Jain earned his MBA in India and moved to Silicon Valley in 1983 at the age of 24 to take up a position with Unisys. He then joined Microsoft, where he spent seven years, In 1996, he left to create Infospace, an online e-mail and phone directory company that he took public. At the height of the dot-com boom, his company was valued at $30 billion. In 2003, he started Inome (formerly named Intelius), an online database and public records company with more than 25 million customers. [34]

In 2013, he was a co-founder of Moon Express, a Mountain View, California-based company that's aiming to send the first commercial robotic spacecraft to the moon next year. In early 2015, it became the first company to successfully test a prototype of a lunar lander at the Kennedy Space Center in Florida. The success of this test, and a series of others that will take place later this year, paves the way for Moon Express to send its lander to the moon in 2016, said Jain, company chairman.

Moon Express conducted its tests with the support of NASA engineers, who are sharing with the company their deep well of lunar know-how. The NASA lunar initiative, known as Catalyst, is designed to spur new commercial U.S. capabilities to reach the moon and to tap into its considerable resources. In addition to Moon Express, NASA is working with Astrobotic Technologies of Pittsburgh, Pennsylvania, and Masten Space Systems of Mojave, California, to develop commercial robotic spacecraft.

Fig. 4.8 Naveen Jain gets his opportunity to experience weightlessness.
(Photo courtesy of ZERO-G Corp.)

Jain said Moon Express also recently signed an agreement to take over Space Launch Complex 36 at Cape Canaveral. The historic launchpad will be used for Moon Express's lander development and flight-test operations. Before it was decommissioned, the launchpad was home to NASA's Atlas-Centaur rocket program and its Surveyor moon landers. His MX-1's maiden moon flight is slated to occur in late 2015 as part of the $40 million Google Lunar XPRIZE, an international challenge to land a robot on the lunar surface, have it travel at least 1650 feet (500 meters), and send data and images back to earth. The first privately funded team to do all of this by the end of 2015 will receive the $20 million grand prize. An additional $20 million is set aside for second place and various special accomplishments and milestones, bringing the prize's total purse to $40 million.

Moon Express is one of 22 teams still in the running to win the grand prize, but its ambitions don't stop there. The company aims to make money flying commercial and government payloads to the moon, and it eventually wants to extract water and other resources from earth's nearest neighbor, both to benefit humanity on its home planet and to help our species extend its footprint out into the solar system. [35]

To underline Jain's commitment to commercial spaceflight, he now serves as a trustee of the XPRIZE Foundation and as a trustee of the Singularity University. As he said to CNBC in 2014, "It's clear that the baton has been passed from the government to the private sector when it comes to space exploration."

The fact is that not every commercial effort to develop new space vehicles is the result of initiatives by space billionaires. The stories in this chapter have covered a range of quite different examples, but together they have convinced the traditional aerospace industries that they too must get into the game or they may well be left behind. Thus we now see the likes of the Boeing Company entering the fray, as discussed in the next chapter.

The Leading Contenders

Table 4.1 sets out our assessment of the top contenders to deploy the first commercial space tourism flights. This is, of course, today's assessment, and could well change in the coming months.

Table 4.1 The space tourism leaders.

Company Name	Vehicle	Technical Approach	Comments	Start of Service
Exos Aerospace (formerly Armadillo Aerospace)	Black Armadillo. STIG launch system.	1 stage. LOX/ethanol engine. (Limited capital investment.) Vertical takeoff and land. (Like the Delta Clipper design.)	New system. Limited tests. No longer key contender due to "hibernation" of Armadillo Aerospace.	Now unknown
Blue Origin	New Shepard launch system. Also building LOX/hydrogen motors for Atlas.	Reusable launch vehicle. Hydrogen peroxide and kerosene fuel. Abort system.	New Shepard is a developmental system. Limited tests.	2015-16?

Company Name	Vehicle	Technical Approach	Comments	Start of Service
Space Adventures (with Myasishchev Design Bureau)	Explorer Spaceplane (C-21) lifted to high altitude by the MX-55 high-altitude launcher plane (HTHL).	Liquid fuel motors HTHL (lifting body with parachute landing).	Based on extension of Russian systems but still a new system.	Current status unknown
SpaceDev (part of Sierra Nevada Corp.)	Dream Chaser	Single hybrid engine (neoprene and NO_2) for sub-orbit. Launch of spaceplane on the side of three large hybrid boosters to reach LEO orbit and ISS.	Both sub-orbital and orbital systems derive from SpaceShipOne but still a new system. Loss of NASA funding major setback.	2017?
SpaceShip Corp. (now wholly owned by Virgin Galactic. See Virgin Galactic.)	SS2 (upgraded version of SpaceShip One with increased cabin size).	Hybrid engine. (aluminum polyimide and NO_2) oxidizer for sub-orbit. Flown to high altitude on a jet-based launcher system.	Undergoing testing program; likely to be first operational system for Virgin Galactic and others. Fatal crash of October 31, 2014, key setback.	2016?

Company Name	Vehicle	Technical Approach	Comments	Start of Service
Virgin Galactic	Fleet of SS2.	See also SpaceShip Corp., above.	Still needs to be subjected to extensive tests following October 31, 2014, crash.	2016-17

And There Are More….

There are, of course, more than just billionaires and well-known entrepreneurs (as well as major aerospace companies) at work in this business. Many of the 50-plus spaceplane developers and space tourism backers are operating out of garages, and, for some, their capital financing plans involve "maxing" out as many credit cards as they can get. Some of them are described in the next chapter. These are in great flux, and many come and go or merge with each other.

Tomorrow's would-be space billionaires are intent on following in the footsteps of Paul Allen, Elon Musk, Jeff Bezos, Robert Bigelow, John Carmack, and Richard Branson. They are diligently toiling away at developing spaceplane designs and technologies, trying to make their dreams a reality. Who knows? They may become the billionaires of tomorrow—or not. Undoubtedly, many will fail, and indeed some have already done so. The current count of failed, folded, or merged projects totals over 25. In the next chapter, however, we will explore those whose dreams and ambitions may not reach to the stars but certainly rise to the stratosphere.

REFERENCES

[1] Anthony, S. "Antares rocket launch failure most likely caused by faulty Russian engine," November 6, 2014, http://www.extremetech.com/extreme/193129-antares-rocket-explodes-during-launch.

[2] Jayakumar, A., "ATK to merge with Orbital Sciences in $5 billion deal," April 29, 2014, http://www.washingtonpost.com/business/capitalbusiness/atk-to-merge-with-orbital-sciences-in-5-billion-deal-spin-off-sports-division/2014/04/29/59dba21a-cfb3-11e3-a6b1-45c4dffb85a6_story.html.

[3] Borenstein, S., Associated Press, May 25, 2012.
[4] Meeting with Bretton Alexander and staff of the CSF in Spring 2010 by IAASS Delegation consisting of Tommaso Sgobba and Joseph Pelton in Washington D.C.
[5] Time.com, "The Space Cowboys," February 22, 2007, http://www.time.com/time/magazine/article/0,9171,1592834,00.html.
[6] Op cit. interview.
[7] "Richard Branson," Wikipedia, http://en.wikipedia.org/wiki/Richard_Branson.
[8] RealityTV, http://realitytv.about.com/od/therebelbillionaire/a/BransonBio.htm.
[9] Virgin Group, http://www.virgin.com/AboutVirgin/RichardBranson/WhosRichardBranson.aspx.
[10] Pelton, J. N., with Marshall, P., *Space Exploration and Astronaut Safety*, 2006 AIAA: Reston, Virginia, Chapter 6.
[11] *Los Angeles Business Journal*, July 23, 2007.
[12] Ibid.
[13] "Paul Allen," Wikipedia, http://en.wikipedia.org/wiki/Paul_Allen.
[14] David, L., Space.com, January 4, 2007.
[15] Space.com, January 4, 2007, http://www.space.com/missionlaunches/070104_bezos_blueorigin_updt.html.
[16] Foust, J., "NASA Extends Commercial Crew Agreement with Blue Origin," *Space News*, November 18, 2014, http://spacenews.com/42584nasa-extends-commercial-crew-agreement-with-blue-origin/#sthash.Dllmgc6S.dpuf.
[17] "United Space Alliance and Blue Origins Announce Partnership," September 17, 2014, http://www.blueorigin.com/media/press_release/united-launch;
Academy of Achievement, http://www.achievement.org/autodoc/page/bez0bio-1.
[18] Pelton, J. N., with Marshall, P., *Space Exploration and Astronaut Safety*, 2006 AIAA: Reston, Virginia; NPR, http://www.npr.org/templates/story/story.php?storyId=5555718.
[19] *Space News*, March 26, 2007, http://www.space.com/spacenews/070326_bigelow_businessmonday.html.
[20] Ibid.
[21] David, L.,"Rules Set for $50 Million 'America's Space Prize,'" *Space News*, November 8, 2004, http://www.space.com/spacenews/businessmonday_bigelow_041108.html.
[22] Arthur C. Clarke Foundation, http://www.clarkefoundation.org.

[23] Howell, E., "SpaceX: First Private Flights to Space Station," Space.com, February 19, 2015, http://www.space.com/18853-spacex.html.
[24] *Space Safety Magazine*, June 20, 2012.
[25] Klotz, I., "SpaceX Launches Space Weather Satellite DSCOVR," Discovery.com, February 11, 2015, http://news.discovery.com/space/private-spaceflight/spacex-launches-space-weather-satellite-dscovr-150211.htm.
[26] Tesla blog, "Nevada Selected As Official Site for Tesla Battery Gigafactory," September 4, 2014, http://www.teslamotors.com/blog/nevada-selected-official-site-tesla-battery-gigafactory.
[27] "Space Tourism: Adventures in Space with Tourism Pioneer Eric Anderson," Space Daily, http://www.spacedaily.com/news/tourism-04a.html.
[28] Ibid.
[29] "Eric Anderson of Space Adventures Speaks at MIT," http://www.spaceadventures.com/press_releases/eric-anderson-of-space-adventures-speaks-at-mit/.
[30] David, L., Space.com, July 27, 2007.
[31] http://www.armadilloaerospace.com/n.x/Armadillo/Home/News?news_id=378.
[32] Parabolic Arc, May 22, 2014.
[33] Zee News, March 12, 2015.
[34] Caminiti, S., CNBC News, April 3, 2014.
[35] Wall, M., Space.com, December 5, 2013.

CHAPTER 5

The Billion-Dollar Corporations and the Other Leading Players

The billionaire entrepreneurs are not alone in pursuing the target of space travel for the masses. Some 50 companies around the world are developing new spacecraft, planning futuristic spaceports, or seeking to offer a range of "space travel" services. Many of these, however, will probably drop out sooner or later—some more likely sooner than later. In fact, about 25 such ventures, some of which were originally XPRIZE hopefuls, have already faded from the scene in just the last few years.

If you want a good mental image of the space tourism "industry" today, just think of the pioneering days of aviation at the start of the 20th century. Then play it fast forward. If you see a motley group of daredevils, wealthy industrialists, inventors, and serious engineers all mixed together, you are not far off the mark.

Some of the aspirants we have already met are billionaires who have their own capital to burn and who have carefully chosen their teams of experts. It turns out billionaires usually become wealthy because they invest their capital and energy wisely rather than foolishly. So far, the billionaire space developers have shown good returns on their aerospace ventures. But what about the established aerospace companies with tremendous technical capabilities, a large collection of patents, and a skilled work force at their command? Have they suddenly become irrelevant? They certainly don't think so.

Many of these giants in the aerospace field have been around a lot longer and will not easily give up their prime positions in the space industry. Boeing, for one, has been in the business of designing aircraft and spacecraft for many decades. The company is certainly far from happy about seemingly being overlooked in the hype generated by Richard Branson and Elon Musk, with their Virgin Galactic and SpaceX ventures. The establishment firms clearly are smarting from the publicity about "upstarts" who claim that they can do everything faster, better, and cheaper than the "guys on the block" with the most experience. Boeing,

Lockheed Martin, Northrop Grumman, and others are now seeking to carve out their own place in the field of commercial space travel.

Boeing Eyes Commercial Spaceflight

For decades, Boeing has been involved in the space programs of NASA but under the conventional approaches accorded to the largest aerospace companies. Now Boeing seems to recognize it has to up its game. Boeing is now busy developing and testing its own spacecraft, the CST-100 Crew Transportation Vehicle. This is shown in Figure 5.1, in a simulation of the capsule separating from the upper stage of an Atlas 5 launcher, which is now planned to support crewed missions to the ISS, beginning as early as 2017.

Boeing and SpaceX, as of September 2014, won out in the NASA competition to develop crewed spacecraft to deliver astronauts to and from the ISS. Initially the competitors included Sierra Nevada, Orbital Sciences, ATK, Blue Origin, and Armadillo Aerospace. The final competition came down to Boeing, SpaceX, and Sierra Nevada. But at this stage, Sierra Nevada's Dream Chaser vehicle lost out. Despite a formal protest from Sierra Nevada, the U.S. General Accounting Office ruled Sierra Nevada had lost fairly and that Boeing and SpaceX should continue. Thus, efforts are now forging ahead to develop the Boeing and SpaceX systems to replace the Russian Soyuz vehicles on which U.S. astronauts now depend. [1] In September 2014, Boeing won a $4.2 billion contract from NASA to develop a commercial crew transportation system, and SpaceX won a $2.6 billion contract to develop its Falcon 9 plus Dragon 2 capsule. These contracts are obviously not exactly comparable. Further, over a billion dollars in additional development contracts will be added into these totals. At least one estimate has concluded that the entire development costs for both ISS transportation systems will be on the order of nearly $9 billion. [2]

Fig. 5.1 The Boeing CST-100 Crew Capsule, which that would carry seven crew members.
(Courtesy of the Boeing Company.)

Exactly what Boeing's ambitions are is still far from clear. Is it to ferry astronauts to the ISS? Is it to carry cargo to space? Or does Boeing aspire to have a role in not only the nascent space tourism business but also in future supersonic and hypersonic flight? Perhaps it is all of the above. [3]

The New York Times reported that Boeing wants to develop the space equivalent of airlines, adding that NASA would then buy seats on those spacecraft to send its astronauts to the ISS. According to John Elbon, program manager of Boeing's crew transportation system, government funding is essential for the emerging space tourism industry to succeed. To date, Boeing has received two preliminary contracts totaling $111 million, and in September 2012, it won a huge new NASA contract that would total $460 million if all of the many specified milestones are finished on time.

Despite all of the significant government support, Boeing has indicated that it might actually not be able to make a profit on this ambitious undertaking and that it is concerned about being able to bring this project to fruition. The objectives of designing a capsule that could withstand 10 repeat missions that include 10 high temperature re-entries and that could sustain up to seven months of in-orbit operations are daunting indeed.

The many dozens of milestones set forth in the NASA contracts are quite difficult, and the costs associated with designing, engineering, manufacturing, and testing the capsule indeed might run to more than a half billion dollars. Thus, Boeing has explored if private partnerships could produce additional revenue streams. [4]

The Boeing Company has, in this respect, already agreed to partnerships with two of the new-generation businesses, Space Adventures and Bigelow Aerospace. In September 2010, Space Adventures announced an agreement to market commercial flights to LEO on the CST-100 spacecraft, which Boeing is planning to launch from Cape Canaveral on Atlas V. The announcement said the company foresees "a market for flying tourists, companies, government agencies and astronauts from other countries to the ISS or other destinations." [5]

Boeing has also partnered with Bigelow Aerospace with the aim of flying passengers to the BA330 inflatable space station, which is now planned to be the first commercial space habitat. Thus far Boeing's efforts seem focused on developing spacecraft capsules, but as will be noted later chapter, this aerospace giant is certainly also considering developing its capabilities in this area. The bottom line here is that Boeing has invested more money in terms of new commercial systems than any other corporation and has won the largest contracts in this area, especially with the $4.2 billion contract from NASA to provide commercial transport to the ISS.

Orbital Sciences Corporation (OSC)

OSC was selected by NASA for the second COTS contract in order to develop the Antares rocket and Cygnus spacecraft. This contract was to develop a transport system to the ISS and to carry out supply missions, with the first launch scheduled during 2013.

This reflects not only a new type of commercial environment but a changed NASA as well. In the past, NASA would have asked for research work to be done and contributed toward this R&D work. Today, under the COTS contract, NASA has made the contract payments contingent on successful cargo deliveries to the ISS as well.

A bright young man named David Thompson was not long out of Harvard Business School when he founded OSC in 1982. In the last quarter century, this company has grown from a small firm dabbling at the edges of the space industry into a billion-dollar business with a full palette of services and products. Today, OSC builds a variety of civil and military satellites for telecommunications, remote sensing, surveillance and scientific missions, and a series of small to medium-size launch vehicles. Its Pegasus and Taurus vehicles now almost dominate the low end of the market. Most recently OSC has used its Taurus II booster as the basis for creating the new Antares launch system as part of the NASA COTS program.

This Virginia-based company was NASA's choice in February 2008 to replace RpK, whose earlier contract under the COTS program was canceled for failing to meet financial milestones. As Thompson explained at a session after the award was given, this was somewhat of a mixed blessing, because this was really not so much a traditional NASA contract award as it was a joint development and service contract under which OSC had to put up its own capital and share the risk with NASA.

Fig. 5.2 David Thompson, founder of OSC. *(Courtesy of OSC.)*

Fig. 5.3 The Cygnus capsule in flight to the ISS. *(Courtesy of OSC.)*

OSC's award from NASA was reportedly worth $171 million to subsidize the R&D needed to build and demonstrate a launch system capable of delivering cargo to the ISS. Unlike the SpaceX Dragon capsule, the Cygnus was designed to burn up on re-entry. OSC's development of its Antares-Cygnus launcher and spacecraft was not only slower than originally hoped, but it has been plagued with difficulties as well. The first launch was over a year late, and the fourth delivery mission to the ISS ended in a launch failure some 15 seconds after lift-off, on October 28, 2014. After the launch failure review, it was concluded that it was due to a turbopump failure in one of the AeroJet Rocketdyne AJ-26 engines, a refurbished Russian NK-33 engine. In any event, these engines had been scheduled for replacement by an improved rocket motor design on the next Antares launch. [6]

Over the years, OSC'S experience with NASA has been a series of ups and downs—figuratively and literally. In August 1996, OSC won a contract for the development of the pilotless X-34 experimental spaceplane under a 50-month contract awarded by NASA. This low-cost demonstrator was to be dropped from a carrier jet plane to then fly up to Mach 6 speeds and to an altitude of 260,000 feet. This contract, like those of so many of NASA's X-series of craft, was canceled in midstream in 2001 for a variety of reasons, the most important likely being the huge cost demands from operating the Space Shuttle and building the ISS.

The X-34 plane still sits rather forlornly at the OSC headquarters site, but now, years later, it serves as an inspiration for future possibilities. In 2010, OSC made a commercial proposal to NASA to develop a spaceplane about one-quarter the size of the Space Shuttle, in response to the CCD2 program. This was for a vertical take-off, horizontal landing (VTHL) vehicle that would be launched on an upgraded Atlas V rocket but land on a runway. It would be able to carry four astronauts. However, this proposal was not selected. [7]

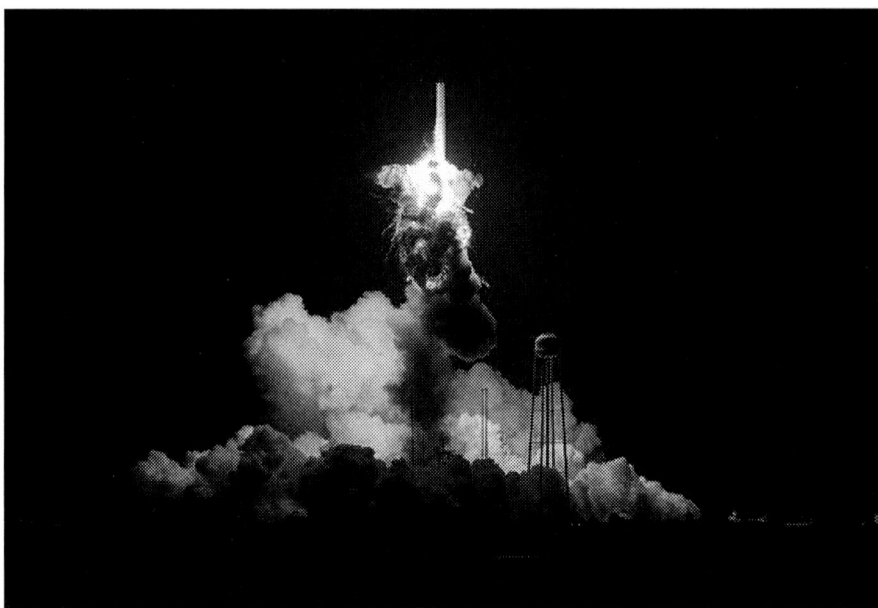

Fig. 5.4 Photo of the Antares rocket explosion on take-off, on October 28, 2014. (Courtesy of OSC.)

The next big thing for OSC was announced in early February 2015, when Alliant-ATK issued a press release revealing that it was merging with and taking over OSC. This announcement indicated that the new name would be Orbital ATK. This ever-expanding organization has grown larger and larger by acquisitions to become a very major player in the U.S. rocket industry over the past two decades. It has a recent new joint technology development agreement with Ad Astra Rocket Company and has taken over AeroJet General rocket engines, Thiokol, ATK, and now OSC. [8]

Orbital ATK

Orbital ATK (formerly Alliant Techsystems), headquartered in Arlington, Virginia, differs greatly from the startup companies seeking to enter the space tourism market. It is part of a multi-billion-dollar advanced weapon and space systems company. It has perhaps been most noted for its Thiokol Division, which developed the solid motor rockets for the Space Shuttle, and AeroJet General, which developed some of the very first rocket engines in the United States. Today the merger of Alliant-ATK with OSC to form Orbital ATK creates a whole new entity that is part aerospace industry giant and part innovative startup with a can-do spirit. Orbital ATK, after its latest mergers and acquisitions, now employs approximately 13,000 people in 22 states. Thus, its business plan and its capabilities are quite different from those of most space billionaire startup companies. [9] Orbital ATK's diverse geographical distribution helps it to have great weight in Congress with its lobbying efforts. Some have suggested that the decision that came out of the Nixon White House and the Office of Management and Budget (OMB) to add a solid rocket booster to the Space Shuttle was just one key example of this geographically diverse aerospace company's political muscle over time. Certainly what is known as the Western Governors Conference intervened in the shuttle decision-making on behalf of ATK. The original design for the shuttle was actually an entirely liquid-fueled rocket system. However, Alliant—now Orbital ATK—has continued to grow and expand in a number of areas, and its various divisions support a wide range of NASA and U.S. Department of Defense (DoD) space activities today, even after the grounding of the Space Shuttle.

In 2007, the company announced the successful completion of the on-pad assembly of its Pathfinder-ALV vehicle. This involved mating the Pathfinder with the ATK-designed launch vehicle, the ALV X-1. The flight of the ALV X-1 was part of ATK's plan to develop a low-cost launch vehicle for the operational responsive space (ORS) market for the DoD and other customers. Possible ORS program requirements include delivering small payloads to LEO in support of DoD missions—such as deploying a dedicated communications satellite to meet specific urgent battlefield requirements in an area lacking telecoms capacity. ORS requirements could also include NASA scientific missions, and commercial and university satellite programs. The rocket carried two NASA research experiments. However, the launch from the Wallops Island facility in August 2008 failed after just 30 seconds. As a result, SpaceX and its Falcon rocket program took over with the U.S. Air Force and other DoD customers from then on.

Alliant also had high hopes for its Liberty rocket and spacecraft system. According to company vice president Kent Rominger, a former astronaut and Liberty's program manager, most of the rocketry and capsule systems have been tested, and a first unmanned test flight was thought possible in 2014.

After that, it was thought that a test flight with a private crew on board could occur as early as 2016. However, this ambitious program was not selected when NASA announced the results of the latest round of contract awards in September 2012. Initially Alliant indicated that the company might continue this development as a private development but then announced that it was "moving on." [10]

NASA also selected ATK to develop a solid rocket motor launcher for the Space Launch Services (SLS) program, which was canceled as part of the Orion Program and its various launcher programs. Thus, over the past five years ATK has experienced a series of start and stop rocket development programs with NASA. This means that most of its contracts are currently with the DoD.

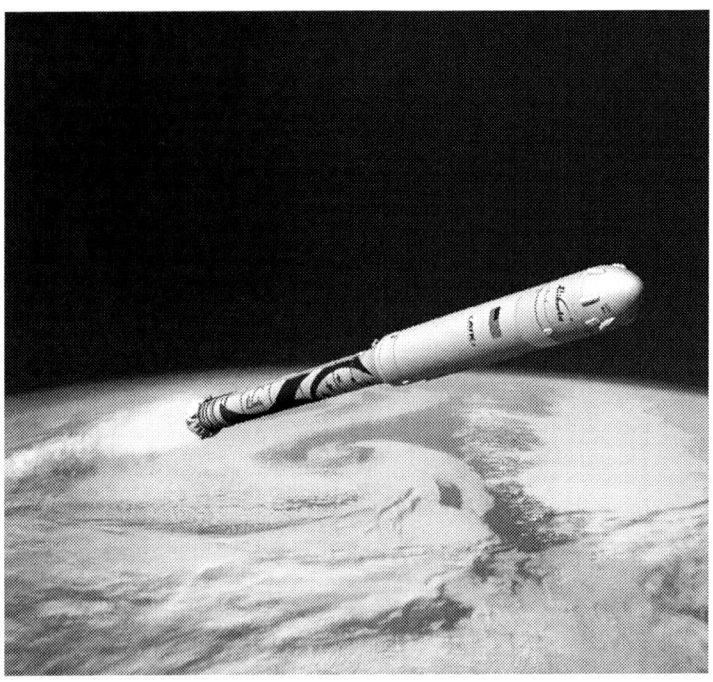

Fig. 5.5 The Liberty rocket, which remained unfunded and is now canceled.
(Courtesy of Alliant-ATK, now Orbital ATK.)

The Alliant development plans in the past few years have included the Liberty rocket and the so-called space-taxi system, which would be able to lift crew to and from the ISS. This design would have included a space capsule that would provide lift capacity for as many as seven passengers to destinations in LEO. This space taxi design would use composite materials and a design based on Alliant's efforts to create a replacement crew vehicle for the now-canceled Orion program. But none of these programs are supported by U.S. government contracts.

Despite ATK's initial ambitious effort to continue developing the Liberty rocket and the so-called space taxi, which could support both NASA and private space tourism missions, the lack of funding constituted a barrier to proceeding. If Boeing felt there were financial challenges with a $460 million contract behind its efforts, these were nothing compared to the challenges that ATK experienced and that undoubtedly led to the latest merger. The largest question at this time is whether the redesigned Antares rocket, with its new rocket motors and the Cygnus capsule, will become the main avenue forward for the newly merged organization. The Antares would give Orbital ATK a new entry in the liquid-fueled launcher arena. However, the company might view its acquisition of OSC as more in the domain of building a smaller class of commercial applications and defense satellites.

Sierra Nevada Corporation

Since the Sierra Nevada Corporation was created in 1963 in Sparks, Nevada, it has been a fast growing business that has operated as a systems integrator in high-technology areas. It is now a $1.5 billion business employing some 2500 people in six different business areas and in 30 locations across 16 states.

In December 2008, the company accelerated its move into the space industry with the acquisition of SpaceDev, founded by the late Jim Benson, one of the pioneers of commercial spaceflight with his Dream Chaser project. In February 2010, NASA awarded SNC $20 million by NASA in phase 1 of its CCD program for the development of the Dream Chaser.

Then, in August 2012, NASA announced new agreements with Sierra Nevada and two other companies to design and develop the next generation of U.S. human spaceflight capabilities—with the objective of launching astronauts into space in the next five years. [11]

This growth profile has recently hit a snag. It was a major blow when NASA decided to go with SpaceX and Boeing for future astronaut flights to the ISS and to pass over Sierra Nevada. And although Sierra Nevada will continue to be a partner in the SS2 development, another setback occurred when Virgin Galactic decided to acquire 100 percent ownership of the SpaceShip Corporation and to stop using neoprene (i.e., the hydroxyl-terminated polybutadiene ([HTPB] rubber-based system) as the solid fuel in SS2 and instead to use a slightly more powerful polyimide plastic fuel.

The Ultimate Dream Chaser

Benson founded SpaceDev in 1997 and never looked back. Indeed, it might be said that it was Benson who started the trend of successful computer and dot-com entrepreneurs moving into the space development arena. After a successful career in the computer industry, he decided to take on the challenge of starting a space commercialization venture, which combined his lifelong interests in science, technology, and astronomy with his successful business experience.

SpaceDev aimed to become the first private-sector enterprise to profitably explore and develop space beyond earth orbit. The company's mission was to help "make space happen" for all of humanity, by developing a comprehensive private space program. Some people start a company. Benson wanted to create a new vision of human development.

Under Benson's guidance, SpaceDev sold a range of innovative space products, services, and solutions to government and commercial enterprises. Its innovations included the design, manufacture, marketing, and operation of micro- and nano-satellites, hybrid orbital maneuvering and orbital transfer vehicles, as well as sub-orbital and orbital propulsion systems for human spaceflight—including the Dream Chaser vehicle. The company developed critical hybrid rocket motor technology and the "throttle-able" rocket engine design based on the innovative combination of laughing gas (nitrous oxide as the oxidizer) and neoprene rubber (as the fuel). This kind of hybrid rocket motor powered SpaceShipOne when it secured the $10 million Ansari XPRIZE in 2004.

During Benson's 10 years with SpaceDev, he served as founder, chairman, chief executive officer, and chief technology officer. He was an outspoken critic of NASA's efforts to design and build a spaceplane (such as the HL-20, the X-33, and follow-on projects).

In 2006, Benson retired from SpaceDev to start a new venture called The Benson Space Company. Shortly before leaving SpaceDev, he was quoted as saying:"I have been waiting for almost fifty years for commercial spaceflight… I have concluded that SpaceDev, through its unbroken string of successful space technology developments, now has the technical capability and know-how, along with our partners … to quickly develop a safe and affordable human spaceflight program, beginning with sub-orbital flights in the near future, and building up to reliable orbital public space transportation hopefully by the end of this decade." [12]

Benson was a founding member of the Personal Spaceflight Federation (now the CSF), and he served on the Board of Directors of the California Space Authority. He founded the non-profit Space Development Institute and introduced the Benson Prize for Amateur Discovery of Near Earth Objects. He was also Vice-Chairman and private sector representative on NASA's National Space Grant Review Panel. He was, he said, "dedicated to opening space for all of humanity."

In 2008, Benson died after a yearlong illness, and shortly afterward the SpaceDev company was acquired by Sierra Nevada and merged with another company in the group, Microstat.

Sierra Nevada Carries on the Good Work

Under the management of Sierra Nevada, the company until quite recently had continued to develop the Dream Chaser as part of NASA's Commercial Crew Development Round 2 (CCDDev-2) program. This is a sub-orbital project, derived from an existing X-Plane concept, as used in SpaceShipOne. It was planned to have an altitude goal of approximately 160 kilometers (about 100 miles), powered by a single, high-performance hybrid rocket motor, and in February 2012, Sierra Nevada announced that the development of the flight test vehicle was "on time and on budget."

Mark Sirangelo, head of Sierra Nevada Space Systems, added,

"SNC is proud to have met its schedule and cost targets in the delivery of our first flight structure as we continue to make preparations for our vehicle's first full-scale flight. The Dream Chaser program is making great strides towards developing a safe and cost-effective space system that will provide our country with the capability to safely transport crew and critical cargo to and from the ISS."

In August 2012, Sierra Nevada was one of three companies to be awarded development funding by NASA under its CCiCap program. The other two were Boeing and SpaceX. The award to Sierra Nevada was worth $212.5 million, and the NASA-stated objective was "to perform tests and work on designs for manned spaceflight between now and May 31, 2014, with the goal of sending crewed demonstration missions into LEO by the middle of the decade."

There was another major step forward in December 2012, when Sierra Nevada announced that it had received a Certification Products Contract from NASA for its Dream Chaser commercial crew orbital transportation system. Craig Gravelle, a senior director of Sierra Nevada, said, "The SNC team is thrilled with the opportunity to work with NASA to certify the Dream Chaser space system for crewed LEO flight."

As noted earlier, however, since that time things have not gone nearly as well for Sierra Nevada. NASA selected Boeing and SpaceX over Sierra Nevada for the commercial crew launch system to the ISS. A formal appeal of this NASA decision was rejected by the U.S. General Accounting Office after a review of the NASA award process. After Sierra Nevada's string of successes and very rapid growth and expansion, these recent developments will undoubtedly set back its growth and meteoric rise.

Fig. 5.6 An artist's impression of the Dream Chaser docked with the ISS – something that will likely not now be achieved. *(Courtesy of Sierra Nevada Corp.)*

The good news is that the Dream Chaser hybrid motor has shown itself to be reliable in its operation. The engine is most notably "throttle-able" by simply turning off the nitrous oxide in the case of a needed emergency shutdown. The problem is that the emissions from this type of engine are many times more toxic than those of liquid-fueled motors. It is not yet certain to what extent the new polyimide plastic fuel (which is substituted for the neoprene rubber fuel in SS2) will reduce the polluting particulates.

In January 2013, Sierra Nevada announced a deal whereby the aerospace giant Lockheed Martin will build the composite structure of the spaceship and assist with certifying the vehicle for human spaceflight.

Jim Crocker, vice president and general manager for civil space at Lockheed Martin Space Systems Co., said the Sierra Nevada partnership will leverage Lockheed Martin's expertise in human spaceflight and composite aerospace structures. Lockheed Martin will assemble the composite structure for the first space-bound Dream Chaser vehicle at NASA's Michoud Assembly Facility in New Orleans. The company will use composites developed for the F-22 and F-35 fighter planes, Crocker said.

The next step for the Dream Chaser is not clear, with Virgin Galactic taking over full control of the SS2 development and changing fuels, and with Sierra Nevada losing the competition to SpaceX and Boeing for the transport system to the ISS. [13]

Lockheed Martin's Prime Focus on Supersonic Flight

The famous bank robber named Willie Sutton was asked at the end of his notorious career why he decided to rob all of those banks. His reply was very logical and to the point: "That's where all the money was."

Lockheed Martin has similarly recognized that after a long history of close involvement in every stage of the NASA-led space industry, commercial spaceflight is now the place to be. The company had taken a more cautious approach to these new developments than others. Lockheed Martin is a part of the United Alliance with Boeing that manufactures the Atlas 5, and it does have the partnership with Sierra Nevada. However, the company has certainly been stung by the Orion vehicle cancellation and has been cautious ever since this traumatic decision.

NASA had originally selected Lockheed Martin as the contractor to build the Orion capsule, which would have been the spacecraft for returning astronauts from the moon as part of the multi-billion-dollar Constellation program. NASA canceled this program when the Obama administration reviewed it, and Lockheed Martin essentially opted out of developing new space hardware. The scaled-down version of the Orion is the CST-100 capsule now being designed and built by Boeing, as described earlier in this chapter.

Now—apart from the Sierra Nevada link, at least for now—Lockheed Martin has seemingly opted to concentrate on designing and building quiet supersonic commercial craft that would fly in the stratosphere. Although the company's business strategy is not publicly revealed, it seems as if its focus on developing craft that could carry millions of passengers in the future is in part bolstered by the idea that space tourism is a small market; on the other hand, the future supersonic and hypersonic transport market could involve a revenue stream that is much, much larger.

Northrop Grumman Is Hiding in Plain Sight

The U.S. aerospace market is today largely dominated by what might be called the big three: Boeing, Lockheed Martin, and Northrop Grumman. If Boeing is making a bid for the new commercial space transportation business, then what are the others doing? Well, Northrop Grumman is actually out there but hiding in plain sight. The $30 billion a year Northrop Grumman bought Scaled Composites, the company that Burt Rutan made famous by designing and building everything from James Bond ultralights to SpaceShipOne and SS2.

Scaled Composites has had a rather complicated life. It was first created in 1982 and was then acquired by Beech Aircraft in 1985. Raytheon, the parent company of Beech, sold ownership back to Rutan in 1988. It was then sold again, this time to Wyman-Gordon, but when Wyman-Gordon was in the process of being purchased, Rutan bought it back again, with Northrop Grumman holding a 40 percent stake, but not a majority control.

This acquisition arrangement had the key stipulation that Northrop Grumman would buy the company outright in mid-year 2007. This arrangement took place as agreed, with the deal actually being consummated on August 20, 2007. Thus, Northrop Grumman was able to complete acquisition of this iconic firm, which had so often been the top innovator in the commercial spaceflight arena. Wisely, Northrop Grumman has kept a low profile and has let Rutan and his team continue to operate in their freewheeling way out in the Mojave Desert.

Northrop Grumman has a very key role in the rapidly emerging space tourism business. It has accomplished this through Scaled Composites, and in turn with the Scaled Composites stake in the SpaceShip Corporation, which is designing and building the WK2 and SS2. Northrop Grumman, of course has other major interests and concerns in this arena. As we will learn later, Northrop Grumman has been working closely with NASA in this arena, and the company has received major R&D awards to develop new commercial craft to fly at supersonic speeds. Northrop Grumman's long-term ambitions were also apparent when it was announced in January 2013 that the company had signed a contract to design the Lunar Lander for the new Golden Spike venture (further details are in Chapter 12). This Colorado-based company is pursuing a plan to return humans to the moon by the end of the decade, and Northrop Grumman is reported to be pushing forward with a study-level review of design options. [14]

Fig. 5.7 One of the design concepts for the Golden Spike Lunar Lander.
(Courtesy of NASASpaceflight.com.)

REFERENCES
[1] Harwood, W., "Boeing, SpaceX on track for commercial crew flights to station in 2017," *Space Flight Now*, January 27, 2015, http://spaceflightnow.com/2015/01/27/boeing-spacex-on-track-for-commercial-crew-flights-to-station-in-2017/.
[2] Ibid.
[3] Sweetman, B., "Why We Don't Have an SST," *Air and Space Magazine*, August 2014, http://www.airspacemag.com/flight-today/Search-for-Quiet-SST-180952125/?no-ist.
[4] Rosenberg, Z., "Flight Not Guaranteed for Boeing's Commercial Crew Capsule," Flight Global, September 17, 2012,

http://www.flightglobal.com/news/articles/flight-not-guaranteed-for-boeings-commercial-crew-capsule-376515/.

[5] "Boeing Saves Room for Tourists in Orbital Space Taxi," Space Adventures, September 18, 2014, http://www.spaceadventures.com/news/boeing-saves-room-for-tourists-in-orbital-space-taxi/.

[6] O'Callagan, J., "Antares explosion caused $20 million worth of damage but the rocket could launch again in 2016 … without Russian engines," Daily Mail.com, December 8, 2014, http://www.dailymail.co.uk/sciencetech/article-2865351/Antares-rocket-explosion-caused-20-million-worth-damage-launch-2016-WITHOUT-Russian-engines.html#ixzz3R62kZy2h.

[7] http://en.wikipedia.org/wiki/Orbital_Sciences_Corporation.

[8] Messier, D., "ATK, Orbital Sciences Announce $5 Billion Merger," Orbital Arc, http://www.parabolicarc.com/2014/04/29/atk-orbital-sciences-announce-5-billion-merger/.

[9] Borenstein, S., AP, May 25, 2012.

[10] "Should SpaceX Fear the New Orbital-ATK?" *Space News*, February 1, 2015. http://www.fool.com/investing/general/2015/02/01/space-news-should-spacex-fear-the-new-orbital-atk.aspx.

[11] http://en.wikipedia.org/wiki/Sierra_Nevada_Corporation.

[12] "SpaceDev Announces Founder James Benson Steps Down as Chairman and CTO; Benson Starts Independent Space Company to Market SpaceDev's Dream Chaser," http://www.businesswire.com/news/home/20060928005366/en/SpaceDev-Announces-Founder-James-Benson-Steps-Chairman#.VYG54vlVhHw.

[13] http://spaceflightnow.com/news/n1301/31dreamchaser/.

[14] "Golden Spike Contract Northrop Grumman for Lunar Lander," http://www.nasaspaceflight.com/2013/01/golden-spike-northrop-grumman-lunar-lander/.

CHAPTER 6

The Challengers and the "Wannabes" in the United States

As described in the previous chapters, the billionaire entrepreneurs are not alone in pursuing the target of space travel for the masses. Some 50 companies around the world have set their sights on these new markets for developing new spacecraft, planning futuristic spaceports or seeking to offer a range of new "space travel" services.

Some of the aspirants we have already met are billionaires with their own capital to burn. Others in the mix are medium-size to large aerospace companies with tremendous technical capabilities at their command. Then there are "would-be" space tourism businesses that are true startups, all mixed in with the seriously wealthy entrepreneurs. Some of the most tenuous of these "rocketeer aspirants" are often operating on shoestring budgets from garages or university lab facilities.

These are the space enthusiasts and dreamers seeking to make their mark on history. They are avid pursuers of space commercialization, and they very often have smart websites, lofty business plans, and great vision statements. They also have a very long shot at becoming 21st-century success stories. Very, very few have the potential to become the latest version of a Hewlett-Packard or Microsoft by springing from obscurity to fame and fortune. Their goals are stratospheric, but their odds of success are longer than those of a plow horse at the Kentucky Derby. Many of them are world-class scientists and engineers who are long on hope and aspirations but short on capital.

Yet if you had tried to assess the chances that three guys, named John Parsons, Ed Forman, and Frank Malina, could have succeeded in launching the Jet Propulsion Lab from virtually nothing nearly 75 years ago, you would have given them probably lesser odds. However, two backyard tinkers without degrees and a Cal Tech graduate student did indeed successfully launch the Jet Propulsion Lab and the Aerojet General solid rocket company.

Mistaking Technical Possibility for Market Opportunity?

The key to understanding the space tourism business is to note that most members of the general public are not willing to risk their lives on "bleeding edge" technology. As our friend Courtney Stadd, formerly of NASA and head of its Department of Space Commercialization, has noted, "Space entrepreneurs … tend to be seduced by the tendency to mistake technical possibility for market opportunity.…" [1]

The sophisticated upscale clientele that is willing to pay $200,000 or more for a sub-orbital flight to the edge of space and back wants security, glitz, and pampering—but most of all security and safety. The would-be spacecraft designers, operating out of makeshift labs and workshop facilities, are perhaps thinking boldly and innovatively like today's Wilbur and Orville Wright. But these "garage tinkerers" have a long shot at providing either true breakthrough technology or reassuring safety and security features. To date, on the order of $3 billion has been invested in the new space tourism industry, and much more will be needed before this industry truly "takes off." We are years and gigadollars away from knowing if this incipient industry will truly succeed or fail. The odds will thus be on those with staying power and deep pockets rather than on those with just a dream. Those with the dream and deep pockets, like Elon Musk, Richard Branson, Jeff Bezos, Paul Allen, and Robert Bigelow, may have the best shot. However, don't count out the likes of Boeing, Lockheed Martin, and Orbital ATK. Today, the early market leaders—Boeing, Blue Origin, Lockheed Martin, Orbital ATK, Sierra Nevada, SpaceX, Virgin Galactic (despite the U.K.-supplied capital), and XCOR— are all essentially based in the United States. Then there is the Stratolauncher project, which is a sort of outlier. But one should not discount technology that is brewing in European companies such as Airbus, Bristol SpacePlanes, and Reaction Engines, Ltd., as well as those in other countries such as China, India, Japan, and Russia.

However, it is probable that fewer than a dozen or so of the ventures described here and listed in Appendix A will ultimately deliver a viable product and offer a reasonably secure service that the high-flying jet set market will embrace. Market validation—and consolidation—is the dominant name of the game.

All of these companies are striving to get there first and to stay the course. But success and the ultimate market lead are certainly not guaranteed to the wealthy entrepreneurs and the best-publicized projects. Many other companies and their ambitious leaders, as described in this chapter, believe they also have the experience, expertise, and determination to play a key role in the development of space tourism. Hope springs eternal, and so forth.

An Open Competitive Laboratory

Various companies are pursuing many technical approaches, and they are fervently seeking the most efficient and safest approach to suborbital flights and, in some cases, orbital flight. Several companies—in the United States and abroad—are indeed seeking to develop craft capable of reaching LEO rather than just the much easier parabolic suborbital flight, which achieves a relatively quick up and down.

Some of these innovators are seeking alternatives to the familiar blast-off used by NASA, ESA, JAXA, ISRO, the Russians, the Chinese, and so on, where the crew and passengers in the spacecraft sit on top of a huge quantity of explosive fuel on the launch pad and make a fiery and thunderous departure to escape the earth's gravity—not exactly a joy ride for "Citizen Astronauts."

This diversity of approach and extended trial periods may lead to duplicative actions and wasted resources, but it also may provide the clearest pathway to success. This open competitive laboratory will allow the best designs to rise to the top—literally and figuratively. We believe this competitive process will ultimately prove invaluable to the development of new and proven passenger craft for the space tourism business. Testing dozens of approaches against one another in a global marketplace seems to be a better idea than prematurely locking onto a single unified design approach. NASA's efforts to develop spaceplanes in the "conventional way"—betting on a particular design up front—have not been a great success. Quite candidly, billions of dollars were largely wasted in this manner with the now defunct X-33, X-34, X-38, and X-43 programs. Table 6.1 illustrates the diversity among the industry leaders and some of the "wannabes." In addition to the companies described earlier, some of the most interesting space tourism and space plane businesses are listed here. (Note that these are the major U.S. players; we will cover the international market in the next chapter.)

Table 6.1 Technical design approaches for U.S. private commercial space systems.

Various Approaches for Accessing Space	Companies Using This Particular Approach
Lighter than air ascender vehicles and ion engines with high-altitude lift systems providing access to LEO	JP Aerospace
Balloon-launched rockets with capsule return to ocean by parachute	PlanetSpace (now defunct)
Vertical takeoff and vertical landing (VTVL)	Exos Aerospace (formerly Armadillo), Blue Origin, Lockheed Martin, Masten Space plus the new SpaceX Grasshopper project
Vertical takeoff and horizontal landing (VTHL) (spaceport)	Aera Space Tours, PlanetSpace (now defunct), SpaceDev,. SpaceX, Sub-Orbital Corp, t/Space, TGV Rocket, and Wickman Spacecraft & Propulsion
VTHL (from ocean site)	Advent Launch Site
Horizontal takeoff and horizontal landing (HTHL)	Andrews, Scaled Composites, the SpaceShip Corporation, Virgin Galactic, and XCOR
Tow launch and horizontal landing	Kelly Space & Technology Inc.
Vertical launch to LEO from spaceport	Alliant-ATK (now Orbital ATK), Inter Orbital Technologies, SpaceHab, UP Aerospace
Launch to LEO from carrier jet drop	Triton Systems, Stratolauncher, Launcher One

XCOR Aerospace

XCOR was formed in 1999 by former members of the Rotary Rocket Company's engine development team and is based at the Mojave Spaceport. After many years of developing and producing safe, reliable rocket engines and rocket-powered vehicles, the company embarked on the development of a viable spaceplane for sub-orbital flights to support a space tourism business.

In 2004, XCOR received a reusable launch vehicle (RLV) mission license from the FAA's Office of Commercial Space Transportation (FAA/AST). The license, which is the first for a commercial RLV that is launched and recovered from the ground, was used to test RLV technologies prior to sub-orbital passenger travel. The launch license does not yet cover passenger operations, though it does allow for revenue-generating payload flights after initial tests are completed.

The company's base at the Mojave Spaceport, licensed by the FAA, was the first inland launch facility to be authorized for commercial launches. According to XCOR's Government Liaison Officer, Randall Clague, "This license covers the full flight test program conducted in a designated test area. A significant feature of the license is that it allows the pilot to do an incremental series of flight tests—without preplanning each trajectory."[2]

Fig. 6.1 XCOR's Lynx Spaceliner. *(Courtesy of XCOR.)*

In addition, the company is working in partnership with Orbital ATK to develop a low-cost LOX/methane rocket propulsion for NASA.

XCOR's president, Jeff Greason, said, "This is a wonderful opportunity for NASA and for XCOR. NASA is reaching out to small businesses, and this contract is an excellent example. Both private industry and the government will benefit from this project, as well as future users of space vehicles." [3]

In March 2008, Greason pulled a rabbit out of the ever-expanding space tourism hat by announcing amid much fanfare his new Lynx Spaceliner. This new vehicle would accommodate only a pilot and a single passenger, who would ride in the co-pilot seat and would climb to an altitude of 37 miles (about 60 kilometers)—well below the accepted definition of outer space. Nevertheless it would still constitute quite an eye-popping ride. (Astronauts are generally defined as having "gone into space" by reaching an altitude of at least 62.5 miles, or 100 kilometers.) However, Greason's "cut rate" version of a spaceliner, with an estimated cost of $10-million (versus $50 million to travel on SS2), is offering much lower passenger fares.

In early 2012, XCOR announced that it had closed on $5 million of equity financing, enough to finish building the two-seater Lynx rocketplane, and that it expected the first test flight of the Lynx to take place before the end of the year. The company also reported that nearly 500 people had signed up for rides at a mere $100,000, compared with the price of around $200,000 for a flight on Virgin Galactic. Of course, the passenger gets far less—a flight of about 30 minutes' duration and at most two minutes of weightlessness—versus a two-hour flight and four minutes of weightlessness. Clearly the more successful entrants there are in the marketplace, the more flight costs will decline. In the case of XCOR and Greason, however, it is not only the costs that have declined but the altitudes as well. Nevertheless, Greason has not given up on true sub-orbital spaceflights as the next objective. The Lynx II, with its more ambitious projected flight up to 103 kilometers with 4.5 minutes of weightlessness, is the next phase of development.

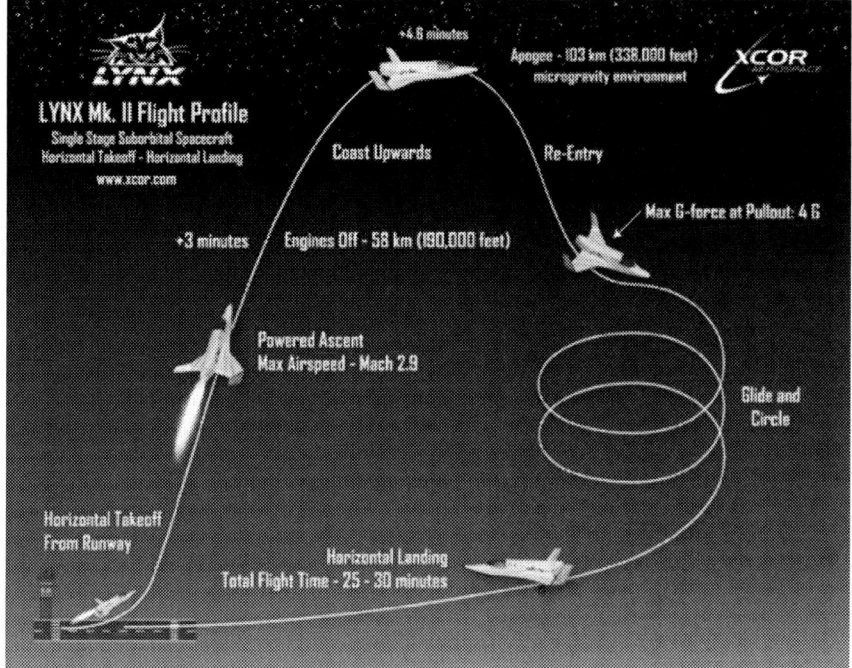

Fig. 6.2 The proposed flight profile for the follow-on Lynx Mark II. *(Courtesy of XCOR.)*

Greason is one of the more outspoken members of the CSF, but he is also one of those guys with the "vision thing." He expects the spaceplane and space tourism businesses to grow and thrive, and to evolve in new and interesting ways. His vision is that there will be vertical evolution, where different firms will develop particular expertise and competence with regard to various subsystems. Thus, some like XCOR will develop propulsion systems, whereas others will design avionics systems, or escape modules, or spaceflight suits. This model, if it proves to be true, could allow a number of today's players to evolve into major components of an overall private spaceflight industry. Greason always talks a good game. Time will tell if his vision is the one that plays out.

PlanetSpace

PlanetSpace, based in Chicago and headed by Dr. Chirinjeev Kathuria, was an ambitious project that like others played the high-risk game of spaceplane development; however, the company went defunct in 2013. The business plan was to team with other more established players to develop a rocket booster and spacecraft that could lift cargo and crew to the ISS and to develop other capabilities such as point-to-point global travel, space tourism, and satellite orbital delivery. Lockheed Martin Space Systems Company was to have developed the orbital transfer vehicles for PlanetSpace, and ATK Launch Systems (now Orbital ATK) was to have led the development of the launch vehicle segment and ground processing systems.

The group also had teamed up earlier with Lockheed Martin on a NASA contract to develop a cargo transfer vehicle system. This project was based on the use of a Lockheed Martin Athena III launcher, the latest in a series of small commercial space launch vehicles.

PlanetSpace, with a Canadian heritage, had started out with the development of the Silver Dart hypersonic glider, which was to operate as an orbital vehicle. This was a two-stage spacecraft, using the Nova launch system, and was designed to double as an unmanned or manned spacecraft. The Silver Dart was based on the earlier FDL-7 design, which is designed to achieve flight speeds as high as Mach 22. It was also planned to have a glide range of 25,000 miles (one earth circumference). Thus it could have been used to support hypersonic transportation from any point to any other point in the world. The state-of-the-art all-metal thermal protection system would have allowed for all-weather flying and was a clear improvement over the ceramic thermal protection system used on the Space Shuttle. The thermal protection system and glide range were designed to create a re-entry vehicle that could not be trapped in space and would theoretically be able to return to base from any orbit around the earth.

In 2009, PlanetSpace lost out in its bid for NASA's $3.5 billion Commercial Resupply Services Contract to service the ISS. After the awards were made to Orbital Sciences and SpaceX, PlanetSpace entered a protest to the U.S. Government Accountability Office (GAO). [4] Earlier, PlanetSpace also worked with the Canadian Arrow company, but that project was closed down in 2010. The end of these two projects was the major reason why the company could not continue and ceased operations in 2013.

Dr. Kathuria is no stranger to privately funded spaceflight nor controversy. Together with the controversial Walt Anderson, he was a founding director of MirCorp, the Russian company that made history on April 4, 2000, when it launched the world's first privately funded manned space program and signed up Dennis Tito through a deal with Space Adventures as the first space tourist (or "Citizen Astronaut"). MirCorp was a joint venture with RSC Energia. This renowned company participated in the launch of the first satellite (Sputnik), the launch that sent the first man to orbit the earth (Yuri Gagarin), and built the Mir Space Station. RSC Energia was also a major partner in the ISS development. However, the MirCorp venture did not proceed due to objections from NASA. The story of MirCorp and its untimely demise is vividly told in the film "Orphans of Apollo," produced by space and telecommunications enthusiast and entrepreneur Michael Potter. [5]

Kelly Space & Technology Inc. (KST)

KST is essentially an R&D enterprise operating out of San Bernadino International airport in California, and it specializes in environmental and dynamic testing systems for the aerospace industry. Its business model is based on having an existing and well-established line of services that provides a revenue stream for its efforts to develop spaceplane capabilities as well as rocket systems. These services, beyond environmental and dynamic testing, include ballistic testing, jet and rocket testing, laser coating removal, burn debridement, biohazard elimination, and wireless products.

KST is also pursuing one of the more innovative concepts for achieving low-cost access to space. With plans to develop a wide array of launch vehicles for both unmanned and manned missions to space, the company has developed and patented a so-called "tow launch" technology, and it started conducting trials in conjunction with the U.S. Air Force as early as 1999.

The system would employ a modified 747 airliner to tow the KST Eclipse Astroliner to at least 20,000 feet (about 6 kilometers) before separation. The Astroliner would then accelerate to speeds of up to 6000 miles/hour (9600 kilometers/hour), which would achieve an altitude of 400,000 feet (112 kilometers). At this point, the second-stage rocket would be released to fire so that the satellite payload could then be injected into LEO. This system is designed to both reduce launch risk and cost per kilogram of payload. [6]

KST is working on a wide range of spacecraft and launch systems to meet any number of needs, including satellite launch systems, ISS re-supply, and crew and cargo transport vehicles. The company's website simulation of how the KST launch system works is a beautiful thing to behold.

Fig. 6.3 The reusable KST Eclipse, towed to an altitude of 6 kilometers before release. *(Courtesy of KST.)*

Now the challenge is to get the real-world machinery to do the same trick. This project (as well as the many other similar projects, such as Stratolauncher and Launcher One)provides a solid rationale for creating an integrated process for a hybrid and integrated approach to air and space launching systems. These developments, however, provide yet additional justification for the increasing calls for the evolution of a combined air and space traffic control and management. As the number of space, protospace, and aviation activities continues to grow, a systematic system of control seems necessary for both safety concerns and international coordination. Few people know that in the case of the *Columbia* shuttle accident, the calculated chance of a collision from the shuttle debris with an in-air aircraft was estimated to be 1 percent.

The increasing prospect of drones, high altitude platform systems (HAPS), dark sky research stations, robotic flight in the so-called protozone, and hypersonic transportation will undoubtedly increase the pressure to create an integrated system for air and spaceflight traffic control and management. One indicator of the increasing interest in this subject is the workshop that the International Civil Aviation Organization convened March 18-20, 2015, in Montreal, Canada, to "learn more about this subject."

JP Aerospace—The "Other Space Program"

JP Aerospace represents a strikingly different and novel approach to the space tourism and space research market. It has designed its unique concept for lighter-than-air systems as a means to augment low-cost access to LEO. Its website describes its efforts as those of a volunteer-based organization that is dedicated to achieving affordable access to space. It is headquartered in Rancho Cordova, California, and its test program has been located at the West Texas Spaceport. [7]

The company is currently developing new technology to fly a balloon (or perhaps more accurately a lighter-than-air craft) to extremely high altitudes. The idea is that from this very high altitude it is possible to deploy an even higher dark sky station. Then from the dark sky station, it would be possible to use ion-engine propelled craft to reach LEO with a special ultra-light-weight high-altitude vehicle. The concept is as simple as what JP Aerospace calls using "the earth's atmosphere as a ladder to space." This project was inspired by the early work of the scientist James Van Allen, who was one of the pioneers in balloon-launched rockets. JP Aerospace engineers have indicated that by using this technology, they might also reduce costs and increase safety in access to LEO. Advocates of this quite logical approach start with the common-sense observation that balloons have carried people and machines to the edge of space for over 70 years. Therefore, with some creative engineering, balloons and highly efficient ion engines can accomplish the task with greater cost efficiency, less pollution, and greater safety than approaches dependent on rockets that use chemical propulsion systems.

For many years, flying an airship directly from the ground to orbit has not been considered practical for a number of technical and aerodynamic reasons. An airship large enough to reach orbit would not survive the winds near the surface of the earth. Conversely, an airship that could fly from the ground to the upper atmosphere would not be light enough to reach space. Thus, the need to reach LEO with a viable lighter-than-air craft that is structurally sound and of proven reliability and safety requires an innovative and unique architecture. This includes staging different types of craft. It also requires augmented propulsive force such as using ion thrusters in the last stage to orbit, when one runs out of atmosphere with lifting power. [8]

JP Aerospace's Innovative Three-Part System

JP Aerospace claims to have overcome these constraints by developing a very innovative three-part system. The first part is an atmospheric airship, the Ascender (not to be confused with other vehicles using this same name). It is designed to be capable of traveling from the surface of the earth to an altitude of 26 miles (42 kilometers). The vehicle is operated by a crew of three and can be configured for cargo or passengers. It is a hybrid vehicle that uses a combination of buoyancy and aerodynamic lift to fly. It is reasonably maneuverable and driven by propellers designed to operate in a near vacuum.

As a first step, the unmanned twin balloon airship flew to an altitude of 95,085 feet in October 2011, four miles higher than any previous airship; this set a new world airship altitude record.

The second part of the system is a sub-orbital space station in protospace, or "sub space." This is designed to be a permanent, crewed facility parked at 42 kilometers and is called a Dark Sky Station (DSS)—"the way station" to space. The DSS is envisioned as the destination of the Ascender, the departure port for the orbital airship, as a protospace habitat for tourists, and as a research station. Initially, the DSS will also be the construction facility for the orbital vehicle.

The third part of the system is the airship/dynamic vehicle that flies directly to orbit. To utilize the few molecules per cubic meter of gas at extreme altitudes, this craft must be of a large volume. The initial test vehicle design, therefore, is 6000 feet (1.82 kilometers) long. The reduced force and speeds of the wind at these super altitudes allow a craft of much greater fragility than the Ascender, which is designed for lower altitudes. This very large and high-volume airship uses its great buoyancy to climb to 200,000 feet (60.6 kilometers), using electric ion propulsion to gradually accelerate. As it accelerates, it dynamically climbs, and over several days it reaches orbital velocity. This approach appears to offer a number of advantages:

- All three parts of the ascension system are either long-term or multiple-reusable in their design.
- The system is scalable to handle larger-scale payloads.
- Large structures can be placed already assembled in orbit using this type of ascent system.
- The ascension system provides a safe and reliable way to access space.
- It offers low-cost access to space for significant payloads.
- It avoids large thermal gradients for re-entry. (This is a significant safety feature.)
- There are no large fuel tanks or rocket motors to explode.
- It is a "green" or "clean" system that is much more environmentally friendly than a conventional spaceplane.

On the other side of the coin, this large cross section could be hit by falling debris or micro-meteorites, or even by ascending rockets and thus its location would need to be "zoned off," and emergency evacuation would be needed in case of such an accident. Once in orbit, the airship truly becomes a spacecraft. With solar/electric propulsion, this type of vehicle could in theory proceed to multiple locations. The Ascender and DSS project has now been in development for over two decades. There have been over 80 real hardware test flights and countless development tests. One of the remarkable aspects of this project is that it is being built essentially with existing technology. However, new polyimide, kapton, or other materials—and the latest in ion engine, solar cell, and fuel cell technology—could be applied to advance the design and performance further.

Remarkably, this new way to space has not required huge capital expenditures to accomplish; each component has its own business application and funding source. Thus the JP Aerospace development program operates as a pay-as-you-go system. Clients seeking an improved high-altitude reconnaissance vehicle provided funding for the atmospheric airship. The DSS development, on the other hand, has been supported by multiple customers in the telecommunications community. JP Aerospace estimates that it is seven years from completing the entire system. This approach is almost completely antithetical to that of rocket propulsion systems, but it should not be counted out as a viable concept for the future. [9]

The Transformational Space Corporation (t/Space)

This company, based in Reston, Virginia, was formed in 2004 to respond to NASA's plans to implement the President's Vision for Space Exploration. The company was one of eight winners in NASA's "Concept Exploration and Refinement" competition to advise the agency on the best architecture for moon-Mars exploration and the best initial design for the Crew Exploration Vehicle (CEV). The effort kicked off in August 2004 with a $3 million contract that was extended in March 2005 with another $3 million.

In 2011, t/Space, working with Scaled Composites, proposed an eight-person crew or cargo recoverable reusable transfer spacecraft to NASA under the CCDev-2 program. The concept spacecraft could launch on a variety of launch vehicles, including the Atlas V, Falcon 9, and Taurus II/Antares but would involve a capsule design based on the Corona, developed by the Air Force; this has flown successfully some 100 times. However, despite all of these innovations, the t/Space proposal did not make the final cut of the four finalists in the $500 million NASA COTS program to create low-cost cargo and passenger vehicles. [10] As noted earlier, Blue Origin, Boeing, Sierra Nevada, and SpaceX were the ultimate winners of this round, with awards ranging from $22 million to $94 million to refine their design concepts. [11]

The t/Space website noted that the design philosophy behind its Crew Transfer Vehicle (CXV) was driven by "Simplicity, Survivability, and Affordability." These principles have guided both the design of the CXV and the earth-to-orbit launch system that was developed for an air launch, in conjunction with the Falcon launch vehicle system, which would serve as the second stage. According to t/Space, its design was optimized to take people into orbit safely and at a price low enough to enable development of a new market for personal spaceflight for people who wish to go beyond sub-orbital flights and actually go into LEO for multiple orbits of the earth. The cost of the t/Space system, if actually successfully implemented, would have cost, in theory, at least three times less than what NASA is currently paying to the Russian Federated Space Agency for a Soyuz launch to the ISS.

Safety considerations in the t/Space design include air launching from an altitude of 25,000 to 30,000 feet, which would provide far safer abort modes during the first few critical seconds of flight. Also, the booster utilized a simple design with fewer failure modes. The engine design has very few moving parts—there are no turbo-pumps or pressurization systems. The capsule shape was designed for safe re-entry, even if all control systems have failed, because it will automatically right itself as it descends into the atmosphere, regardless of its initial orientation. The capsule's thermal protection system was conceived as a double layer of SIRCA tiles, developed by the NASA Ames Research Center; the tiles shield the vehicle and crew from the heat of re-entry. The capsule's water landing uses proven systems from the Apollo program. Many believe today, however, that new metallic thermal protection systems are better suited for future missions.

And Some of the Rest….

Still other companies have entered into the private space business in recent years. These are very hopeful enterprises that nevertheless are generally considered to be "longer-odds wannabes." These companies include the following:

Advent Launch Services was formed in Houston, Texas, in 1999, but the Advent concept started long before that. As part of the Future Programs Office at NASA's Johnson Space Center, several individuals were studying ways to make the shuttle program more cost efficient. At the time (1990), NASA was considering boosters using liquid propellants to replace the solid propellant boosters for the shuttle system. The initial cost analysis of the shuttle system prompted the team's engineers to consider features for a launcher concept that could achieve minimum cost.

The Advent concept is a spacecraft that launches vertically from water and lands horizontally like a seaplane. It is a winged rocket designed to glide down to the ocean surface for a safe, controlled landing. It would, in theory, be safer than an airliner because it is mechanically much simpler, having fewer components, a shorter run time, and very robust mechanical parts. In fact, the Advent vehicle requires only seven simple "on" or "off" signals to fully control the propulsion system. The guidance system uses redundant control surfaces on the trailing edges of the wing for propulsion efficiency and steering, atmospheric re-entry, and aerodynamic control for gliding and landing. [12]

However, in 2008 the company's plans suffered a setback when an explosion destroyed the rocket engine designed for its prototype spacecraft. No one was hurt in the blast, which was caused by a build-up of methaneoxygen fumes in the engine's combustion chamber. Advent Launch Services of Houston, Texas, and Advent Orbital Services of Austin, Texas, are still active and still seeking to develop the Mayflower launch system. [13]

Andrews Space, Inc., located in the Seattle, Washington, area, has been in business only since 1999. Its goal has been to become a catalyst in the commercialization, exploration, and development of space. The company has developed a range of capacities in terms of small satellite buses, avionics, and torque systems, and it has demonstrated the capability to serve as an affordable integrator of aerospace systems and as a developer of advanced space technologies. It was one of six finalists for the NASA COTS contracts. Most recently Andrews Space has been awarded a $250 million-ceiling, eight-year contract by the U.S. Air Force for ground and flight experiments and demonstrations in support of the Air Force's proposed Reusable Booster System. Andrews is also developing the Spaceflight Secondary Payload System (SSPS), a secondary payload system, and the SHERPA in-space tug for Spaceflight Inc. In addition, Andrews has a range of spacecraft and launch system development efforts including the Small Agile Tactical Spacecraft for the U.S. Army SMDC. Andrews is one of several companies that has developed new space capabilities through funding that has come primarily from the Air Force rather than from NASA. Andrews has a variety of other products to support its business revenues, such as a SCOUT visible imaging microsatellite, which is capable of 1.0 meter visible imagery, and the Sentry bus platforms for a range of missions that begin with cube sats and scale up to nanosatellites, with the Sentry 4000 bus being the largest of these platforms. [14]

Interorbital Systems Corporation (IOS) is an American aerospace company based in Mojave, California. It was founded in 1996 by Roderick and Randa Millirond, who also co-founded Trans Lunar Research. The company developed a line of launch vehicles aimed at winning America's Space Prize. Interorbital Systems was also a competitor in the Ansari XPRIZE. The company successfully tested a number of rocket engines in the thrust range of 500 to 5000 foot pounds (2 to 22 kilo Newtons).

NEPTUNE launch vehicles by IOS are optimized for cube sat launches. Due to the modular nature of the launch propulsion system, everything from "tubesats" to 1-, 2-, and 3-unit cube sats can be launched by these vehicles. Currently over two dozens of these NEPTUNE launches are manifested—often for university-based research projects.

These NEPTUNE Modular Series rockets are based on concepts advanced by the OTRAG and rocket designer Lutz Kayser as proposed in the 1970s. These small rockets are designed for minimum cost and maximum reliability. In order to cut cost and improve performance, the NEPTUNE rockets do not include wings or turbopumps in the powerplant, nor do they have the architecture of IOS rockets. Since the NEPTUNE Modular Series launch vehicles are designed to be deployed from a private island launch site or from the open ocean, IOS promotional literature indicates that launch costs can be radically reduced (compared to standard spaceport fees), and launch scheduling will be based on customer demand (not on placement in a spaceport's launch rotation). The IOS modular rocket system is an evolved version of a similar system that was under development by OTRAG in the 1970s. Kayser, the former head of the OTRAG team, is an important consultant on the IOS project.

Payload capacity can be varied by increasing or decreasing the number of CPMs and varying the burn times of each CPM in the array. There can be much larger configurations, and IOS is currently registered to compete for the Google Lunar XPRIZE. IOS projects that the so-called N36 can carry a payload of 36 kilograms to the moon's surface, and the N36B is projected to be able to carry a 100-kilogram payload to the lunar surface. IOS's prototypes are all designed to be amphibious (capable of launch from land or sea). Although IOS currently operates from Mojave Spaceport, it plans to develop and implement a method for launching directly from the water without a launch platform. The vehicle is towed into position by a 33-meter (110-foot) crew boat. Then from the ocean position, a ballast is suspended from the submerged tail section to provide proper orientation and stability during launch. According to IOS, sea launches are inherently less dangerous than land launches because they reduce the chance of damage to people or property in the event of a failed launch. As a result, it is easier and less expensive to obtain a license for sea operations. In July 2007, IOS announced that Tim Reed, a Midwestern businessman and adventure traveler, was the first to purchase a ticket for a week-long orbital expedition aboard the five-passenger IOS NEPTUNE Spaceliner. The company claims that the ocean-launched NEPTUNE will be the first manned orbital launch vehicle built totally without government funding. Reed bought the first of IOS's $250,000 "promotional fare" spaceliner tickets, which come with a complete rebate of the purchase price, redeemable two years after the crewmember's flight. After these advance-purchase tickets are sold, IOS say that the price for an orbital expedition with IOS will revert to a standard fare of $2 million per passenger. [15] Crewmembers will undergo a rigorous 30-day training program before the sea-based launch out of the Pacific Ocean off the coast of California. The novel design of the NEPTUNE, a stage-and-a-half-to-orbit (SAAHTO) vehicle, allows the rocket's two spherical 21-foot-diameter liquid oxygen tanks to serve as a spacious habitat for the five-person crew during their seven-day stay in orbit. A capsule return with ocean splashdown and recovery completes their space adventure. In February 2010, IOS stated that the company was commencing the construction of the launch pad for its NEPTUNE 30 spacecraft. Since that time, IOS seems to have refocused its attention on developing the N36 and N36B launch vehicles as part of the lunar XPRIZE competition.

Masten Space Systems is a Mojave, California-based rocket company that is developing a line of reusable VTVL spacecraft, and related rocket propulsion hardware. Masten announced its intention to compete against Armadillo Aerospace in the Lunar Lander Analog Challenge, sponsored by NASA and Northrop Grumman. In October 2009, its rather crude prototype, Xombie, looked more like a lab mockup than a flight craft, and it actually came in a distant second to Armadillo Aerospace, although it still won a prize of $150,000. Then two weeks later, in the "Level 2" Lunar Lander Challenge, Masten's Xoie system beat Armadillo to win a $1 million first prize—with an average accuracy of about 190 mm on two landings. [16]

Masten's XA (eXtreme Altitude) line of sub-orbital (VTVL) vehicles represents a family of incrementally developed vehicles starting with technology demonstrators, and leading to commercially operated manned and unmanned sub-orbital launch vehicles. The XA series of sub-orbital vehicles could also help to prove—or reject—some of the technologies necessary for future orbital launch vehicles, lunar landers, and other spacecraft. The first vehicle in its line of VTOL spacecraft is the XA-0.1 (or Xaero). This vehicle probably can best be viewed as a prototype demonstrator for the testing of engines, controls, and other systems for future vehicles. It uses four liquid oxygen/isopropyl alcohol rocket engines, each of which can produce up to 500 pounds of thrust at full throttle.

The company suffered a setback in September 2011, when its Xaero sub-orbital launch and landing vehicle crashed during a test flight at the Mojave spaceport in California. The craft was coming in to land after a flight to above 3000 feet when it experienced oscillations during its powered descent. This led to a loss of control, and the craft was destroyed in the ensuing crash. The cause of the failure has not been confirmed; however, a stuck throttle valve is considered the most likely cause. A second Xaero rocket has been constructed to replace the crashed craft.

The colorful slogan for the Masten Space Systems is "Just Gas 'em and Go!" So far the company still has a long way to go. [17]

TGV Rockets was founded in Maryland in April 1997 by Patrick Bahn. He developed the company with the founding management team of Dr. Earl Renaud and Kent Ewing. TGV received an initial design study contract for a reusable launch vehicle with the U.S. Naval Research Laboratory in Washington D.C. in 2003 and has expanded this line of business since that time. Its prime customers have been the Air Force, the Naval Center for Space Technology, the U.S. Navy Research Labs, and the Oklahoma Center for Advanced Science and Technology. The company now essentially operates from its Norman, Oklahoma, location and has grown revenues to over $5 million in services.

TGV has completed the design, development, and testing of large liquid propellant rocket engine systems through test firings at NASA's rocket testing facilities. It has demonstrated new types of injector and combustion chamber design and variable throttle operation. It has also conducted detailed heat transfer measurements, computer aided simulations of test firings, and low-power-level ignition. TGV's expertise is largely centered in the design, analysis, prototyping, testing, and validation of the design of complete propulsion and reaction control systems.

TGV's pockets are far from deep, although its ambitions remain high. [18] Dr. Renaud, the Chief Operating Officer, has said that the company saw its future in developing hardware rather than just providing engineering services: "TGV wants to be perceived by industry and government as a viable, low-cost, agile alternative to the established larger aerospace companies."

In 2007, the company began ground testing its TGV-RT30 reusable, throttle-able rocket engine at NASA's Stennis Space Center. Then, in 2010, it announced that it had become the exclusive agent in the United States for test development and marketing of CHASE-10, a South Korean methane-liquid oxygen rocket engine. Again, TGV may not become a dominant supplier, but it may be able to win a viable niche in the evolving commercial spaceflight market. [19]

Triton Systems is based in Houston, Texas, ad stems from the combination of several individual consulting operations that began in the Houston area in the mid-1990s. Triton was incorporated in Texas in July 2004 and in August 2005 opened its quite modest Houston office a short distance from the Johnson Space Center. This is quite a small company that claims to have expertise consisting of aerospace specialists in flight mechanics, trajectory optimization, computer software, aerodynamics, and other related disciplines.

Triton's president and chairman, Wes Kelly, is an Air Force veteran and aerospace engineer with three decades of experience in industry. He is the developer of Triton Stellar-J launch vehicle concepts. The papers posted on Triton's website are interesting, but the last posted paper was in 2008. This suggests that the company is largely now acting as a consultant to the Johnson Space Center and is not currently developing the Stellar-J vehicle.[20]

UP Aerospace of Hartford, Connecticut, describes itself as "the world's premier supplier of low-cost space access." It launched its first unmanned SpaceLoft XL vehicle from the White Sands Missile Range and by 2012 had conducted 10 successful launches from New Mexico's Spaceport America.

The SpaceLoft XL vehicle can launch up to 110 pounds (50 kilograms) of scientific, educational, and entrepreneurial payloads into space, with an altitude capability of up to 140 miles (225 kilometers). It gained a high degree of notoriety with its April 28, 2007, launch of the ashes remains of James "Scotty" Doohan ("Scotty" from "Star Trek"), Mercury astronaut L. Gordon Cooper, and a number of others. [21] The SpaceLoft XL does not have a human-rated launch capability, and current business plans apparently are aimed primarily at launching only educational and scientific payloads. Its payload capacity is limited to (25 centimeters x 2.2 meters) (10 inches x 7 feet), and so a human payload would be more than a bit difficult for most passengers. Longer-term plans for a manned capability are nevertheless likely still being considered. [22]

Wickman Spacecraft and Propulsion (WSPC) was founded in 1981 in Casper, Wyoming, by John Wickman. The company supported Thiokol (now Orbital ATK) on the post-*Challenger* solid rocket booster field and nozzle joint redesign. It also worked with the United Technologies Chemical Systems division on the Titan recovery program. In the late 1980s, the founder teamed with Dr. Adolf Oberth to move the company into research and development.

In the 1990s, the mission of the company was expanded to include manufacturing. This goal led WSPC to explore new areas of propulsion, launch vehicle construction, and lunar base design and construction. Building on its previous work, WSPC has undertaken to develop new solid rocket propellant technology. WSPC was the first company to develop the technology for using low-cost commercial-grade materials in solid rocket motors and launch vehicles. The company claims that its "cutting-edge technologies" are now enabling it to achieve low-cost access to space with an innovative small launch vehicle and sounding rockets. Two remaining problems with many solid fuel rocket systems are that they create "dirty particulates" in the upper atmosphere, and they cannot be throttled on and off for safety reasons.

In time, new types of rocket engines that can burn materials found on the moon, Mars, and outer planets will be needed. WSPC has invented the rocket propulsion technology of mixing metal powders found on the moon into liquid oxygen and making a successful rocket engine using it. This knowledge enabled WSPC to successfully develop a rocket engine that burns carbon dioxide and magnesium powder, and this engine could conceivably be used for future Mars missions. Working on several NASA development contracts, the group has built and tested a turbojet engine that could eventually operate on Mars for durations of 30 minutes, or perhaps longer. [23]

In the summer of 2011, the company began work on Phase II of this project for NASA. In addition to its small launch vehicle launch capabilities, WSPC is also designing a high-altitude Sharp Spaceplane for the Air Force. [24]

Conclusions

The great diversity of design and engineering concepts reflected by all of the above companies is a great strength for the emerging space tourism and commercial space market. However, there is also danger in this diversity of approach. This new and emerging industry is, to an extent, only as strong as its weakest link. Small start-up enterprises with slogans like "Gas 'em Up and Go" suggest a level of safe development and technological sophistication that simply does not exist. Such safe, reliable, and low-cost access will not be achieved for many years to come.

In short, false expectations and false confidence in our ability to achieve mass travel to space are real dangers. This heady optimism went up in flames to an extent with the October 31, 2014, crash of the SS2. This fatal crash has stirred caution in the minds of would-be fliers. Overselling the safety, level of development, and maturity of this new and fragile space tourism business must be avoided.

If this new enterprise is to succeed, effective regulatory oversight and careful legislative guidelines are crucial. Patricia Grace Smith, who headed the FAA's Office of Space Commercialization until 2008 (and then joined the board of SpaceDev), emphasized that major risks are out there and cannot be ignored. "Licensing of spaceplanes is not anything we will be able to do soon," she said. [25] She also noted that the first thing they ask is for consumers to read and sign the waiver statement that unequivocally states that space tourism flights are "dangerous." George Nield, Ms. Smith's successor at the Office of Space Commercialization, has continued to express similar words of caution.

Clearly the various spaceplane companies in the United States represent a heady stew of creativity that helps to expose innovative ways to access space as well as reveal many dead ends. Entrepreneurial skills and competitive designs will certainly drive down costs over time. Competing designs and companies not only lowers cost and prices but also create a "creativity crucible" that ultimately highlights the best ideas. If we can survive the next 10 years of spaceplane and LEO rocket development and shakeout without a major disaster, the chances are that a space tourism industry—and perhaps much more—can truly emerge.

REFERENCES

[1] Space Planes and Space Tourism, GW University, SACRI, 2007.
[2] XCOR press release, April 23, 2004.
[3] Kaufman, M., "Another Firm Joins the Commercial Space Race," *Washington Post*, p. A-3, March 30, 2008.
[4] http://www.nasaspaceflight.com/2009/01/planetspace-officially-protest-nasas-crs-selection/.
[5] "Orphans of Apollo," http://orphansofapollo.wordpress.com/.
[6] Kelly Space & Technology, http://www.kellyspace.com.
[7] http://en.wikipedia.org/wiki/JP_Aerospace.
[8] http://en.wikipedia.org/wiki/JP_Aerospace.
[9] "NASA Awards Next Set of Commercial Crew Development Agreements," Release 102 (2011), http://www.nasa.gov/home/hqnews/2011/apr/HQ_11-102_CCDev2.html.
[10] Advent Launch Services,– http://www.adventlaunchservices.com/our_concept.html.
[11] "CRX Crew Transfer Vehicle," *Encyclopedia Astronautica*, http://www.astronautix.com/craft/cxv.htm.
[12] Advent Launch Services business profile, http://www.corporationwiki.com/Texas/Houston/advent-launch-services-llc/34632983.aspx.
[13] Andrews Space, https://en.wikipedia.org/wiki/Andrews_Space.
[14] "Advent Launch Services Methane-Fueled Rocket Engine Explodes," http://www.aero-news.net/index.cfm?do=main.textpost&id=db95e1c1-f136-4cf4-9afb-54e079248f40#.
[15] Interorbital, http://www.interorbital.com/.
[16] http://en.wikipedia.org/wiki/Masten_Space_Systems.
[17] Masten Space Systems, http://www.masten-space.com.
[18] TGV Rockets Inc., http://www.tgv-rockets.com/.
[19] http://www.space.com/5284-tgv-suborbital-reusable-rocket-design.html.
[20] Triton System, http://www.stellar-j.com/.
[21] UP Aerospace, http://www.upaerospace.com/.
[22] http://www.parabolicarc.com/2012/04/05/up-aerospace-conducts-tenth-launch-from-spaceport-america/.
[23] http://www.wickmanspacecraft.com/marstvc.html.
[24] Wickman Spacecraft & Propulsion Co., http://www.wickmanspacecraft.com/sharp.html.
[25] Interview with Patricia Grace Smith at Center for Strategic and International Studies (CSIS), Washington D.C., April 2007.

CHAPTER 7

The International Scene

The space tourism business is rapidly expanding. It is on the verge of becoming a true global enterprise. In the next 10 years, spaceports and space vehicle programs will circle the globe. Many of these new space tourism initiatives are commercial and entrepreneurial in nature. But in many countries, including China, India, France, Germany, and Russia, there are projects backed by governmental agencies or even funded by the military. Even in countries like France and Germany, where a private aerospace company is taking the lead, the national space agencies (CNES of France and DLR of Germany) are providing significant support.

In the United States, there is generally great enthusiasm for commercial space initiatives. Private ambitions and heroic individual efforts are clearly in vogue. In America, there is a historical feeling, remaining from past centuries and dating back to the settling of the Old West, that creative enterprise wins the day. This is sometimes simply called "Yankee Ingenuity." Americans tend to believe in rags-to-riches entrepreneurs as in the famous "Horatio Alger stories," where a poor boy from the slums could rise to great wealth and power. American literature is dotted by many heroes whose hard work and creativity let them rise to fame and fortune. Today's real life examples, like Jeff Bezos, Paul Allen, Bill Gates, and John Carmack, are thus seen as "prototypical" Americans, destined to succeed against the odds.

Ever since the exciting maiden flight of SpaceShipOne, the American public has been rooting for the private aerospace designers. The average man or woman in the street is likely to believe that these daring and entrepreneurial innovators will be able to produce a viable spaceplane better, faster, and cheaper than those at a federal agency like NASA will.

Where Private Spaceflight Is Still Science Fiction

In other parts of the world, however, the predominant vision is dramatically different. The deregulation and privatization of traditionally state-run activities elsewhere have, for the most part, been much slower than in the United States. And in this environment, it is not too surprising that the citizenry in many countries tends to believe that governments should undertake such advanced space and high-tech research missions. In many parts of the world, private space initiatives still seem like Buck Rogers science fiction. Although progress is now being made, this attitude is still found, even in parts of Europe.

Russia is a unique case unto itself. Here, the changes after the breakup of the rigidly controlled and totalitarian Soviet Union created a headlong rush toward capital markets where and when available. The most agile of the new entrepreneurs found ways to achieve privatization within the structure of formerly state-owned assets. This has been particularly true of the technologically sophisticated but under-capitalized space launcher industry.

In China and India, the government strictly oversees and controls the aerospace industry, and in the Republic of Korea there is a close symbiotic relationship between the public and private sectors. But public or private, an array of spaceplanes, spaceports, and other facilities for testing and training is mushrooming around the world. The game is afoot in Russia and in Europe—including at least France, Germany, Switzerland, and the U.K. There are also ambitious space programs in China, Canada, India, Australia, Israel, and the Republic of Korea. Spaceport initiatives can be found in other locations, such as Singapore, the Middle East, Sweden, and Spain. Although fewer than half of these various initiatives are strictly private commercial space enterprises, private money and commercial intellectual capital lie behind many of them. Each country's space initiative represents a unique story, with a varied but always interesting cast of characters.

Europe Plays Catch-up

In Europe, there is now a proactive relationship between private space ventures and the various official space agencies. All of the players are seeking to promote the rapid development of this new industry. In terms of technology, regulatory frameworks, testing, and training and simulation facilities and spaceports, Europe is currently behind the United States in the space tourism enterprise, but it is sincerely trying to play catch-up. (Virgin Galactic, Sir Richard Branson's enterprise, of course, is an entirely different species—neither American nor European—and perhaps, as such, the leader of the pack.)

Only a decade ago, topics such as space tourism and private space ventures were considered to be science fiction in Europe—somewhere on the outer fringe of thinking. Today, this activity is seen as more and more realistic and achievable. Winning the Ansari XPRIZE in 2004 was a great boost to space tourism efforts in the United States, but in Europe it strongly suggested they had made a strategic "boo boo" by not providing greater governmental support for the private sector. More than one political leader in Europe has asked whether the lackadaisical approach to space tourism over the previous decade might have been another muffed opportunity on the high-tech frontier of new commercial opportunities.

In 1996, for example, the faculty at the International Space University in Strasburg overruled a proposal by students and ISU founder Peter Diamandis to undertake its final class project on this topic. But the European faculty, at that time, saw space tourism and space commercialization as too *outré* for a serious academic research project.

ESA Recognizes the Space Tourism Market

However, by 2004, ESA had initiated a number of studies that showed the potential for developing the concept of commercial human spaceflight. Then, in July 2006, it announced a program involving private companies working on the development of crewed space vehicles for the space tourism market. This initiative by ESA was seen as a two-way process that will result in new and interesting inputs into its own launch technology program. European space officials have expressed the objective of establishing closer and ongoing links between ESA and the emerging new space tourism industry. [1]

In 2011, ESA announced a step forward in its plans for an unmanned orbital mission to be launched from Europe's spaceport at Kourou in French Guiana. This Intermediate Experimental Vehicle (IXV) was intended to reach an altitude of 450 kilometers and collect a large amount of data during various parts of its hypersonic and supersonic flight. Originally, the initial flight was planned for 2013, but this was delayed about 18 months, as is common in such development programs.

[2] The IXV was designed to test the framework of the Future Launchers Preparatory Program (FLPP). The first flight, on February 11, 2015, was successful and was completed in a nearly two-hour-long test flight that proceeded according to plan, including its touchdown by parachute in the Pacific Ocean. NGL Prime SpA was the lead company in the IXV's development. This vehicle represents the evolved technology from the earlier projects, such as the Phoenix program, Pre-X, ESA's Atmospheric Re-entry Experimental Vehicle, and ESA's Atmospheric Re-entry Demonstrator (ARD), which flew in 1998. The next step in the development will be a project called Programme for Reusable In-orbit Demonstrator in Europe, or PRIDE. [3]

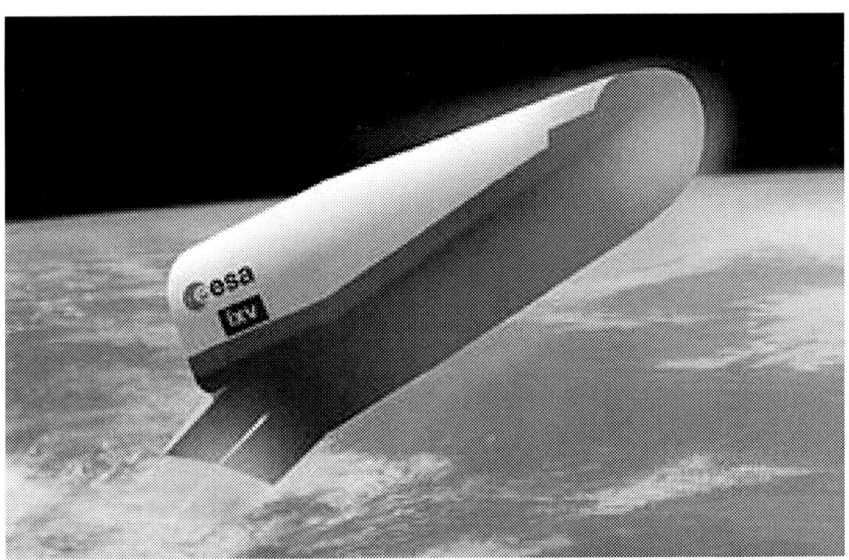

Fig. 7.1 The IXV, which flew successfully in February 2015.
(Courtesy of ESA.)

Meanwhile, ESA is also exploring what types of launch systems it will need to deploy to follow on from the successful French-led Ariane-5 launch vehicle. However, in some ways this multi-national agency is behind the national development programs of spaceplanes, and these programs have already been underway for some time in the U.K., France, and Germany.

ESA is also involved as a sponsor of Germany's Enterprise

Program, being undertaken by the TALIS Institut and DLR. In this case, the German development is also working in tandem with the Swiss Propulsion Lab. Both the Enterprise and the IXV/PRIDE efforts are aimed at developing an RLV in the form of a spaceplane.

Europe Partners with Russia

ESA is also undertaking a joint development launch program with Russia's Roskosmos agency for an Advanced Crew Transportation System (ACTS). One ultimate objective is a manned Mars mission. As part of this plan, ESA embarked on a simulated 520-day experiment during 2011; it involved a six-man crew in a mock spaceship to study the physiological and psychological effects of an 18-month round trip to Mars. [4] Also, at the European Astronaut Center in Cologne, Germany, there is a continuing program of astronaut selection, training, and medical support, and at least one European each year has taken part in missions to the ISS on board the Space Shuttle or Soyuz spacecraft.

In the spaceplane sector, ESA is also exploring the market for space tourism from a business perspective and has undertaken a significant project to help private companies get their space tourism ventures off the ground. In 2006, it launched a "Survey of European Privately Funded Vehicles for Commercial Human Space Flight." The stated purpose of this ESA study was to assess the feasibility of various businesses' mission concepts. A number of private companies already involved in planning for space tourism were invited to submit their plans to ESA, and experts in the General Studies Program selected a final three for further study. Each selected company is to receive 150,000 euros to develop their plans further. In the past, this type of hybrid, public-private undertaking would have been considered very European in nature. But NASA's COTS—funded to the tune of nearly half a billion dollars in research effort— trumps the European measure by a wide margin. In short, in the field of spaceplane development and flight to LEO, the dominant path forward these days seems to be public-private partnerships.

The survey aims to review the spacecraft designs and mission profiles critically, ensuring they are technically and financially feasible. Another key element in the ESA program is an assessment of safety standards and requirements that are seen as necessary to make space tourism viable.

Meanwhile, under its FLPP program, ESA appears to be intent to build on the national initiatives already under way, especially in France and Germany. One of the particular challenges is satisfying the stringent economic requirements of the space tourism market, which requires safe, reliable, and cost-effective vehicles.

Commercial Space Transportation from Airbus

Airbus is now the official brand name of the group that also owns the EADS consortium (including its Astrium subsidiary), which is the largest developer of satellites and rockets in Europe. Airbus thus competes with Boeing to supply aircraft to the world's airlines as well as to design and manufacture spacecraft and rockets. Although it is largely a Franco-German group, Airbus has important plants across Europe and in Britain as well. As such, it is likely to be the dominant European entry into this high-risk and capital-intensive commercial spacecraft business.

Airbus/EADS Space Transportation/Astrium, based in Toulouse, France, has been carrying out serious studies and new technology demonstrators for a spaceplane that could, at a later stage, be incorporated into the overall FLPP program. Initial flight tests of its Phoenix project were first carried out in 2004. These tests were designed to explore the final approach and landing of an RLV. Another demonstrator, Pre-X, is the intermediate experimental atmospheric re-entry vehicle to be developed as part of FLPP. Pre-X successfully passed its System Requirement Review at the end of 2005. As noted above, these development projects fed technology into the ESA IXV project. [5]

In 2007, François Auque, the executive president of EADS Astrium, gave a first glimpse of possible ambitions. He stated, "Space tourism is already emerging through sub-orbital flights, and could evolve towards orbital flights if the initial business development is successful … these flights can evolve in parallel with exploration … private sponsored contests might be organized to undertake such missions as planetary rover races, solar sail races and the like." [6]

A European Spaceplane Is Unveiled

Then, at the Paris Air Show in June 2007, Astrium, now known as Airbus-Astrium, unveiled its revolutionary new vehicle for space tourism. This spacecraft, the size of a business jet, is designed to carry four passengers 100 kilometers up into space, giving them more than three minutes of "zero G," or weightlessness. Guests at the Paris event were shown a full-sized mock-up of the forward section of the revolutionary craft including its cabin, designed by Marc Newson. [7]

Fig. 7.2 An artist's impression of the Airbus-Astrium spaceplane.
(Courtesy of Airbus-EADS-Astrium.)

According to Astrium, the passenger-carrying craft will take off and land conventionally from a standard airport using its jet engines. However, once it is airborne at an altitude of about 12 kilometers, the rocket engines will be ignited to give sufficient acceleration to reach an altitude of 100 kilometers. In only 80 seconds, the craft will have climbed to an altitude of 60 kilometers. For passenger comfort and safety, it features highly innovative seats that balance themselves to minimize the effects of acceleration and deceleration. The rocket propulsion system is shut down as the ship's inertia carries it to over 100 kilometers, where the passengers will join the very few to experience zero gravity in space. The pilot will then control the craft using small rocket thrusters to hover weightlessly for three minutes and to provide passengers with a spectacular view of earth. After slowing down during descent, the jet engines are restarted for a normal landing at a standard airfield. The entire trip will last approximately an hour and a half.

However, in June 2011, Auque said more money was needed for the project to be launched commercially. He announced a partnership with a consortium of Singapore industries that will build a small-scale demonstrator of the spaceplane. Their ambition, he said, was to compete with Virgin Galactic in the emerging high-end space tourism market. [8] EADS is committed to developing a plane that will fly at speeds up to and above 3000 miles per hour and to travel up to 100 kilometers into space at a cost per passenger of around $200,000. This commitment suggests that Virgin Galactic and the other space tourism operators may indeed have stiff competition from Europe in due course. Perhaps more significantly, Airbus, as the owner of Astrium/EADS, can be expected to be thinking about more than 30- to 60-minute trips to nowhere in their business planning. High-altitude flights to and from actual cities for corporate executives must clearly be a part of the longer-term planning process. [9]

Germany's Project Enterprise

Project Enterprise is a joint venture of the German Space Agency DLR, the German-based TALIS Institute (itself an alliance of five research laboratories and companies), the Swiss Propulsion Laboratory, and VEGA, a multi-national aerospace consulting firm. This project is also being carried out with the sponsorship and technical support of ESA. Although there is official participation by the Swiss and German governments and ESA, the development of the Enterprise spaceplane is essentially being privately funded. The Enterprise spaceplane, in theory, could be privately operated. Spaceflight operations for Project Enterprise are currently planned for a European spaceport.

The Enterprise spaceplane is intended to eventually emulate a commercial aviation charter operation. The vehicle as initially envisioned was intended to take off like a conventional aircraft and then perform a vertical climb to take it above the atmosphere, reaching a top speed of more than Mach 3. Burnout of the liquid rocket engines will occur at an altitude of 50 to 80 kilometers. The vehicle will then follow a ballistic ascent path to reach a maximum flight altitude of 100 to 130 kilometers. After burnout, the passengers will experience weightlessness for a period of –five to eight minutes, depending on the altitude reached. The craft will then re-enter the lower atmosphere for landing back on the original runway. The altitude reached will be sufficient for passengers to see from the North Pole to the Baltic Sea, including all of Central Europe, the United Kingdom, the Alps, and the northern Mediterranean Sea. [10]

The funding for this project has apparently been cut back, and as an alternative it seems that joint development activities with the Sierra Nevada Corporation in the United States with regard to its Dream Chaser spaceplane are being explored. In January 2014, a press conference was held "to announce the global expansion of its Dream Chaser Space System based on recently finalized technical understandings to identify areas of collaboration with the European Space Agency (ESA) and the German Aerospace Center (DLR)." [11]

The United Kingdom

Excalibur Almaz is a hugely ambitious private spaceflight company that plans to orbit manned spacecraft by using modernized TKS space capsules and Almaz space stations, derived from the formerly secret Soviet space program. Missions will support orbital space tourism and provide test beds for experiments in a microgravity environment. The group (based in the Isle of Man, in the U.K., with offices in Houston and Moscow) is following "a lightweight and efficient business mode, by owning its spacecraft but contracting expert services, including refurbishment, launch, control, and recovery. In 2011, the company exhibited its reusable return vehicle (RRV) and claimed that it was the only existing reusable spacecraft that had flown multiple times. [12]

The company's founders include the space law expert Art Dula and the space commercialization veteran Buckner Hightower. As the CEO, Hightower brings many years of experience of working with aerospace organizations in Russia. Before joining Excalibur Almaz in 2005, he was VP and CFO of Space Commerce Corp, which formed the first Soviet-American joint venture in aerospace. [13] Another senior executive is Colonel Vladimir Titov, a highly experienced cosmonaut who established a record by spending a year on the ISS. In June 2012,

Dula announced that the company is planning to fly the first mission to the moon since Apollo 17 in 1972. He said,"This is not space tourism; it is a real scientific expedition. They would go further than anybody has gone before in space." [14]

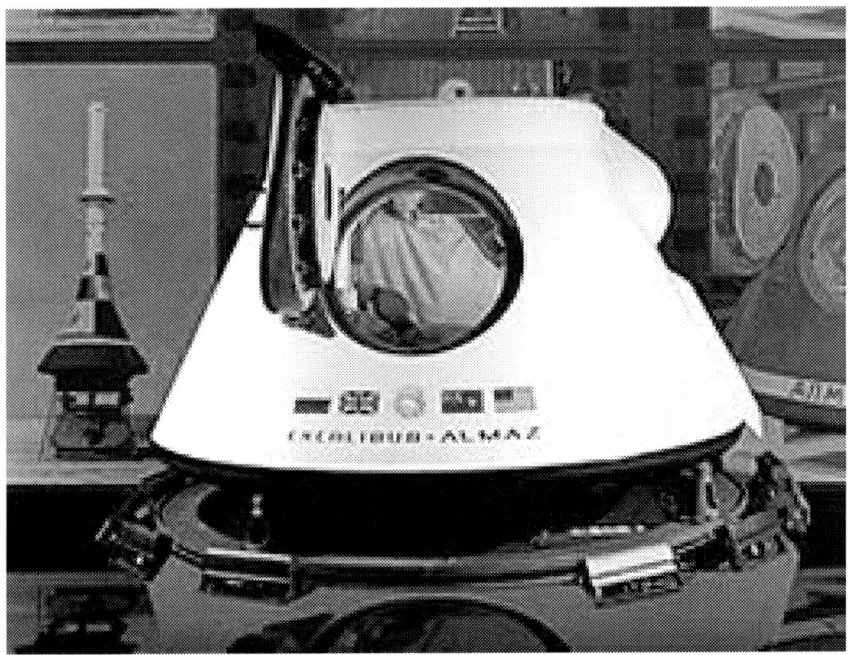

Fig. 7.3 The Excalibur Almaz two-person capsule.
(Courtesy of Excalibur Almaz.)

Not surprisingly, Excalibur Almaz also entered into a partnership with the United States-based Space Adventures in marketing its venture, and in a joint press announcement, the companies made it clear that there are as yet no plans to actually land on the moon. The plan is to launch the shuttle craft from Kazakhstan to an orbiting space station, which would then use its thrusters to achieve an orbit 1000 kilometers above the moon. Dula added, "We want to have the same kind of tradition that Britain had in the 16th and 17th centuries, when its explorers went to the ends of the earth seeking knowledge and information and bringing back wealth." These ambitious plans, however, have been recently cast in doubt due to suits by major investors. [15]

The TKS-derived space capsules that have been proposed for a lunar orbiter mission resembled in many ways the American Gemini capsules from the NASA manned space program. Unlike the two-person Gemini, however, they were designed to be reusable and could carry three passengers or operate autonomously. Recently, though, all has not gone well with this project. Major investors have sued Excalibur Almaz leadership and claimed that the existing equipment cannot perform an actual lunar mission. Until these lawsuits are resolved, the status of this project remains in substantial doubt.

Reaction Engines Ltd, a long-established U.K.-based company, has received the go-ahead from the U.K. and European space agencies in 2011 for its Skylon spaceplane, which is being developed at its facility in Oxfordshire, England. This is an unmanned horizontal take off and landing (HOTOL) spaceplane that can launch into LEO after taking off from a conventional runway. It is described as "a self-contained, single stage, all-in-one reusable space vehicle with no expensive booster rockets or external fuel tanks." The vehicle's hybrid SABRE engines use liquid hydrogen combined with oxygen from the atmosphere at altitudes up to 26 kilometers and speeds of up to Mach 5, before switching to on-board fuel for the final rocket-powered stage of ascent into LEO. The Skylon is intended to cut the costs involved with commercial activity in space, delivering payloads of up to 15 tons including satellites, equipment, and even people into orbit at costs much lower than with vehicles that use expensive conventional rockets. [16]

This development is intended to lead to the so-called A2 hypersonic jet, which might be able to fly at speeds of Mach 5 and carry up to 300 passengers. Alan Bond, managing director of Research Engines, told the BBC in April 2012 that engine tests were under way. He added, "We can reduce the world to four hours, the maximum time it will take to go anywhere. It also gives us an aircraft that can go into space, replacing all the expendable rockets we use today." [17]

In an earlier interview with the U.K.'s *Guardian* newspaper, Bond said, "The A2 is designed to leave Brussels international airport, fly quietly and subsonically out into the north Atlantic at Mach 0.9 before reaching Mach 5 across the North Pole and heading over the Pacific to Australia. The flight time from Brussels to Australia, allowing for air traffic control, would be four hours, 40 minutes. It sounds incredible by today's standards, but I don't see why future generations can't make day trips to Australasia." [18]

Fig. 7.4 Off to Australia in four hours, 40 minutes! The planned A2 hypersonic jet. *(Courtesy of Reaction Engines Ltd..)*

One of the most significant parts of the design is the same hydrogen-fueled engine that has been developed over several years by Bond and his team. It would avoid either nitrogen oxides or carbon oxides being spewed into the fragile upper atmosphere. The key element here, therefore, is the increasing interest in developing hypersonic transport craft to ultimately move people around the globe rather than just taking a quick jaunt up to peek into space. However, the design is still at the conceptual stage, and it is estimated that it might be as long as 20 years away from commercial market.

Bristol SpacePlanes was created in 1991 by David Ashford and has been involved in a series of space development projects in Britain ever since. In 2010, Ashford was appointed Team Leader for the preparation of a technology road map for a U.K. small satellite launcher as part of a U.K. Space Agency initiative.

Meanwhile, his group has continued to work on its Ascender

vehicle, which was originally planned to compete for the Ansari XPRIZE and is designed to carry a crew of two pilots plus two passengers to an altitude of 100 kilometers. It takes off on a conventional airstrip and climbs to an altitude of 8 kilometers like a normal jet aircraft, and then it ignites a rocket engine to climb to an altitude of 64 kilometers and a maximum speed of Mach 2.8. It then coasts on a steep parabolic path to a height of 100 kilometers before descending for a horizontal landing. The steep ascent and descent on the sharply defined parabolic path means that the time of weightlessness will be only two minutes.

The Ascender design is based on an earlier spaceplane study carried out for the ESA by four U.K. aerospace companies. The design is in many ways a derivative concept from the UX-15 rocket plane, and the thrust levels are actually quite comparable. The span of the Ascender is 7.9 meters and it is 13.7 meters long.

The key to the Ascender's operation for a space tourism flight is the near vertical trajectory that achieves the altitude of 100 kilometers, even though the rocket motor thrust terminates at 64 kilometers. The pullout from the returning glide path occurs at an altitude of 46 kilometers and at its maximum return speed. On the return, the pilot pulls out and gradually flattens the parabolic path to reduce the diving speed using the craft's jet engines. The conventional jet engine then takes over and allows a conventional jet aircraft landing just some 30 minutes after takeoff. The Ascender spaceplane is thus as much a jet aircraft as it is a space rocket. Bristol SpacePlanes claims that with the Ascender space frame and proven rocket engines, the vehicle could operate several flights a day with a relatively short turnaround between flights.

Progress to date has included tethered engine testing. In addition, the company has continued to work on a long-range development plan for a new capability called the Spacecab, and it is then moving on ultimately to the Spacebus. This would be an expanded and higher-passenger-capacity spaceplane, with two pilots and the ability to carry six passengers or a full crew for the ISS. Alternatively, it could carry a small satellite for orbital insertion. The Spacecab would be lifted to the upper atmosphere by a supersonic carrier jet designed to fly at speeds of Mach 2. After release by the carrier jet, the Spacecab would reach speeds of Mach 4.

The Spacebus concept is a much larger vehicle, able to carry 50 passengers to an LEO "space hotel." The first-stage launcher system would be powered by four high-thrust ram-jet engines that resemble the design used in the German Zanger (but the system avoids the need for air-breathing hypersonic scram jet technology). The launcher, with the docked Spacebus aboard, would climb to 24 kilometers and reach a speed of Mach 2. The Spacebus would then detach and fly into outer space at Mach 4. The later versions of this craft are envisioned as being able to achieve LEO and then to dock with a space station or "space hotel."

Bristol SpacePlanes says that it intends to achieve safety standards similar to those of commercial airliners. The company also estimates that the Spacebus would have operating costs equivalent to about four times those of a Boeing 747. On this basis, it projects that in coming decades the cost to orbit and return could be as low as $5000 per seat per flight (in 2015 dollars). Although this analysis is highly speculative, even a price of twice this amount would seem to be economically viable, if the independent marketing analysis conducted by the U.S. Futron Corporation is even close to estimating the size of the potential market at variously assumed pricing points. As has been seen time and time again, however, early market projections for totally new services can go seriously wrong.

Projecting demand against future time frames, future prices, and future service options with any degree of accuracy can be dicey at best. Then, when one starts to make appropriate adjustments for inflation, the number of competing companies offering space tourism services, and other key variables, one can quite easily miss the mark between profitability and market failure. The latest Futron study makes a number of key assumptions that would rather significantly impact any space tourism company's business plan. [19]

Spacefleet Ltd., created in 2004, is another private U.K. company to design and construct vehicles for the space tourism industry.

Fig. 7.5 Spacefleet's EARL project.
(Artist's impression courtesy of Spacefleet.)

In February 2013, Spacefleet announced its EARL project, a sub-orbital reusable launch vehicle that will carry payloads up to 100 kilograms and reach an apogee of not less than 100 kilometers. This spacecraft features remote piloting and is designed to carry out research functions and then return to a runway landing.

Preliminary studies have been carried out on the vehicle hull design, and a consortium of companies, based in the U.K., the Czech Republic, Ukraine, and Georgia, has been formed to build it. The company is targeting the launch market for scientific and engineering research payloads, and it has estimated a market of about 60 launches per year worldwide. EARL, it says, will be cost-competitive because for such launches it is reusable. [20]

Starchaser Industries, based in Manchester, England, originally announced that the launch of its Nova II rocket would take place in 2012 or 2013, when it would have been the largest rocket launched from the U.K. This was part of its plan for developing two concepts with space tourism as the objective. The first is a three-person reusable capsule called Thunderstar, which will be used for sub-orbital trips to a height of 100 kilometers. The second is a vertically launched eight-seat sub-orbital craft that could be upgraded for orbital applications. Both the capsule and spaceplane design feature pressurized environments where the occupants will be protected by Russian Sokol-derived spacesuits. Both will also be fitted with a launch escape system. Starchaser says that taking these features into account, the survivability from a catastrophic failure of hardware will be "close to 100 percent and significantly higher than other contemporary space tourism systems."As with many other companies, the ambitious launch dates have not been met, and Steve Bennett, the founder and chief executive of Starchaser, says on his website, "Our overall design philosophy has been to take things one step at a time. We design it, build it, test it and then learn from it. Then we can be sure of the next step."

Virgin Galactic, described in Chapter 2, is also nominally a U.K. company, but this is a special breed apart. It is neither U.K., nor U.S., nor Asian in concept and execution. But this very high-profile space tourism enterprise, funded by Sir Richard Branson's own tremendous wealth, is planetary in its vision.

Sweden could become the first European base for Virgin Galactic's passenger flights (to fly up to see the Aurora Borealis up close and personal). At a press briefing in April 2008 at Sweden's Esrange launch site, Sven Grahn, senior adviser to the Swedish Space Corp., which operates the facility in the northern Swedish town of Kiruna, said Esrange's record as a site for launches of sub-orbital rockets has established a regulatory regime in Sweden to cover third-party liability that also could apply to Virgin Galactic. To reduce the value-added tax that would be levied on Virgin Galactic operations, Grahn said the Swedish Space Corp. is investigating whether the space-tourism activity could be fitted into the same low-tax regime that covers the operations of hot-air balloons. [21]

Virgin Galactic has also indicated that it might consider operations from the U.K., using facilities at existing military bases either in Cornwall, in southwest England, or at Lossiemouth, in Scotland. [22]

Russian Space Initiatives—The "Anything Goes" Model

The widely divergent space initiatives in Russia could perhaps be best described as a free-for-all in which everyone is seeking their own part of the prize. A more kindly interpretation is to say that a public-private hybrid model is now emerging. Russian aerospace companies are working hand-in-glove with the official Russian Federated Space Agency (Roskosmos) and a host of international partners (especially from Europe, but also the United States) to develop a wide range of options.

The current geopolitical rift between the United States and Europe on one hand and Russia on the other, which is currently being played out in the Ukraine, serves as a significant barrier to cooperative space projects. The so-called "hybrid warfare" that Russia is waging in Eastern Ukraine has not yet severed cooperation on such initiatives such as the ISS, but unless a true ceasefire and a political settlement can be reached in coming months, it seems that new cooperative space initiatives will become impossible to pursue.

For several years, Roskosmos has been saying—over NASA's predictable objections—that it is happy to fly "Citizen Astronauts" to space just as long as wannabe space tourists are equally elated to pony up $20 to $25 million for each ride to space. However, these opportunities for "Citizen Astronauts" were restricted when the U.S. Space Shuttle was retired in 2011 and the Russian Soyuz became the only remaining link with the ISS.

RKK Energia Space Missile Corporation, in conjunction with Roskosmos, proposed a cooperative development with the European aerospace industry and ESA in 2006. This new Russian initiative was launched with great enthusiasm, on the 45th anniversary of the first-ever manned spaceflight by Yuri Gagarin. The initiative would be to develop the reusable Kliper Spaceship, which could, in about a decade, operate to and from the ISS. The concept was to develop by about 2015 a system that could go to the ISS and, in time, be part of a lunar transportation system in years to follow. This new spaceship would launch from both Baikonur and the Russian Northern cosmodrome at Plesetsk.

By 2011, Roskosmos was able to demonstrate a mock-up of the new spaceship at an air show at Zhukovsky near Moscow. The head of

RKK Energia, Vitaly Lopota, said that trials would begin in 2015 and that several versions of the spaceship would eventually be built: one designed for earth orbit, another for lunar orbits, and a third "to remove outdated spacecraft from orbit." [23] The Kliper system was conceived as a new capability that could integrate into the existing space infrastructure that includes the Soyuz-2 and Soyuz-3.

Finally, an upgrade of the Soyuz launchers will support the Kliper project for a six-person spacecraft to send astronauts into orbit and return them to earth on a recurring basis. Upgrades of the system might support missions to earth orbit and, in time, potentially, to the moon or Mars. Separate cargo pods could also be launched on a separate rocket, so that both the Kliper and cargo pods could then be towed to the ISS.

The anticipated funding from European partners and the ESA, however, dried up, and thus the Kliper project as first envisioned has been canceled. Nevertheless, it might be reactivated at a future date with a less ambitious design. ESA Director General Jean-Jacques Dordain explained the European decision on Kliper as follows: "It is not a question of member states for and member states against. I think the decision could not be taken for reasons that are not linked to Kliper itself. The decision could not be taken because of budgetary restraints." [24]

Myasishchev Design Bureau (MDB) has a long history of aerospace development for both advanced aircraft and space vehicles. In particular, it is famous for its strategic bombers: the 3M, the M-4, and the M-50. It was also responsible for developing the cabin for the Buran orbiter, an earlier Russian project in the 1990s, which made a series of test flights.

In the space tourism area, MDB was involved in 2006 with upgrading the Cosmopolitan XXI high-altitude jet into an operational Explorer Space Plane (the C-21) in partnership with the U.S.-based Space Adventures company. The C-21 will also operate from the MX-55 high-altitude launcher plane to create a horizontal take-off and vertical parachute landing (HTVL) spaceplane system. In addition, MDB is developing the same capability, namely the C-21, with the MX-55 high-altitude launcher plane, for another U.S. client, the Suborbital

Corporation. The second stage will be the C-21, a rocket-powered lifting body with parachute landing.

As discussed in Chapter 1, Space Adventures pioneered the space tourism business by teaming up with Russian partners, and especially Roskosmos, to deliver "Citizen Astronauts" to orbit. But by 2012, these activities appeared to be on the back burner while Eric Anderson, president of Space Adventures, pursued other parts of his varied offerings. Time and time again over the past two decades, Anderson has shown his prowess in developing not only a Plan A but also a Plan B for his company's strategic planning process.

The Chinese Space Program

China reached a milestone in its space history in June 2012 with the successful launch of three astronauts to the orbiting Tiangong laboratory, a 60-ton craft (the ISS is 400 tons). The trio included China's first female astronaut, Liu Yang, and the astronauts were scheduled to carry out a program of scientific experiments during the seven-day mission. Their Shenzhou-9 spacecraft was launched on a Long March rocket from the Jianquan spaceport on the edge of the Gobi Desert. [25]

In October 2003, China launched its first piloted spaceflight into earth orbit. Blasting off from Jianquan atop a Long March 2F rocket, a Chinese astronaut named Yang Liwei went into orbit around the earth aboard the Shenzhou-5 spacecraft, according to the official Xinhua News Agency. Chinese President Hu Jintao was at the launch site to witness the launch in person and said, "The party and the people will never forget those who have set up the outstanding merit in the space industry for the motherland, the people and the nation."

As a result, China became only the third nation on earth capable of independently launching its citizens into orbit. The former Soviet Union was first in 1961, followed by the United States in 1962. The space capsule, whose more modern design is largely based on the Russian Soyuz spacecraft, made 14 orbits and remained in space for about 21 hours before executing re-entry and a parachute landing onto Chinese soil. Experts commented at the time that China's space infrastructure, its array of launchers, its space industry, and now a piloted space mission placed the Chinese above even the Japanese, in terms of demonstrated space capabilities, and put them in the same category as the United States and Russia.

In early 2007, China's State Council approved the country's 11th five-year plan on space development. The plan covered 2006 to 2010, with manned spaceflight, lunar exploration, new launch vehicles, and high-resolution earth observation being areas of priority. The manned space program made steady progress, and in September 2011, the Chinese launched Tiangong-1, the first module to test their plans for a space station. A month later, they launched the unmanned Shenzhou-8 capsule, which docked twice with the Tiangong-1. Following on from the manned mission in 2012, more flights are scheduled over the next year or two, and a second spacelab, Tiangong-2 (some 20 tons and 14.4 meters in length), is also planned. The first stage of a permanent space station is scheduled for 2015, with completion by 2020. [26]

Programs Have Rocketed Ahead in China

China's economic, political, and technological programs have rocketed forward in figurative and literal ways over the past decade. When one of the authors was dean of the International Space University and attended the International Astronautical Federation meeting in Beijing, Vice Minister Li Pen gave a briefing to the assembled heads of various space agencies from around the world. He outlined current and planned space initiatives that China would undertake in the next two decades. He talked about communications and other application satellites, new and improved launchers, and even trips to the moon and beyond. He spoke in great depth about all of these programs and answered questions—all without any notes and with ready knowledge about both space applications and manned space programs. This was not a high-level political leader reading from a script, but a person dedicated to a mission.

China has clearly been dedicated to becoming a space power of the first rank for decades, and the results of this effort are now apparent. Since that time, China has, if anything, redoubled its efforts. Its astronauts have flown safely into earth orbit. A Chinese engineered and manufactured communications satellite has been launched to meet the domestic telecommunications needs of Nigeria. Cooperative projects have been undertaken with Brazil. Numerous applications and scientific satellites have been launched to meet localized needs in meteorology, TV, communications, remote sensing, and more. A robotically deployed project to build an observatory on the moon in cooperation with Italy and other European partners has been announced, but not actively pursued since. Lunar exploration by Chinese astronauts is no longer a vision but a serious program.

The China Academy of Launch Vehicle Technology (CALT), which develops launchers for China's space program, unveiled its road map for RLVs at the International Astronautical Congress in Valencia, Spain, in September 2006. This long-term, 20-year, Chinese launcher development plan envisaged three generations of RLVs. The first is a partially reusable two-stage-to-orbit (TSTO) launcher. This is to be followed by a fully reusable vehicle with VTHL for the second stage. The third launcher in this longer-term plan is for a fully reusable aircraft-like single-stage-to-orbit (SSTO) vehicle with HTHL. CALT research and development center manager Yong Yang, in his presentation at the International Astronautical Conference, stated, "We don't have an official schedule for the program, but the first generation could be developed in 15 years." [27]

In the short to medium term, meanwhile, China's strategy for launcher development involves cryogenic liquid oxygen/liquid hydrogen expendable launchers using a common booster architecture. The cryogenic propellants will replace the solid and other toxic fuels used in the country's existing expendable rockets.

The Chinese focus in space also includes space applications. They have a strong series of development programs that range over satellite navigation systems, remote sensing, and space communications. The interlinking of these technologies is designed to achieve what the Chinese are now calling a spatial information super highway. This will include a network of communication and broadcasting, earth resource, meteorological, navigation, and scientific experiment satellites, and so on. China hopes to involve the private sector to sell these technologies and systems to a global market and to develop the space service sector. [28]

Not too surprisingly, the officially sanctioned Chinese space development program is oriented toward development capabilities to support governmental programs, rather than to pursue commercial space tourism programs. As the space tourism business develops, however, it is possible that some of the Chinese CZ and Long March vehicles will be adapted to commercial purposes, but there are no announced initiatives in these areas at this time.

Canada's Commercial Space Program

Canadian Arrow Corporation, a privately funded rocket and space travel project based in London, Ontario, was leading the way until 2010. Led by Geoff Sheerin, Dan McKibbon, and Chris Corke, it had been considered one of the top three candidates for the XPRIZE competition before losing out to Burt Rutan in October 2004. Its two-stage vehicle was a 16.5-meter-long (54-feet-long) three-person sub-orbital rocket, with the second stage doubling as an escape system. It was designed with four main parachutes to slow the second stage for a splashdown on water at approximately 30 feet/second.

Canadian Arrow received enough encouragement from the Canadian government and private investors to continue its quest to be a leader in the new space tourism business. During 2006, the company joined forces with PlanetSpace, based in Chicago, and recruited a team of test astronauts for its spacecraft program. In August 2006, PlanetSpace/Canadian Arrow announced an agreement with the province of Nova Scotia for a 300-acre launch site and training center at Cape Breton. This agreement included financial incentives from the Nova Scotia government, with plans to transition the launch vehicle testing operation from the Meaford military installation in Ontario to Cape Breton. [29]

Plans for the new Cape Breton site called for a massive orbital launch facility that could eventually be on a scale comparable to NASA's operations at Cape Kennedy. "We're basically building a private manned space program for Canada," said Dr. Chirinjeev Kathuria, the head of PlanetSpace. [30] However, in November 2010, Sheerin, the president of Canadian Arrow, stated the company is unlikely to fly a Canadian Arrow rocket as a space tourism vehicle, and the project appeared to have been closed down.

da Vinci Project spacecraft was another Canadian initiative that was born of the 2004 XPRIZE competition. This project envisioned launching a small 18-meter-long craft to a height of some 60,000 feet via a balloon. The craft would then disconnect from the balloon and rocket to 100 kilometers. In 2006, Brian Feeney, the Canadian founder of the da Vinci Project, launched DreamSpace™, a new brand aimed at the space tourism market. The objective is to create a line of successively larger spacecraft that can provide personal spaceflights to an ever-increasing tourism base.

The technical team behind DreamSpace comes from the original da Vinci Project's team of aerospace engineers and project personnel. This team introduced the XF1 as the first of a planned line of multiple passenger spacecraft. The one-person technology demonstrator will be used to constantly broaden the spaceflight envelope, in preparation for the DreamSpace goal of space tourism. Technical details for this new ship were released at the 2006 XPRIZE Cup event at Las Cruces, New Mexico.

The XF1 employs a new liquid rocket propulsion system and aircraft turbine engine. The objective of the da Vinci Project is to develop a winged space aircraft design that can operate out of an airport-style spaceport with airline efficiency. Such objectives and visionary aims are easy to put on a website or proclaim at a news conference, but they are hard to deliver in practice.

In 2010, Feeney gave a progress report in which he said that the DreamSpace prototype could fly "as early as next year." However, he also said that he would need $2.5 million in venture capital funding to build the prototype and another $12.5 million to build two three-person spacecraft. He added that "in a young industry that's not proven, you're always running up against the 'Can you demonstrate to the venture capital community that you can manage the risk and get to the marketplace?'" [31] The needed funding was, however, not forthcoming, and, as with many other bold entrepreneurial projects, the development of DreamSpace has apparently ceased.

Australian Projects

Australia was involved in early space research and rocket development. Indeed, it was the third country to have a satellite launched—from its Woomera Range in the 1960s. However, it abandoned space activities as a national government enterprise some years ago on the basis that space capability is "not essential" to its strategic longer-term mission. The current Australian national strategy is to encourage private enterprise to develop launch capabilities instead of developing a national rocket program or governmental-operated launch facility.

This policy decision led to a variety of different activities undertaken by private institutes, universities, and research organizations. These initiatives include those of the Australian Space Research Institute (ASRI), the Australian Rocket Program (AusRoc), the Cooperative Research Centre for Satellite Systems, and the Queensland University of Technology. The AusRoc program, sponsored by ASRI and corporate and private donations, has developed the AusRoc 1, 2, and 2.5 sounding rocket programs. Efforts are also currently underway to develop the AusRoc 3 and 4 programs. The AusRoc 4 is intended to launch a small satellite of 35 kilograms into LEO.

Virtually all of these launcher developments are centered on the launch facility in Woomera, and it appears possible that a new Australian enterprise will evolve out of these activities. The range is also used by the Australian Defense Force for trials of missiles and aircraft weapons. Access is also leased to foreign militaries and private companies for testing weapons systems, rockets, and drone aircraft. Both RpK and DTI Associates signed agreements to use the Woomera launch facility for their manned spaceflight projects, but neither of these U.S.-based companies is now proceeding with their earlier plans.

At the end of 2009, up to 10 different tests were occurring at the Woomera range daily, and future bookings have been made as far in advance as 2023. This has prompted the Australian government to allocate $500 million to update systems and infrastructure at the range. [32]

Initiatives in India

The Indian Space Research Organization (ISRO) was created in 1969 and reached a series of important milestones as it developed its indigenous capabilities. Its first satellite was launched by the Soviet Union in 1975, and in 1980 it achieved its first satellite launch using an

Indian-made launch vehicle. An impressive program of launches has followed for commercial, military, and scientific users.

Meanwhile, ISRO was also developing a human spaceflight program at the Vikram Sarabhai Space Center for an RLV program (), together with new expendable launch vehicles. The Geosynchronous Satellite Launch Vehicle Mark-III (GSLV-III) will enable India to become less dependent on foreign rockets for heavy lifting.

The reusable capsule will carry a crew of three. An astronaut training center was opened in Bangalore in 2012, and the first unmanned flight of the RLV was planned for 2013, with the first manned mission to follow in 2015. This seven-day flight in LEO at an altitude of 400 kilometers will make India the fourth nation to carry out a wholly indigenous manned spaceflight, after the United States, Russia, and China. [33]

In fact, the launch of the unmanned vehicle took place in December 2014, reaching an altitude of 78 miles, and the crew capsule was successfully recovered at sea. Work is now proceeding for the first launch with two or three astronauts, but no date has yet been announced.

Longer-term plans include an upgraded version of the RLV, which will incorporate a rendezvous and docking capability. According to space scientists, India's reusable vehicle project will have a three-fold advantage: lower-cost launches, more frequent missions, and more international customers for launch services. Currently, ISRO's cost to place a kilogram of payload in orbit ranges between $12,000 and $15,000. Projected costs for the new launch capability might reduce these costs by as much as 10 times, to the level of $1200 to $1500. [34]

The Israeli Spaceplane

Even Israel attempted to be a part of the new space tourism industry as the fever to go into space spread worldwide. The Israel Space Agency (ISA) and IL Aerospace Technologies (ILAT) had plans in the past decade for the Negev-5 hybrid spacecraft, to be launched from the ground using a hot-air balloon as a first stage. After being dropped from the balloon at an altitude of 10 kilometers (6.25 miles), the spacecraft would fire a solid rocket motor for 96 seconds. At engine burnout, the vehicle would be traveling at Mach 3.4 and the crew would experience weightlessness for four minutes, reaching a maximum altitude of 120 kilometers before descending to an ocean landing by parachute. At an altitude of 5 kilometers, drogue and main chutes would deploy sequentially, and the main chute would slow the capsule down to 7 meters/second (about 24 ft/second) before splashdown.

However, in recent years there have been no reports of further progress, and it would appear that other priorities have overtaken the ISA, with almost all activity in the space sector coming from the defense industry. [35]

Republic of Korea

The Republic of Korea has been increasingly a player in the aerospace field over the past two decades. South Korea has made a solid move into the world market for automobiles and has essentially said if we can do cars, then why not rockets too. Samsung, LG Corporation, and Hyundai have all developed space capabilities, as well as governmental research labs. Projects have ranged from the Hyundai Corporation serving as a system integrator (for the Globalstar mobile satellite system) down to microsatellites undertaken by South Korean universities.

To date, activities in space have been limited to developing and building scientific and meteorological satellites, which have been launched by Japanese and Russian rockets. In a 2012 interview, Dr. Seung-Jo Kim, president of the Korean Aerospace Research Institute (KARI), outlined its ambitions to develop its own launch capability. He said Korea had recently opened its new launch site and developed its first launch vehicle, the Korean Space Launch Vehicle (KSLR). The next step is to launch a domestic communications satellite.

Korea took its first step into manned spaceflight in 2008, when So-Yeon-Yi flew on the Soyuz spacecraft to spend 10 days on the ISS. Now, according to Dr. Seung-Jo Kim, the Korean Air Force is considering the establishment of an astronaut training center. [36]

Argentina: Why Can't We Also Get into the Space Business?

Argentina's so-called Gauchito "Little Cowboy" project sought to develop a manned sub-orbital flight vehicle. This project, however, was discontinued in 2004. It was originally undertaken in response to the

XPRIZE contest. The vehicle was designed to use a hybrid rocket with a powered vertical take-off and parachute descent to a water landing. It was conceived by Pablo de León and Associates of Buenos Aires.

De León is a man of some vision. He is the co-founder and president of the Argentine Association for Space Technology (AATE) and a graduate of the International Space University. The Gauchito was envisioned as a clustered design made up of four hybrid engines producing a maximum acceleration of 3.5G. The engines would burn until the vehicle reached an altitude of 34 kilometers, followed by a coast to 120 kilometers. After five minutes of weightlessness, the pilot was to orient the thermal shield for re-entry. De León and Associates flew a subscale test capsule to 30 kilometers and conducted one-third-scale rocket test firings, but a half-scale rocket test firing resulted in a failure. Testing of flight suits and other critical systems was also conducted, but this program was discontinued after the Ansari XPRIZE was successfully claimed in 2004.

International Space Hotel Initiatives

Most international efforts have been aimed at developing spaceports or space vehicles to support the space tourism industry. But in 2008, Galactic Suites of Barcelona, Spain, announced the incredibly ambitious $3 billion initiative to deploy a space hotel. This initiative reportedly had backing from Spain, the United States, and other countries and aims to charge $4 million for a space stay. This project has a long way to go to catch up with the Bigelow initiative in the United States, but the approach Galactic Suites is taking to use a cluster of "cells" for space tourists is clearly much different than the Transhab technology approach.

Conclusions

The international dimensions of the space tourism business can only continue to grow. The demand for space travel is certainly global. The market is not restricted to any nationality, creed, or belief system. Spaceports (described in Chapter 8) may dominate the first wave of globalism, but as can be seen by the many projects described above, efforts to develop spaceplane technology and to undertake needed research will continue to expand around the world as well. Several recurring themes can be seen in this global review. These include the various ways that private enterprise and governments cooperate, collaborate, and even compete—all at the same time. In addition, areas of strategic concern involve possible defense or military use of spaceplane technology as well as issues relating to intellectual property and the U.S. ITAR regulations. The most likely space tourism ventures to succeed may very well be those that can raise capital from a global market and tap into a wide range of scientific and engineering knowledge from around the world. Virgin Galactic and Space Adventures are clearly designed as multi-national ventures that are following such a multi-national model.

Finally, there are unresolved issues about how spaceplane flights will be regulated as well as who will control and oversee the rising number of spaceports. There appear to be at least some unresolved international and national legislative and policy issues, despite actions undertaken by the U.S. Congress and the FAA to establish an initial framework for such flights in America. Prime among these issues is whether the International Civil Aviation Organization (ICAO) or some other body is in charge of international safety standards. During the three-day "Aerospace Symposium," held from March 18 to 20, 2015, ICAO brought together experts from around the world to consider the safety, regulatory, and traffic control issues related to space tourism and planned space transportation systems. Despite this initiative, it is far from clear what international regulatory systems will evolve for space and near-space (protozone) air traffic control and management.

Just as the diversity of technical approaches creates a global laboratory against which one might test the best spaceplane technology, the international community appears to be forming a sort of "international policy framework" against which space tourism safety and regulatory controls will be tested as well. With literally thousands of "Citizen Astronauts" on the verge of flying at least into sub-orbital space within the next few years, some of these safety and regulatory control issues will need to be answered sooner rather than later.

Further, the pending entry of Airbus/EADS/Astrium as well as U.S. aerospace companies into the future high-altitude transport business realm suggests that space tourism may yet give birth to high-altitude, supersonic air transport for corporate executives and other high-rolling jet setters. This has several implications. The most serious of these may be the air pollution factor for the stratosphere and how this new industry may give rise to issues that go beyond global warming. Destruction of the ozone layer in terms of genetic mutation and cancer might prove, in the nearer term at least, to be even deadlier to humans than the rise of carbon pollutants. This is an international issue that deserves paramount attention. It is ironic that Sir Richard Branson, who has offered top prize money to address global warming issues, may be the father of a new industry that could be a source of deadly atmospheric pollution. Go figure.

REFERENCES
[1] ESA General Studies Programme, http://www.esa.int/esapub/br/br263/br263.pdf.
[2] *The Register*, June 22, 2011.
[3] "ESA Experimental Spaceplane Completes Research Flight," February 11, 2015, http://www.esa.int/Our_Activities/Launchers/IXV/ESA_experimental_spaceplane_completes_research_flight.
[4] *Discovery News*, August 2011.
[5] Phoenix Reusable Launch Vehicle, http://www.spacefuture.com/archive/phoenix_m_a_small_ssto_launch_vehicle_for_commercial_space_transport_missions.shtml.
[6] Taher, A., "Europe Joins Space Tourism Race," *Times of London*, June 10, 2007.
[7] EADS/Astrium press release, June 13, 2007.
[8] http://www.flightglobal.com/news/articles/paris-eads-astrium-looking-for-funds-to-launch-spaceplane-358289/.
[9] Taher, A., "Europe Joins Space Tourism Race," *Times of London*, June 10, 2007.
[10] http://www.youtube.com/watch?v=oeA084fPgrc.
[11] "Sierra Nevada to Outline Cooperation with ESA, DLR on Dream Chaser," January 7, 2014, http://www.parabolicarc.com/2014/01/07/sierra-nevada-outline-cooperation-esa-dlr-dream-chaser/#sthash.djOOorJ9.dpuf.
[12] "Excalibur Almaz—Now What?" NASA Watch, nasawatch.com/archives/2012/07/excalibur-almaz-2.html.
[13] http://www.parabolicarc.com/2011/05/11/excalibur-almaz-usa-appoints-

buckner-hightowner-ceo/.
[14] *Daily Telegraph*, June 20, 2012.
[15] "NASA and Excalibur Almaz Inc. Complete Space Act Agreement," http://www.nasa.gov/exploration/commercial/crew/eai_spaceact_complete.html. Also see http://www.parabolicarc.com/2012/09/15/investor-sues-art-dula-excalibur-almaz-for-alleged-fraud/.
[16] http://news.discovery.com/space/skylon-spaceplane-budget-approval-110526.html.
[17] *The Guardian*, February 7, 2008.
[18] http://travel.aol.co.uk/2012/04/29/Skylon-spaceplane-can-fly-anywhere-in-four-hou/.
[19] Bristol SpacePlanes Ltd., http://www.bristolspaceplanes.com/projects/spacebus.shtml.
[20] http://spacefleet.co.uk/info.htm.
[21] SpaceNews.com, April 1, 2008.
[22] Hoffman, C., "Now Boarding," *Wired*, pp. 152-153, June 2007.
[23] http://en.rian.ru/russia/20110407/163417171.html.
[24] "Europe unites over space budget," *Nature*, May 12, 2005.
[25] http://www.bbc.co.uk/news/science-environment-18458544.
[26] http://en.wikipedia.org/wiki/Chinese_space_program.
[27] Coppinger, R., "First RLV by 2020?" Flight Global, October 17, 2006, http://www.flightglobal.com/articles/2006/10/17/210002/first-rlv-by-2020.html.
[28] Bhonsle, R.K.,2007 NewsBlaze, *Daily News*.
[29] "Cape Breton joins space race: N.S. signs private launch facility deal," *Toronto Star*, August 16, 2006.
[30] PlanetSpace, http://www.planetspace.org.
[31] http://www.cbc.ca/news/technology/story/2010/11/12/spaceflight-x-prize-feeney-da-vinci.html.
[32] http://en.wikipedia.org/wiki/Woomera_Test_Range.
[33] http://en.wikipedia.org/wiki/Indian_Space_Research_Organisation.
[34] http://en.wikipedia.org/wiki/Geosynchronous_Satellite_Launch_Vehicle_Mk_III
[35] http://www.haaretz.com/print-edition/features/israel-falling-behind-in-space-research-warns-outgoing-agency-chief-1.390151.
[36] http://www.jaxa.jp/article/interview/2012/vol70/p2_e.html.

CHAPTER 8

The Spaceport Stampede

The prospect of private spaceflight has led to almost a glut of so-called spaceports in the United States and around the world. By 2015, more than 50 launch sites in 37 countries were listed as certified or intended spaceports. [1] But pride of place as the first new facility constructed specifically for "Citizen astronauts" is Spaceport America in New Mexico, opened by Sir Richard Branson of Virgin Galactic in 2012.

The reported cost of the spaceport is $209 million, and at the official ceremony, as Branson stood alongside his SS2, he said, "The building is absolutely magnificent. It is literally out of this world and that's what we are aiming at creating. We are ticking the final boxes on the way to space." [2]

Fig. 8.1 New Mexico Governor Bill Richardson with Richard Branson at the 2012 opening of Spaceport America. *(Courtesy of Virgin Galactic.)*

Elsewhere in the United States, there are seven federally operated spaceport facilities: the Cape Canaveral Air Force Station; Edwards Air Force Base; NASA's Kennedy Space Center; the Kwajalein Island tracking and telemetry facilities; Vandenberg Air Force Base; the Wallops Island, Virginia, facility; and the White Sands Missile Range. In addition six fully federally licensed spaceport facilities have considerable capability and are available to support private space vehicle programs. These include the California Spaceport, the Kodiak (Alaska) Launch Complex, the Poker Flats Alaska Launch Site (which is just for sounding rockets and unmanned polar orbiting satellites, not intended to be licensed for spaceplane operations), the Mid-Atlantic Regional Spaceport (MARS), the spaceport operated by the Florida Space Authority at Cape Canaveral, and the Spaceport America facility located near Upham, New Mexico.

The List Goes On and On

Then, add to these the growing number of regional facilities in the United States. These include Brazoria County, Texas; the Willacy County, Texas, facility; Spaceport Alabama in Baldwin County, Alabama; Spaceport Washington in Grant County, Washington; the West Texas Spaceport in Pecos County, Texas; and the Wisconsin Spaceport in Sheboygan, Wisconsin. Recently proposals have come forward for operations in Colorado, Hawaii, Indiana, and Wyoming. Beyond the United States, numerous launch facilities are operated by the ESA and by CNES of France. China has three state-run facilities, as detailed later, and other launch sites are operated by the national space agencies of the Russian Federation, the Ukraine, Japan, and India. And then there are new spaceport projects in several other countries including Australia, Canada, Israel, Malaysia, Singapore, Spain, the U.K., and the United Arab Emirates (detailed later as well). Even in the unlikely event that *all* the planned and potential new space tourism operators are successful in meeting their projected timetables and passenger forecasts, however, it is difficult to see how most of these new facilities could operate profitably. The thought process of many of those developing, promoting, or backing these projects must be that of "loss leaders." These backers seem to be thinking that there is money to be made in other ways. For example, some American states see a spaceport as a job magnet for high-paying "green jobs" that will come as aerospace, computer, and other high-tech companies flock to be associated with this exotic new business. State economic development units are eager to not be left behind and to present themselves as forward thinkers and "high flyers" in the new frontier of space. These states—which include Alabama, Alaska, California, Colorado, Florida, Hawaii, Indiana, New Mexico, Oklahoma, Texas, Virginia (in cooperation with Maryland), Washington, Wisconsin, and Wyoming—have not only supported the creation of one or more spaceports but in several cases have enacted new legislation to bring state rules in line with federal laws concerning liability claims and public safety rules.

The Theme Park Model

Others have a more elaborate concept of what a spaceport complex is all about. Space Adventures, the leader in space tourism to date, has focused its planning on four sites—in Florida, Las Vegas (the location from where the ZERO-G parabolic flights operate), the United Arab Emirates, and Singapore. The two U.S. sites Florida and Nevada) are in top tourist destinations that are already associated with vacationers who go there not only to relax, but also to be entertained, let off steam, and experience something new and different. At a typical theme park, a family of four will spend hundreds, if not thousands, of dollars within the amusement facility. In terms of a business model, it is probably important that at the same time they will probably spend thousands of dollars at local hotels, restaurants, and retail shops.

As an "entertainment enterprise," the spaceport business case becomes ever more complex and challenging. Hollywood has actually had to re-invent its business model several times over in the past few decades. Movie studios such as Universal make less money on blockbuster movies like "Jurassic Park" and "Lost World" than on the associated rides and associated hotel and restaurant businesses at their amusement parks.

Consider the basic economics. Someone going to a movie spends about $10 to $15 (or perhaps pays $4 for a rental). Someone buying a "Lost World" computer game might spend $60. But someone going to an amusement park, staying at a hotel, and eating at restaurants, paying for parking, and shopping for a new swimsuit and souvenirs may easily drop a thousand dollars or more. You do the math. The movie studios already have. The successful "space park/spaceport" operators will try to do the same—that is, to turn the space ride experience into a reliable and large revenue stream.

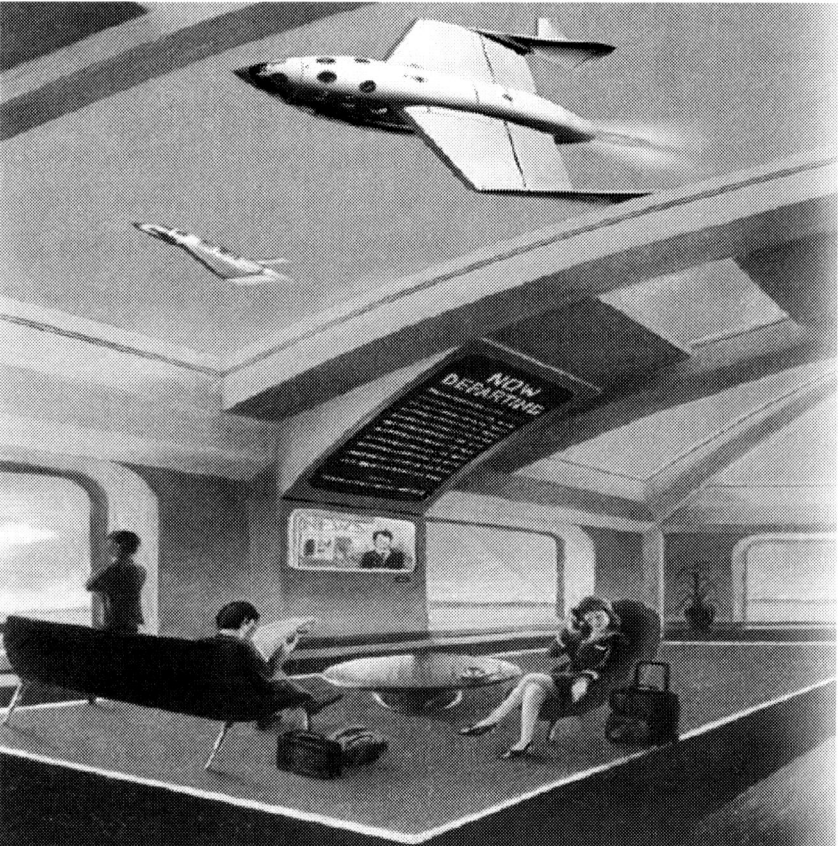

Fig. 8.2 An artist's conception of "Citizen Astronauts" waiting to fly into space. *(Courtesy of Futron.)*

The Disney Model and Much More

Eric Anderson of Space Adventures, no dummy when it comes to inventing new entrepreneurial business models, has already given this issue some serious thought. He feels that he might be able to franchise high-tech spaceports that are a new type of 21st-century amusement park. Two facilities are already under serious planning, in Singapore and the United Arab Emirates, where potential partners have some very deep pockets. These will look and feel like a totally new animal—part air and spaceport, part Disney World Epcot Center, part game arcade, part hotel, part restaurant, and part shopping mall.

These new spaceports will have lots of interesting features. The elaborate and capital-intensive facilities could easily include space simulation facilities, space training and fitness centers, as well as take off and landing terminals for spaceflights. But this is just to get started.

Closely associated with the spaceport might well be a space amusement park with rides and experiences focused on the future and space, with lots of corporate tie-ins and product placement opportunities. Although the high-end space tourists are expected to pay $200,000 to $250,000 per ride and then spend perhaps another $50,000 on "extras," a much larger "gate" of family tourists and fun-seeking vacationers could be attracted to a glittering spaceport for some low-altitude rest and relaxation and a few thrills that will cost a lot less. What are these possibilities?

Possible options are already percolating through the fertile minds of XPRIZE executives such as Peter Diamandis. You can take one of Diamandis's ZERO-G flights for under $4,000. This high arc in the sky can let fun seekers experience about 40 seconds of weightlessness. This can be extended to minutes of floating in space if you sign up for multiple parabolic arcs, such as what Professor Stephen Hawking took. ZERO-G is closely aligned with Space Adventures and operates from both Florida and Las Vegas. But far more than parabolic rides in a jet are available. There are the XPRIZE space demonstrations and potential "space races" such as those already planned to be staged in Las Cruces, New Mexico. The NASA Centennial prize competitions are already part of the "rocket to the future" fun.

Fig. 8.3 An artist's impression of space races in Las Cruces, New Mexico.
(Courtesy of XPRIZE Foundation.)

Setting Safety Rules for Spaceports-plus

Meanwhile the FAA is figuring out the safety controls that need to apply to all of these activities. The regulators are trying to see how best to combine serious, highly explosive rocket systems on one hand with tourism, entertainment, and performances geared to mass market audiences on the other. There is a comparison with NASCAR racing, which brings excitement and close-up views of high-octane race cars locked in dangerous combat. Protective barriers and other precautions offer reasonable protection to the grandstand audiences. Pit crews, race officials, firefighters, and other racetrack workers have exposure to crashes and other risks; but fatalities, other than among the race drivers themselves, have been few and far between. The FAA officials are not sure whether NASA and Air Force guidelines for launch operations are the best point of departure for "space races and shows" or whether NASCAR races should be the starting point for regulating safety. To date the FAA has been trying to set safety guidelines that leverage off those for air shows as well as the best regulations from past air and space programs.

This much is clear: Various "spaceport developers" are seriously planning to expand the facilities into space-related leisure and recreation centers along the lines of Epcot or even Disney World. The list of possible offerings only continues to expand:

- Air and space museums
- Elaborate educational and scientific exhibits
- Thrill-a-minute simulated spaceflights for the public
- SpaceX, NASA, and other "challenge prize competitions" to test or prove the capability of new state-of-the-art lunar and Mars landers, new rocket planes, and scram jets demonstrations
- Parabolic flights to achieve weightlessness
- Rides in extremely high-altitude jets
- High-G centripetal force centrifuge rides and astronaut training activities
- Potentially, space rocket races within "marked courses" in the sky

Spaceports or Recycled Airports

As already mentioned, a number of U.S. States that are bent on economic development and seeking a new high-tech cachet have latched onto what they see as a very appealing idea. Their plan is to spruce up discontinued airfields into shiny new facilities. They hope to turn their down-at-the-heel airport Cinderellas into shining spaceport princesses.

Already a burgeoning—and perhaps disturbing—number of decommissioned U.S. Air Force bases and shut-down airports are being re-invented as spaceport facilities. This, at first sight, seems a great idea. It provides a new use for abandoned or under-utilized locations. Further, these locations already have a great deal of expensive infrastructure in place, and it gives a high-tech flair to proclaim it as a spaceport. Advocates see the opportunity to attract cutting-edge industry, new jobs, and an infusion of scientific talent. The enthusiastic economic projections that were made in the market study in New Mexico have caught the eye of politicians in a number of other states.

There is, unfortunately, a flaw in the thinking. This is simply that only a very limited number of space tourism companies will succeed. The proliferation of spaceports will soon outnumber the potentially viable spaceplane ventures, perhaps five to one. As Burt Rutan has bluntly observed, a spaceport without any flights is probably going to fail.

The transformation of old airports into spaceports is one thing, but making them successful enterprises is another. Renovating these facilities and transforming them into high-tech, exciting "amusement parks" is a gigantic undertaking. Just to calibrate such an effort, the Walt Disney Corporation's creation of a new park outside Tokyo—more than a decade ago—involved a capital investment of $4 billion. The creation of a spaceport and a "space park" takes time and requires environmental impact statements, extensive planning and engineering, and potentially billions of dollars of investment over a period of many years. Such an effort involves not only planning, building, and staffing the "space amusement park facilities" but also creating the rest of the infrastructure—from hotels and restaurants to water and sewer systems, and so on.

Regulations and Licensing by the FAA

Since 1996, FAA/AST has been responsible for licensing the operations of non-federal launch sites in the United States. These new spaceports are expected to serve both commercial and, potentially, government payload operators. According to the FAA/AST 2007 report "Commercial Space Transportation Developments and Concepts, Vehicles, Technologies, and Spaceports," about $200 million had been invested into non-federal spaceports across the United States. In the past eight years, this figure has expanded to over a billion dollars with new investments in Alaska, California, New Mexico, Florida, Virginia, Maryland, Texas, Oklahoma, and Florida for upgrades to existing commercial airports or for the creation of totally new spaceports. And there are many more waiting in the wings.

In short, capitalization associated with these sites continues to grow rapidly in response to new initiatives. In the United States, current activity to establish or upgrade spaceports is primarily funded by the individual states, augmented by private sponsorship and some federal government support. It is estimated by the FAA/AST that an average of about $3 million is being spent yearly at each facility to operate these established and licensed spaceports. Again these operating cost estimates appear to be conservative, and these figures will continue to increase for both U.S. and overseas spaceports. [3] Table 8.1 shows the 14 states that have federal, non-federal, and proposed spaceports. Although the majority of licensed launch activity still occurs at U.S. federal ranges, much of the future launch and landing activity may originate from private or state-operated spaceports. Of the growing number of non-federal launch sites licensed by FAA/AST, six are co-located in conjunction with federal launch sites: the California Spaceport at Vandenberg AFB; the Mojave Space Port, near Edwards Air Force Base; the FSA spaceport at Cape Canaveral; the MARS spaceport (originally called the Virginia Space Flight Center), at the Wallops Flight Facility, Virginia, but now also supported by Maryland; Spaceport America in New Mexico, near the White Sands federal facility; and a licensed spaceport in Kodiak, Alaska. The Poker Flats site in Alaska has been used for sounding rockets and some other previous launches under the supervision of the University of Alaska and NASA, but it is not currently licensed by the FAA as a spaceport to launch into orbit. The launch site owned by the space entrepreneur Jeff Bezos in Texas has been used for experimental tests but has not yet been fully licensed.

Table 8.1 United States-based spaceports—federal, commercial, and proposed.
(Derived from FAA/AST records.)

State of Operation	Federal	Non-Federal Facility (Licensed)	Not Yet FAA Licensed
Alabama			X
Alaska		X	X (Poker Flats intends to remain a sounding rocket test facility)
California	X	X, X	
Colorado			X
Florida	X	X	
Hawaii			X
Indiana			X

State of Operation	Federal	Non-Federal Facility (Licensed)	Not Yet FAA Licensed
Kwajalein (Marshall Islands)	X		
New Mexico	X	X	
Oklahoma		X	
Texas		X	X,X
Virginia/Maryland	X	X	
Washington			X
Wisconsin			X
Wyoming			X

Today, other spaceports have been proposed for Alabama, Colorado, Hawaii, Indiana, Texas, Wisconsin, and Wyoming.

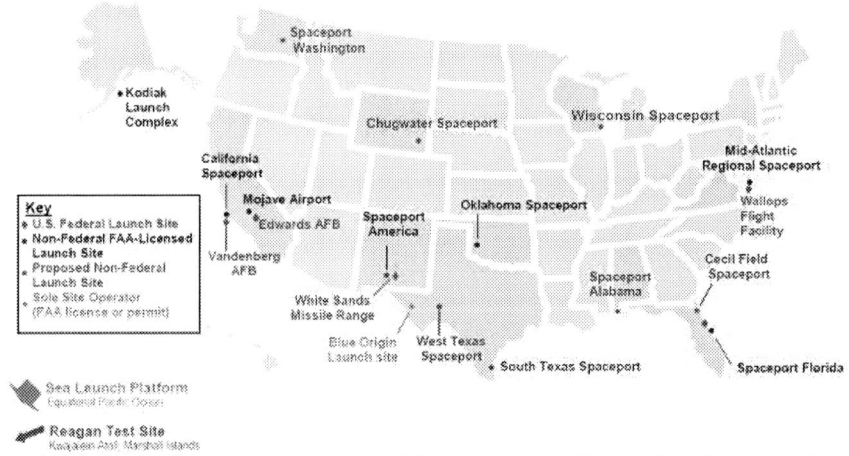

Fig. 8.4 U.S. federal and commercial spaceports—licensed and proposed.
(Map courtesy of FAA/AST.)

In the United States alone, political and financial muscle is at work to install new spaceports in a number of states. Texas, in its tradition of being the biggest in all its ventures, has three spaceport projects in various stages of planning and licensing. These states and a growing number of others are busy pushing tax credits, free land and facilities, and other incentives to get developers to build a shiny new spaceport in their backyard. This proliferation of spaceports can only lead to a number of stresses and strains, if not serious problems. One obvious observation is that this number of spaceports is going to stress the capabilities of the FAA/AST staff to complete an in-depth inspection of all the facilities before they can be licensed, and then inspected and re-licensed on a five-year schedule.

Meanwhile, on the world scene, in addition to the existing space agency launch sites in Russia, China, India, and French Guiana, there are the other commercial or hybrid projects mentioned earlier. There is the existing Woomera site in Australia, and in Scotland, a spaceport has been touted for a former military airfield. The Minister of Aviation in the U.K. has recently announced plans to create a commercial spaceport in the U.K. by 2018 at one of three possible coastal sites: Scotland, Wales, or England.

There are also plans to build spaceports in Canada; one in Barcelona, Spain; and the two projects, one in the United Arab Emirates and one in Singapore, in partnership with Space Adventures. There have also been suggested locations for new commercial spaceport facilities such as the Woomera site in South Australia, as well as in Canada, Malaysia, Spain, Sweden, Israel, Brazil, and Argentina. Of course, there are always other options such as sea launch and the high-altitude launch from carrier aircraft or balloon. It is worth noting, however, that in the past a number of spaceports have been proposed, only to falter for various reasons. A spaceport involves a lot more than putting down a launch pad or a landing strip. The experience with the Burns Flat Spaceport in Oklahoma is just one case study of high expectations, followed by disappointment.

Full-Service Transportation
Overall, there could easily be some three dozen operational spaceports spread out around the planet. Most of these are government owned and operated, but many of the newer projects are private initiatives. However, the promise of scheduled spaceliners blasting off with ticketed passengers is still just a promise. In 2015, we are still talking about experimental flights where passengers have to sign away all of their rights to sue the government and launch companies. In short,

these "experimental passengers" blasting off from a spaceport must sign a statement to the effect that they are aware they may not come back.

"One might look at a spaceport as an innovative, new century version of what you remember airports first looked like," observed Patricia Grace Smith, formerly the head of the FAA's Office of Space Commercialization. "They will be a gathering place for people to learn and witness, for the first time, the capabilities and benefits of space." [4]

Smith pointed to the New Mexico Spaceport America as "a full-service transportation entity for space. So you'll be able to go and take sub-orbital rides and experience zero-gravity, but also become educated and aware of all the various aspects of space."

Eric Anderson, president of Space Adventures, remarked, "Countries around the world are only just realizing the enormous commercial possibilities of space tourism." The market potential for sub-orbital spaceflights alone, Anderson has suggested, might grow to $1 billion annually. If one truly plans and builds "space parks" along with the spaceports, that number could rise to many billions. Yet Anderson's hyperbole must be taken with a grain of salt in light of the most recent Tauri Group market studies, which show gross revenue figures as low as a half billion dollars (U.S.) for the entire first 10 years of operation for the space tourism flight operators.

Polarization of Spaceport Providers

Derek Webber, director of Spaceport Associates in Washington D.C., has taken a hard look at spaceport types. He makes the case that it is probably not a workable plan to attempt to cover all markets with a single spaceport.

"There is emerging a polarization of spaceport providers," Webber observed.
"Throughout the world, the already established government spaceports are likely to continue to provide expendable launch vehicle services to government, military, and some commercial users. Meanwhile, new commercial spaceports are emerging that will focus primarily on space tourism—both sub-orbital and orbital—and will thereby support the development of the reusable launch vehicle mode of spaceflight." [5] According to Webber, it seems unlikely that a single, all-inclusive type of

spaceport that can satisfactorily handle all the diverse kinds of spaceport business will emerge. This observation must also be taken with some caution, in that Webber is in the business of promoting new spaceports. In the spaceport business almost everyone involved has a separate view of how this will all work. Some believe that the government must operate launch safety range operations and be able to push a button to "terminate" a flight out of control. Others believe that private enterprise will take over everything and work much like a private airport. Indeed, as detailed earlier, many of the proposals for spaceports turn out to be recycled airports and former military air bases. The CSF, at least in the past, has had no compunction in telling government officials who might slow down their flight to the skies to stay out of their way. This has been carried out by Commercial Space Transportation Advisory Committee (COMSTAC), lobbyists, and other forms of communications.

One of the drivers of how this all plays out is the quickly emerging issue of what is called Space Traffic Management. As regular aircraft traffic, spaceplane traffic, and space launches increasingly interact, the argument is that a centralized control is needed. Some have suggested that the ICAO, the United Nations specialized agency in the field of aviation, should assume this role to make sure that a catastrophic accident does not occur. Certainly safety advocates such as the International Association for the Advancement of Space Safety (IAASS) and the International Space Safety Foundation (ISSF) have begun to champion serious study of a single coordinative agency with regulatory authority that extends at least from the earth's surface to LEO. [6]

Currently there are new efforts to try to sort out the best way forward. These include the ICAO Space Symposium, held March 18-20, 2015, in Montreal, Canada, to bring decision makers, operators, and academics together to discuss these issues, and the McGill University Global Space Governance Study, which will address the regulation and control of commercial space travel and possible regulation of the so-called "protozone" for air and space traffic management and control. The so-called protozone is that area above commercial space (21 kilometers) and up to the start of outer space as often considered (the Von Karman Line, or 100 kilometers in altitude).

Fields of Dreams and the Fantasy Traffic Models

A more cautious and realistic perspective came from Thomas Matula of the School of Business at the University of Houston-Victoria. Anybody engaged in the new spaceport boom should learn the lessons from the first one, he said. Speaking at the International Space Development Conference in May 2006, Matula explained that the first wave of spaceports occurred in 1989 and 1999. Those "fields of dreams,, he said, were stirred up by such government projects as the Delta Clipper-Experimental (DC-X), the failed NASA/industry SSTO VentureStar program, and the privately backed Kistler rocket.

Eventually reality set in, Matula stated. Often spaceports focused on single firms, so when the firm failed so did the spaceport. Those backing spaceports didn't ask the "hard" questions, he said, like what is the *real* demand for launch services? Also, did the launch firms really have viable business models? And are the proposed launch vehicles technically feasible?

Matula said his counsel to spaceport proponents is that they must craft a realistic business model. He suggested that most spaceports should at first serve as business incubators, not transportation facilities. The Mojave Air and Spaceport, which supports Rutan's various enterprises, plus a growing number of other aerospace and spaceplane ventures, are practical examples of "incubator" models. The problem is that this "model" has also proved to have its dangers. After the accident with the rocket test explosion that killed three people and injured three others, the U.S. government safety group known as OSHA imposed a large fine for unsafe practices on Rutan's company, Scaled Composites. In Mojave, the distinction between Scaled Composites, XCOR, and the Mojave Spaceport is sometimes hard to distinguish. However, as will be noted later, XCOR is moving to the Midland, Texas, spaceport facility. [7]

Thus, Matula counsels most spaceport projects to start small and expand as needed, leveraging existing facilities before building new ones. He emphasized the need to ask hard questions about markets, revenues, and the viability of launch firms sooner rather than later. Of course, not everyone was listening. The mega project in New Mexico was backed in a promotional way by Virgin Galactic, but it was paid for by the State of New Mexico's tax payers, who invested nearly a quarter of a billion dollars of state bonds in this new state-of-the-art facility. Clearly, Space Adventures also has the bit between its teeth, and the company and its partners believe that its entertainment-oriented business model can and will work. Large tax incentives are also being offered by other states. So far, the returns on such tax incentives or public bond issues have been modest to nil.

FAA-Licensed Commercial Spaceports in the United States

As of mid-2015, the FAA Office of Commercial Space Transportation had officially licensed only eight commercial spaceports in the United States (Table 8.2). [8]

Table 8.2 Officially licensed FAA spaceports and license expiration dates.

License	Spaceport Authority	Name of Facility	Location by U.S. State	License and Expiration Date
LSO 09-012 (Rev 1)	Jacksonville Aviation Authority	Cecil Field	Florida	January 10, 2020
LSO 14-015	Midland International Airport	Midland International Airport	Texas	September 14, 2019
LSO 04-009 (Rev 1)	Mojave Air & Spaceport	Mojave Air & Spaceport	California	June 16, 2019
LSO 08-011	New Mexico Spaceflight Authority	Spaceport America	New Mexico	December 14, 2018
LSO 03-008 (Rev 2)	Alaska Aerospace Development Corporation	Kodiak Launch Complex	Alaska	September 23, 2018
LSO 02-007A (Rev 2)	Virginia Commercial Space Flight Authority	Wallops Flight Facility	Virginia	December 18, 2017

License	Spaceport Authority	Name of Facility	Location by U.S. State	License and Expiration Date
LSO 01-005	Spaceport Systems International	Vandenberg Air Force Base	California	September 18, 2016
LSO 06-010	Oklahoma Space Industry Development Authority	Burns Flat, Oklahoma	Oklahoma	June 11, 2016

Mojave Spaceport Made the Headlines

Rutan, the designer of the SpaceShipOne and SS2—and other exotic aircraft—is the prime reason that Mojave Air and Spaceport in California has gained a good deal of fame in the space commercialization world. This facility, near the famous Edwards Air Force Base, where Chuck Yeager first broke the sound barrier, was the first inland commercial launch site to be licensed by the FAA. The license was awarded on June 17, 2004. This allowed the facility to support sub-orbital launches of reusable launch vehicles, including the White Knight carrier vehicle, SpaceShipOne and SS2, and the XCOR Lynx flights.

It was there that on the 100th anniversary of the Wright Brothers' first powered flight, December 17, 2003, Rutan's Scaled Composites flew SpaceShipOne, which broke the speed record for the first manned supersonic flight by a craft developed privately by a small company. SpaceShipOne then flew from Mojave, past the boundary of space on its way to meeting the Ansari XPRIZE qualifications.

Rutan has predicted big things ahead for the Mojave Spaceport. As reported in the *Mojave Desert News*, Rutan said that "significant infrastructure" will be erected at Mojave to handle the space tourism business, including new spaceliner assembly facilities. He went on to say, "Oddly, spaceports are popping up but they have nothing to fly, with investments spurred by very little information. It's almost humorous to watch the worldwide battle of the spaceports ... they're everywhere." [9]

The original Mojave Airport was established in 1935 and serves as a Civilian Flight Test Center, as the location of the National Test Pilot School, and as a base for modifying major military jets and civilian aircraft. One of the three runways was extended as a critical element of the spaceport's expansion program aimed at the recovery of horizontal landing RLVs. The Mojave Airport is also home to several industrial operations, such as BAe Systems, Fiberset, Scaled Composites, AVTEL, XCOR Aerospace, OSC, IOS, and General Electric. XCOR Aerospace has performed flight tests at this facility, including multiple tests of its Lynx Spaceplane and its EZ-Rocket, which was successfully demonstrated in October 2005 at the Countdown to the XPRIZE Cup event in Las Cruces, New Mexico.

New Mexico Chosen by Virgin Galactic

All the buzz and pizzazz about spaceports is not just smoke and mirrors for customer brochures. Back in 2008, Virgin Galactic unveiled the architectural plans for its Spaceport America in New Mexico, which was officially opened three years later. They say that the structure, developed by a team of U.S. and British architects and designers, symbolizes the world's first purpose-built commercial spaceport. The 100,000-square-foot (9290-square-meter) facility will serve as the primary operating base for Branson's Virgin Galactic sub-orbital services, including the fleet of two WK2 carrier aircraft and five SS2 spacecraft now under construction by Scaled Composites in Mojave, California. It will also provide the headquarters for the New Mexico Spaceport Authority. This facility thus became the second inland site for a spaceport.

This development followed on from an announcement first made late in 2006 by New Mexico Governor Bill Richardson and by Branson to start operations from the Southwest Regional Spaceport (renamed as America's Spaceport). The chosen location is some 45 miles north of Las Cruces and 30 miles east of Truth or Consequences, near Upham, New Mexico. This site was selected due to its low population density, uncongested airspace, and high elevation.

According to New Mexico's planners, their eagerness to build the Southwest Regional Spaceport was driven by several money-making activities, such as the following:

- The emerging commercial space tourism sector, including operations of Virgin Galactic
- NASA contracts for ISS commercial cargo and crew re-supply services
- Proposed low-altitude racing competitions, such as those sponsored by the Rocket Racing League Evolving demand for low-cost human-rated reusable launch vehicles and rocket-powered racing aircraft

According to Space.com, state officials have estimated that the economic impact on the region could be sizable. One probably wildly optimistic estimate has suggested that the spaceport could add in excess of $750 million in total revenues and perhaps 3500 new jobs in the region by 2020. This "guesstimate" included all commercial space transportation services, related manufacturing and services activities, plus tourism-related visitor spending. [10]

MARS

MARS at Wallops Island is one of four U.S. spaceports with the current capability to launch rockets into orbit, and during 2013, it expected to begin supply missions to the ISS. These would be carried out by OSC, using its Antares rocket to carry the Cygnus spacecraft under the COTS agreement with NASA.

The spaceport began its life with the creation of the Center for Commercial Space Infrastructure by Old Dominion University, in Virginia. The university established this entity to create a commercial space research and operations facilities within the state, and it has been a prime mover in the development of the commercial launch infrastructure at Wallops Island, Virginia. In 1995, the organization became the Virginia Commercial Space Flight Authority and focused its efforts on promoting the growth of aerospace business in the region while also developing a commercial launch capability.

In July 2003 a bi-state agreement was reached between Virginia and Maryland to allow cooperation between the two states for future development and operations, and to promote the further development of the launch facility. Maryland now provides funding, and the name was changed from the Virginia Space Flight Center to MARS. The FAA's launch site operator's license for MARS was renewed in December 2002 and has been renewed twice again for five year increments. The facility is designed to provide what its backers characterize as "one-stop shopping" for space-launch facilities and services for commercial, government, scientific, and academic users, although its targeted market is largely geared toward smaller-payload missions such as the Taurus II and Antares rockets.

In 1997, the original group signed an agreement with NASA to use the Wallops Island facilities in support of commercial launches. This 30-year agreement includes access to NASA's payload integration, launch operations, and monitoring facilities on a non-interference, cost-reimbursement basis. There is currently a partnership agreement with DynSpace Corporation, a Computer Sciences Corporation company, of Reston, Virginia, to operate the spaceport. The State of Virginia, however, maintains ownership of the spaceport's assets.

MARS has two launch pads at Wallops Island. The first, 0B, was designed as a "universal launch pad," capable of supporting a variety of small and medium-size expendable launch vehicles (ELVs). From its location on the Atlantic coast, MARS can accommodate a wide range of orbital inclinations and launch azimuths. The facility also provides an extensive array of services, including the provision of supplies and consumables to support launch operations, facility scheduling, maintenance, inspection to ensure timely and safe ground processing and launch operations, and coordination with NASA on behalf of its customers.

The successful launch of the four-stage Minotaur I rocket by MARS in December 2006 was the beginning of a more intensive effort to encourage private commercial launch firms, such as SpaceX and OSC, to use this facility. This set the Wallops spaceport on a trajectory to send spacecraft to re-supply the ISS. With this aim in view, the spaceport began a $500,000, federally funded study to determine if it was a suitable location for orbital taxi missions to the ISS. MARS is also ready to consider human sub-orbital flights. [11]

The launch failure of the Antares rocket, which was unsuccessfully launched from the MARS spaceport in October 2014, however, has cast somewhat of a pall over these spaceport operations. If future Antares launches with the new U.S.-designed and manufactured rocket engines are successful, this bad memory will fade over time.

And on the West Coast

California Spaceport is co-located with the Vandenberg launch facility, operated by the U.S. Air Force. This facility was the first commercial spaceport to be licensed by the FAA/AST, on September 19, 1996. In June 2001, the spaceport's license was renewed for another five years, and further renewals were granted in 2006 and 2011. This facility offers commercial launch and payload processing services and is operated and managed by Spaceport Systems International (SSI), a limited partnership of ITT Federal Service Corporation. It is located on the central California coast and can leverage off the infrastructure developed to support the Vandenberg launch facility operated by the U.S. government. SSI signed a 25-year lease in 1995 for the 0.44-square-kilometer (0.17-square-mile) site. The California Spaceport can support launches with azimuths ranging from 220 degrees to 165 degrees. Construction of the commercial facility began in 1995 and was completed in 1999. The design concept provides the launch pad with the flexibility to accommodate a variety of launch systems, including low-polar-orbit inclinations. Although the facility is configured to support solid-propellant vehicles, plans are underway to equip the launch facility with the support systems required by liquid-fueled boosters. [12]

Alaska's Kodiak Launch Complex

The Alaska State Legislature passed legislation in 1991 to create the Alaska Aerospace Development Corporation (AADC). It was structured as a public company to develop aerospace-related economic, technical, and educational opportunities for the State of Alaska. In 2000, the AADC completed the $40 million, two-year construction of the launch complex at Narrow Cape on Kodiak Island. It was the first licensed launch site not co-located with a federal facility and the first new U.S. launch site built since the 1960s. Owned by the State of Alaska and operated by the AADC, the Kodiak Launch Complex received initial funding from the U.S. Air Force, U.S. Army, NASA, the State of Alaska, and private firms. However, today, it is self-sustaining through launch revenues and receives no state funding.

Kodiak is designed to serve several markets. These include military launches, government and commercial telecommunications satellites, remote sensing, and space science payloads weighing up to 1000 kilograms (2200 pounds). These payloads can be delivered into LEOs, polar orbits, and highly elliptical orbits. The first launch was a sub-orbital vehicle, Atmospheric Interceptor Technology 1, built by OSC for the U.S. Air Force in November 1998. A second launch followed in September 1999. Further launches included a Quick Reaction Launch Vehicle, a joint NASA-Lockheed Martin Astronautics mission on an Athena 1 (the first orbital launch from Kodiak), in September 2001, and later, a Strategic Target System vehicle was launched.

In February, 2005, the Missile Defense Agency (MDA) launched the IFT-14 target missile, one of several rockets from Kodiak to test the U.S. missile defense system. The AADC is also supporting development of ground station facilities near Fairbanks, Alaska, in cooperation with several commercial remote-sensing companies.

The high-latitude location of Alaska makes the Fairbanks site specially favorable for polar-orbiting satellites, which typically pass above Fairbanks several times daily. [13]

Alaska's Poker Flats Launch Site

This facility is owned and operated by the University of Alaska. It has largely been used for experimental launches, including those of sounding rockets and polar orbital flight satellites. To date, it has not been discussed as a site for commercial space tourism flights because of its isolated location and relatively sparse facilities and equipment.

Spaceport Florida

In 1989, the State of Florida established the Spaceport Florida Authority and then renamed it the FSA in January 2002. This entity is authorized by the state government to act just like an airport authority, to oversee the space launch industry and space-related economic, industrial, research, and academic activities. The FSA occupies and operates space transportation-related facilities at the Cape Canaveral Air Force Station, owned by the U.S. Air Force. FAA/AST first issued a license for spaceport operations on May 22, 1997, and renewed the license in 2002 and 2007 for additional five-year terms.

Under an arrangement between the federal government and FSA, excess facilities at Cape Canaveral have been licensed to the FSA for use by commercial launch service providers on a dual-use, non-interference basis. To date, FSA has invested over $500 million in new space industry infrastructure development in Florida. This includes an RLV support complex (adjacent to the Shuttle Landing Facility at Kennedy Space Center), and a new space operations support complex. FSA is planning to develop an innovative, flexible, and cost-friendly commercial spaceport to attract commercial launch companies to the state. This is also intended to accommodate the growing need for rapid response launch vehicles and the launching of smaller payloads for government, commercial, and academic users.

A 2005 analysis by FSA included a market assessment of the number and types of launch vehicles that could possibly use such a facility, and concluded that a new commercial spaceport is feasible from both a market and technical standpoint. The conclusion reached by this study was that a Florida commercial spaceport would primarily benefit from the sub-orbital space tourism market—at least from an economic and usage level. It was estimated from this study that such a facility would generate increased economic activity, earnings, and jobs, and would raise Florida's profile as a space state.

FSA is investigating the possibility of having specific Florida airports apply for an FAA Launch Operators license to support horizontally launched spacecraft. They believe this will attract space tourism companies such as Virgin Galactic to use this technology. (In this particular case, however, Virgin Galactic has already signed a contract with the New Mexico Spaceport facility.) Several statewide airports have shown great interest in participating in space tourism, but so far none has taken final action and sought FAA licensing. [14]

Midland, Texas, Spaceport

In September 2014 the Midland, Texas, Airport received its official license from the FAA/AST as a spaceport. This licensing process, which took an extended period of time, represented the key step in the relocation of the XCOR Corporations from their Mojave facility in California to Texas. The XCOR research facility and headquarters will now be moved to the Texas facility, which is still its final preparation. As of April 2015 XCOR's website still listed the Mojave, California, address for its headquarters.

XCOR's current generation LYNX spaceplane will, in fact, not actually fly into outer space; rather, for a price of $95,000 plus taxes, it will make a parabolic flight up to some 37 miles (about 60 kilometers). Although this is not outer space, it will be high enough to be in the dark sky of space and to see the curvature of the globe. Although this move by XCOR to the Midland facility was announced in 2012, the licensing of the local airport as a spaceport took over two years, and the retrofit of the hangar for XCOR research activities was somewhat delayed pending the regulatory approvals. Although Midland, Texas, is not at a coastal location, the spaceport was judged to be an appropriate location from which XCOR flights could take place. The Midland facility is now licensed through September 14, 2019. [15]

Oklahoma Burns Flat Spaceport

The Oklahoma Burns Flat Spaceport is perhaps the most isolated of any of the currently licensed spaceport facilities. It is about 250 kilometers away from Amarillo, Texas, (to the west) and about the same distance to Oklahoma City (to the east).

Fig. 8.5 The 4000-meter-long runway at the Burns Flat Spaceport.
(Courtesy of Burns Flat Spaceport.)

Since the facility was licensed in 2006, the Oklahoma Space Industry Development Authority (OSIDA) has been particularly aggressive in seeking spaceplane companies to establish residency at this rather isolated facility by offering many tax and other incentives to get operators to use this site. To date, Rocketplane Corporation is the only company to have come to Oklahoma and indicated its intent to use this location. Rocketplane, however, declared bankruptcy in 2009 after having benefited from $18 million in Oklahoma tax credits. Efforts were also made to recruit XCOR to come to this site, but as noted above, this company has decided to operate from the Midland, Texas, spaceport. Thus, no active spaceplane operator is currently seeking to operate from this site.

The runway at the Burns Flat facility, the third-longest in the United States, was once part of the former Clinton-Sherman Air Force Base, which closed in 1969. This gigantic runway is well over 4000 meters (13,500 feet) long and is about 100 meters (300 feet) wide. It also has an additional 300 meters of asphalt overrun on either end. The runway continues to play a key role in Altus and Vance Air Force Base operations, with 30,000-plus flight operations by the military each year. The runway is being still used to recruit spaceplane companies as well as those wishing to design, test, and manufacture drones. Nevertheless, today this is a spaceport without a spaceplane operator. Once-ambitious plans to create a space-based theme amusement park in this remote part of Oklahoma have now faded away. [16]

Non-Federal U.S. Spaceports Separate from Federal Sites

Several states now plan to develop private spaceports that are intended to provide multiple launch and landing services. These proposed private spaceports have several common features. One unusual aspect of these projects is that rather than being coastal sites, whereby aborted missions can be easily accommodated, there are now several inland facilities, and the number of such sites is growing. Initially, the Mojave spaceport was the only U.S. FAA-licensed spaceport with an inland location rather than an ocean-based abort capability. The Spaceport America facility is located next to the very isolated White Sands Federal Facility. and the Burns Flat, Oklahoma, and Midland,

Texas, facilities are located in areas with very low density development.

In short, non-coastal spaceport facilities are generally designed to have sufficient space and area to respond to launch or landing mishaps and are often co-located with larger federal sites or former airport facilities.

Assessment of U.S. Spaceport Safety

The licensing of spaceport facilities will likely remain on a case-by-case basis for some time to come. This is, of course, just what is being done for the licensing of spaceplane and rocket launches because the number of these spaceplanes, launches, and facilities remains reasonably small and because the nature of the launch operations at each site remains quite different. Certainly, FAA inspectors must address a number of key questions and issues in the case of each spaceport. The answers to these questions will then trigger further investigation and inspections. Different launch operations, different propulsion systems, and different ground support requirements will clearly alter the nature and the stringency of the licensing process that follows. Key questions for spaceport licensing, according to FAA/AST safety inspectors, include the following:

- **Abort and mission mishap processes.** The licensing process must address the degree to which the launch site or spaceport has a large and securable perimeter area (and/or ocean frontage) that allows launch aborts or landing mishaps to be addressed with a minimum of safety risk to ground crew and surrounding residents and infrastructure. Land-based spaceports and ocean-adjacent sites give rise to different assessment issues. (Requirements as to who oversees and executes launch range safety and whether there are termination capabilities still need to be addressed.)
- **Take-off and landing processes for spaceplanes or RLVs.** Vertical take-off, horizontal take-off, jet carrier, jet-towed launcher systems, and balloon-based launch systems all lead to different types of safety questions and regulatory concerns. Systems that involve ocean-based launches and balloon-based launches where the crew is remotely located (or largely remotely located) from the launch create a

significantly different environment from one where the launch take-off is from a spaceport with a significant number of employees and ground crew. This is particularly the case where a vertical rocket launch and/or landing are involved.

- **Escape capability.** One of the key issues is the degree to which there is a launch-to-land capability for crew and passengers to escape via a separate module or capsule. A key question concerns the landing provisions of the escape system and whether it can make a runway landing or a parachute or parafoil terrestrial landing, or whether a splashdown landing is required.
- **Re-entry mode and the thermal protection system (TPS).** A critical part of any spaceplane sub-orbital mission or de-orbiting from LEO is adequate thermal protection and the nature of required aero-braking, parabolic arcs, and pull-out profiles. In this respect, safety margins against engine malfunctions are particularly key, and mission profiles should ensure that any accident would prevent a crash landing over urban or heavily populated areas and that the craft could land in glide mode.
- **Basic infrastructure.** Any fully licensed spaceport will need to be well equipped with testing facilities; fully backed up communications, power, and security provisions; and other critical infrastructure. Fortunately, because many spaceports are augmentations of modern and fully equipped airports, most of this key infrastructure is already there. In today's uncertain world, security barriers, monitored access for all personnel and customers, parking, transportation infrastructure, and so on are needed.
- **Testing, assembly, training, and simulation facilities.** The spaceport should be well equipped with testing, assembly, flight training, and simulation facilities. Alternatively, access to these facilities should be provided at other convenient and accessible locations. In light of the accident at the Mojave testing facilities, clear safety standards need to be established and enforced for rocket and hazardous material qualification. Training facilities need to be overseen for adequacy and safety as well.

These are only some of the basic questions that FAA inspectors consider in licensing private spaceports. They will also consider the degree to which the operator of the facility is fully financially viable, bonded, and insured by a responsible governmental entity or private equity company against accidental loss and liability claims. In cases where issues of full viability or completeness of spaceport infrastructure remain, it might be advisable to grant only a provisional license, pending resolution of any remaining questions or pending successful performance. Fortunately, a great deal of experience has been gained with regard to licensing federal launch centers over the years, and even commercial spaceports have, in some instances, been inspected and licensed for periods that now exceed 10 years, although five years is the usual licensing period. The recent trend toward rapid expansion of commercial spaceports suggests that vigilant inspection and licensing procedures need to remain in place.

International Spaceport Projects

A number of national launch sites have been owned and operated around the world for a number of years, and commercial spaceport projects seem to have mushroomed in the past year. Clearly, it is appealing to many countries to have at least one spaceport in their country to create an image of being modern and a part of a totally new and "sexy" industry. The appeal is sufficiently great that some governments have even offered financial incentives to locate a spaceport within its borders, just as some city and state governments in the United States have offered tax and other incentives.

Many new initiatives to create commercial spaceports around the world have been announced. These include projects in Canada, as described below. The U.S.-based Space Adventures company is involved in spaceport developments in the United Arab Emirates and Singapore. And in Australia, the Woomera launch center is being modernized and upgraded to attract U.S. operators such as the DTI Corporation. (The U.S. company RpK was to have been the prime client for Woomera, but this particular company is now out of the business.) Other possibilities have been noted above, but because it is uncertain which spaceports will actually be built, only some examples are provided below. In 10 years, however, over a dozen spaceports may well be operating in the United States and over a dozen around the world.

We believe that the four international spaceport initiatives described below indicate the range of ventures that might be anticipated in coming years.

Canada's Spaceport Ranges

The Canadian Forces Meaford Range facility near Cape Rich, just off Georgian Bay and near Barrie, was selected by the Canadian Arrow/PlanetSpace operation as an initial test site and proving ground for its manned sub-orbital spacecraft. It was considered a prime spot for a private space launch facility because of its restricted airspace and waterways. The Cape Rich peninsula stretches 2.5 miles (4 kilometers) out into Georgian Bay and thus a launch there is almost equivalent to a barge launch. This Canadian military site is 70 square kilometers (44 square miles). For safety purposes, initial test launch operations have been held at the end of a peninsula spit.

This site has hosted engine test firings and escape system shakedowns for the Canadian Arrow launch vehicle. Canadian Meaford Range officials agreed to allow Canadian Arrow to use the base's facilities as needed for these initial tests. The project received authorization from the Canadian transportation agency, Transport Canada, for test flights that go beyond the current engine test firings, but it now seems that actual test and operational flights will be relocated to Cape Breton in Nova Scotia.

The latest development in the planning for permanent facilities for the Canadian Arrow/PlanetSpace launch operations, however, involves a new location in Nova Scotia. Arrangements have been made for the spaceport to occupy 1 square kilometer (300 acres) of land on Cape Breton, with the Nova Scotia government also providing financial and tax incentives. In addition to testing facilities, launch pads, and manufacturing facilities, there will be a new state-of-the-art training facility for astronauts. All of these arrangements are now contingent on the future success of the PlanetSpace venture and its carrying through with the agreement with its Canadian partners. [17]

The United Arab Emirates and Singapore Projects

The highly entrepreneurial Space Adventures company intends to offer space rides at new spaceport locations in the United States and in the United Arab Emirates and Singapore. In February 2006, the CEO, Eric Anderson, announced plans for these two new spaceports near major airports, saying,. "These sites in the United Arab Emirates and in Singapore are just the initial steps [for private spaceflight]."

According to Space Adventures' press releases, an agreement is in place to construct the first of these, at a cost of $265 million, at the Ras Al-Khaimah International Airport in the northernmost of the seven emirates that constitute the United Arab Emirates. Sheikh Saud Bin Saqr Al Qasimi, crown prince of Ras Al-Khaimah, said in a statement in February 2006 that they also looked forward to expanding operations elsewhere. Anderson, in announcing the project, said Crown Prince Al Qasimi had been extremely supportive and had personally invested $30 million.

Space Adventures has also reported that it is working with a consortium of investors in Singapore to develop Spaceport Singapore, a facility that will offer not only sub-orbital spaceflights but also astronaut training, parabolic flights to simulate weightlessness, and other high-altitude attractions. Current plans call for this $115 million complex to be built near Singapore's Changi International Airport.

Lim Neo Chian, chief of the Singapore Tourism Board, expressed strong support of the project when it was announced in 2006 by saying, "With the proposed Spaceport Singapore, we now stand at the threshold of an unprecedented opportunity to launch into space practically from our own backyard." [18]

In Space Adventures' announcement of these two spaceport projects, the company indicated that these two initiatives do not rule out an American spaceport in the future, although several unrelated projects are already well underway in the United States. The ZERO-G parabolic rides to achieve weightlessness originate in Las Vegas and Florida, and these locations would seem to be prime candidates for U.S. sub-orbital flights.

Fig. 8.6 An architectural rendering of the proposed spaceport in Singapore.
(Courtesy of Space Adventures.)

Woomera, Australia

The Woomera launch facility dates back to the 1960s. This was the site the U.K. selected for its launch center when it was developing the Blue Streak rocket as a part of the European effort to develop an integrated multi-stage rocket for the European Launcher Development Organization (ELDO). This facility was being updated and modernized, at a cost of some $80 million, to support the operations of DTI and RpK (now defunct).

The clear atmospheric conditions that exist in Woomera allow launches at virtually all times, and the vast open areas that surround the launch site allow the recovering of staged rockets from the K-1 launch. Australia is currently adopting a new set of space policies for the future, and the outcome of these studies will likely impact the future of the Woomera facility.

U.K. Locations

The powerful Institute of Directors published a report in 2012, and the report warned that "thousands of jobs will go to other countries" if Britain does not take the initiative to create a land-based spaceport.

The report's author, Dan Lewis, said, "We are just on the verge of space tourism moving from what people might think of as Star Trek fantasy to the start of something more realistic." [19]

Two military air stations are now under final considered as locations for new facilities to service commercial spaceflights as an interim step to the creation of a separate commercial facility. The two interim locations are the Royal Air Force (RAF) base at St. Mawgan, near Newquay in Cornwall, and the RAF base at Lossiemouth, near the town of Elgin in Scotland. Lossiemouth has the advantage of not having controlled air space issues that could hamper St. Mawgan, where restrictions exist because of flight paths above Cornwall that serve aircraft traveling to and from London and Continental Europe across the Atlantic.

The latest information on commercial spaceport planning came in March 2015, when U.K. Aviation Minister Robert Goodwill explained that a temporary spaceport would be established at one of the two military air bases, but that this would be followed with a completely new commercial spaceport in a selected coastal location. Goodwill stated, "I want Britain to lead the way in commercial spaceflight.... Establishing a spaceport will ensure we are at the forefront of this exciting new technology." He indicated that the commercial spaceport might be established as early as the end of 2018. [20]

Fig. 8.7 Conceptual rendering of the newly planned U.K. commercial spaceport. *(Courtesy of the U.K. Ministry of Aviation.)*

As long ago as 2007, the Science and Technology Select Committee of the British Parliament warned the government that it must provide the U.K. with a coherent space strategy if it is not to be left behind by other countries. The committee said the lack of government support for early-stage technology development already places the U.K. at a disadvantage.

A spokesman for the committee said, "On space tourism, such as those ventures planned by Richard Branson's Virgin Galactic, the committee takes a different view from the government and is excited by the potential afforded by sub-orbital travel and the rise of space tourism industry. The MPs say they do not believe that it should be the responsibility of the government to fund this work but developments in this area should be encouraged through appropriate regulation." In response to this urging, the U.K. government has now embarked on a very proactive role to create a true commercial spaceport and to do so on an accelerated basis.

There will likely be several other spaceport facilities in Europe to support the European-developed spaceplanes. A site in Sweden for sub-orbital flights to see the Northern Lights has been mentioned in press releases, and sites in other parts of Europe are in the pipeline to support the spaceplanes now being developed in France, Spain, and Germany.

Nationally Operated Governmental Launch Sites around the World

To complete the global picture, governmental agencies develop and operate a number of launch sites around the world.
Many of these launch facilities are, in fact, operated by national defense-related agencies, and until the advent of space commercialization, there were only about two dozen major space launch sites around the world. Some of these, like Woomera in Australia and the commercial sites co-located with U.S. federal launch centers described earlier, may well play a role in the future of private spaceflight.

What is clear is that a lack of spaceports and launch facilities is in no way likely to be a barrier to the development of commercial space transportation systems. Other issues and problems constitute far greater problems. Indeed, many such launch and spaceport facilities seem to be popping up everywhere. If there is a concern, it is that too many of these facilities will be developed both in the United States and around the world. The perhaps-misguided hope seems to be that new commercial spaceports will attract new business, new jobs, and new tax revenues to areas with lagging economies. This may be the case in some areas, but this cannot be the case for many dozens of spaceports when there will be so few flight opportunities for some time to come.

REFERENCES
[1] http://en.wikipedia.org/wiki/List_of_rocket_launch_sites.
[2] http://www.dailymail.co.uk/sciencetech/article-2050328/Spaceport-America-Richard-Branson-opens-209m-space-terminal-New-Mexico.html.
[3] "2006 Commercial Space Transportation Developments and Concepts: Vehicles, Technologies and Spaceports," FAA/AST, Washington D.C., January 2006.
[4] David, L., Space.com, May 17, 2006.
[5] Spaceport Associates, http://www.spaceportassociates.com.
[6] Sgobba, T., Jakhu, R., et al, "An ICAO for Space?" IAASS, 2011.
[7] "Midland International Airport Receives Historic Federal Aviation Administration Spaceport License Approval," September 17, 2014, http://www.xcor.com/press/2014/14-09-17_midland_airport_receives_spaceport_license.html.
[8] "Active Licenses," FAA Office of Commercial Space Transportation, http://www.faa.gov/data_research/commercial_space_data/licenses/.
[9] Space.com, http://www.space.com/businesstechnology/060517tech_spaceport.html.
[10] Space.com, http://www.space.com/missionlaunches/050913_nm_spaceport.html.
[11] Virginia Commercial Space Flight Authority and Mid-Atlantic Regional Spaceport, http://marsspaceport.com/.
[12] "Calspace Facilities Overview," http://calspace.com/Home/Facilities_Overview.html.
[13] Kodiak Launch Complex Overview, http://www.akaerospace.com/klc_overview.html.
[14] Florida Spaceport System Plan, http://www.spaceflorida.gov/docs/spaceport-ops/florida-spaceport-systems-plan-2013_final.pdf?sfvrsn=2.
[15] Midland, Texas, Spaceport, http://www.photostospace.com/spaceports/midland-texas-spaceport.
[16] Ellis, R., "Officials defend spaceport efforts," NewsOK, January 21, 2013. http://newsok.com/officials-defend-spaceport-efforts/article/3747447. Also see "Rocketplane Global Still Not Ready for Liftoff,"

http://www.nbcnews.com/id/21019308/ns/technology_and_science-space/t/rocketplane-global-still-not-ready-liftoff/#.VP4UHfnF_OE.

[17] Malik, T., "Private Space flight Group Selects Canadian Launch Site," Space.com, June 5, 2005, http://www.space.com/missionlaunches/050602_planetspace_launchsite.html.

[18] Malik, T., "Sub-orbital Fleet to Carry Tourists into Orbit in Style," February 22, 2006, http://www.space.com/2074-suborbital-rocketship-fleet-carry-tourists-spaceward-style.html.

[19] *Daily Telegraph*, May 22, 2012.

[20] McCormick, R., "British government approves the first spaceport in Europe," March 4, 2015, http://www.theverge.com/2015/3/4/8146729/british-government-approves-the-first-spaceport-in-europe.

CHAPTER 9

How Safe Is Private Space Travel?

To encourage "Citizen Astronauts" to venture into space in significant numbers, it will be essential to create a sense of safety. Over time, this sense of reliability must increase to the level experienced in the commercial aviation industry. The safety factor is especially key if space tourism is to branch off into an entirely new commercial venture, namely hypersonic aviation transportation. This might be commercial flights for executives or even for regular passengers. ESA is backing concepts such as the A-2 spaceplane, which could fly 300 "space passengers" across the globe in five hours; the Swiss Space Systems (S-3) spaceplane developments, and the Skylon spaceplane development by Reaction Engines Ltd. JAXA, the Japanese Space Agency, has also been concerned with aeronautical research since it absorbed the Japanese National Aeronautics Laboratory (NAL). It is currently testing liquid hydrogen and liquid oxygen prototype spaceplanes that now fly at Mach 2 and in the next phase will fly at Mach 4. In the United States, there is Sierra Nevada's Dream Chaser, as well as development projects by Boeing, Lockheed Martin, and the various projects that have originated with Burt Rutan's Scaled Composites. Then there is the so-called Quiet Super Sonic business jet for CEOs and high-flying billionaires—hypersonic transport designed for executives with fewer than a dozen high flyers aboard.

To achieve this new attitude towards the safety of spaceflight, the history of NASA's two tragic shuttle disasters will need to be consigned to history—in the same way as major airline crashes that have occurred around the world have eventually faded away. Today, anyone who looks seriously at the history of human spaceflight and related vital statistics would note that 1 percent of all such launches have ended with fatalities and, even more disturbing, 4 percent of all astronauts and cosmonauts have met their end on space missions. This high death toll statistic arises because multiple people are flying on most space missions and highly trained astronauts and cosmonauts usually fly multiple missions.

Currently the safety statistics on human spaceflight are, at best, a

99 percent success rate, and the typical current aspiration is for 99.9 percent successful launches and descents without a fatality—or one accident per 1000 flights. If we were to apply this goal against the 50,000 flights a day that take place in commercial aviation, however, that would translate into 50 fatal aircraft crashes a day. Such a record would close down commercial aviation for a detailed investigation as to why such a horrendous number of accidents had occurred. In addition, the safety challenges for "normal" air transportation are not static. The current projections that ICAO has carefully assembled are that there will be some 87,000 commercial aviation flights a day by 2040—just some 25 years hence. On the ICAO website, you can see air traffic flow charts that show the growth of traffic and air routes for 2020, 2030, and 2040; the projections for conventional traffic are alarming. Essentially all of the United States and Europe is painted in red as air traffic, air traffic routes, and movements are projected to increase each year. This raises concern not only for air traffic control and management and the interaction between air traffic and vehicles accessing space and protospace, but also the environmental impact of this many flights into the stratosphere. [1] The ICAO air traffic flow projections show this remarkable increase in aircraft movements even though we are seeing an ongoing increase in the passenger-carrying capacities of aircraft such as the A380, which can now carry 544 passengers. [2]

Clearly we have a long way to go in space safety to come even remotely close to aviation safety. One can only hope that commercial projects with corporate survival in mind will be able to improve reliability and success by many orders of magnitude in the months and years ahead.

The pioneering passengers who sign up to travel on spaceplanes in the new, shining world of space tourism and, in time, hypersonic travel must be reassured that both the new space tourism industry and government regulators are focused on safety issues as their prime concern. Today, it must be clear that governmental regulators are not focused on the safety of those that opt to ride on experimental aircraft and seek to go into space at high risk, but rather to protect people in aircraft and people and infrastructure on the ground from crashes or accidents associated with these spaceplanes.

U.S. Government Regulations Are "Relatively Light"

In the United States, the FAA's rules for commercial spaceflight are generally considered to be relatively light as far as regulations go. Applicants for experimental permits need to provide a description of their program, a flight test plan, documentation showing the operational safety of the spacecraft, and a plan for response in case of a mishap. The U.S. government has made it clear that it still considers these flights experimental and that it is not planning to "certify" spacecraft for some time to come. This framework was first established during Congressional debate back in 2004 and under a law that took effect in 2005. The latest bill approved by the U.S. Congress continues to extend this experimental licensing concept through 2015. The odds are that Congress will again opt to extend the experimental licensing program, which has been the basis of regulation for the past decade.

It is significant that the government extended the liability coverage that extends to this service for only one year at a time. Unless this liability coverage is extended for a longer period of time, those industries involved in space tourism may have to scramble to find needed new risk coverage from the commercial insurance market.

The FAA's current rules call for launch vehicle operators to provide certain safety-related information and to identify what an operator must do to conduct a licensed launch with a human on board. The operators must also inform passengers of the risks of space travel in some detail. They also ask all passengers to sign a waiver to hold the U.S. government harmless if there is an accident.

Physical Examinations "Recommended but Not Required"

Despite the length of the proposed rules, the requirements are not particularly onerous. The operator of a space tourism business must follow only these requirements:

- There must be a pilot with an FAA pilot certificate (Class Two only), and each crew member needs a medical certificate issued within a year of the flight.
- There must be training of the crew and pilot to ensure that the

vehicle would not do harm to the public, even if abandoned.
- Operators of such systems must inform all participants of the associated risks and must have them sign a consent form.
- Physical examinations for passengers would be recommended but not required unless a "clear public safety need is identified."

The FAA has indicated that each review process would be on a case-by-case basis. This applies to both the licensing of "manned launch systems" intending to provide space tourists with access to space and the licensing of spaceports. This is because each program and spaceport facility tends to be highly individualized. For this reason, comprehensive guidelines are not possible. Although a growing number of spaceports have now been approved or are seeking licenses, it is useful to examine those spaceports that are executive members or associate members of the CSF to see which are currently most active in the field.

An indication of government policy came in 2006 from then-FAA Administrator Marion Blakey, who said, "... from the government's perspective, our official policy is this ... to embrace the private sector's daring spirit and clever ingenuity. And yes, you better believe that includes space tourism. We are in the business of encouraging and enabling the private sector. We develop regulations to make this high-risk business as safe as possible...And we make sure potential passengers are properly informed and are willing to accept the risks that remain. And then? Well, then we'll step aside ... get out of your way ... and let you do what you do best: innovate." [3]

Safety Leads to Reliability and Sustainability

The U.S. government's proactive role in terms of being an advocate for the new space tourism business was again underlined in a *Wall Street Journal* interview that ran with the headline "How Safe Is the Race to Send Tourists into Space?" It was in the course of an interview with Patricia Grace Smith, then head of the FAA Office of Space Commercialization, that the pro-industry role was unequivocally set forth. She said, "I believe that to work with industry to ensure safety is an appropriate governmental role. We are happy to play it given the fact that our industry is clearly on the same page. In giving us responsibility for regulating and promoting private human spaceflight, Congress directed us to protect the uninvolved public. The spaceflight participants, or passengers, travel at their own risk, once they have been provided the safety record and data related to the vehicle they are flying on. The spaceflight companies that we are working with are clearly focused on safety ... and believe strongly that safety leads to reliability, which leads to more business and sustainability. So do we." [4]

Fig. 9.1 Patricia Grace Smith, former FAA administrator. *(Courtesy of FAA.)*

The FAA holds an annual conference on commercial space travel, and at that conference a few years back there was a key panel entitled "When is a launch vehicle ready to carry passengers?" This panel included top industry leaders with representatives from Virgin Galactic, XCOR, and RpK, among others. The members of this panel all agreed at least on on one sentiment. They unanimously expressed the view that "the 'vehicle' will fly when it's safe to fly." This concept has apparently proved the key pacing item in that the dates for the first space tourism flights have been delayed and delayed again. Some predicted the first flights would come in 2008 or 2009. Now it seems that the first flights by Virgin Galactic will not now come until perhaps 2016 or later.

As noted in the previous chapter, the law that gave the FAA experimental licensing authority for such flights expired in 2012 and thus had to be extended through 2015. It appears likely that new legislation will extend the experimental licensing yet again. This nearly a decade of extension should not be viewed as a bad thing, but rather as a good thing. The positive conclusion to be reached is that the industry will postpone these experimental spaceplane flights until they really feel they have a vehicle that is as safe to fly as possible.

When Peter Diamandis was asked about the future of commercial space travel, he said, "I would actually love to see some of these companies strive for larger breakthroughs, but they will probably not do this in the early days because they need to strive for safety and a reliable revenue stream." [5]

In another interview, Alex Tai, at the time Virgin Galactic's chief operating officer, said he believed the space tourism industry could bounce back from a disaster. He explained that the industry would need to act carefully and responsibly, with passengers being fully warned of risks before take-off. If customers are properly briefed on the potential risks and dangers of spaceflight, there is debate as to whether they could find a viable basis on which to file a lawsuit and win in case of an accident. "God forbid it should happen on the first flight. Hopefully it's many, many years out," Tai added. [6]

Some from the legal profession who have taken on liability suits in the past are not so sure. Right now, the one specific thing that is happening is that the states that are hosting spaceports are revising their state laws to line up with federal regulations in terms of liability provisions.

Since the Obama Administration took office in 2009, the sense of advocacy for the new space tourism by the U.S. government and by the FAA has become somewhat more muted, but the support still seems to be there—in both Congress and the Executive Branch. Dr. George C. Nield, who was deputy to his predecessor, Patricia Smith, heads the office today with considerable competency. With his team at FAA/AST, he has continued to provide cautious oversight and to emphasize safety while promoting this new industry.

When Congress enacted the law that extended the FAA/AST role to oversee commercial space tourism launches and to promote this activity at the same time, Congress also considered the report by a non-profit organization. This study, conducted by the Aerospace Corporation, together with George Washington University and MIT, answered a number of precise questions that Congress had previously posed about the potential conflict between "safety oversight" on one hand, and industry promotion on the other. These were difficult questions to answer. Congress had taken away this dual role for the aviation industry after the crash of Air Florida Flight 90 into the Potomac River on January 13, 1982. The consensus within Congress was that the airline industry was firmly established and that the FAA should thus devote itself entirely to safety regulation. [7]

The final report to Congress, "Analysis of Human Space Flight Safety—Report to Congress," was completed in the fall of 2008. It noted that there was an iron clad wall between the unit that undertook the promotional activities of the FAA and those units that were addressing safety regulation for the new commercial spaceplane activities. The report recounted the history of the FAA's role regarding aviation and suggested that this issue continue to be reviewed in coming years, but there were not inherent difficulties of these dual roles continuing. Despite this report's recommendation, there was considerable debate about this part of the report within the study team. [8]

One of the key initiatives has been to establish a Center of Excellence for Commercial Space Transportation for 2010 through 2020. Participating in this are the University of Colorado—Boulder, Stanford University, University of New Mexico, New Mexico Institute of Mining and Technology, University of Texas—Medical Branch, Florida State University, University of Florida, University of Central Florida, and Florida Institute of Technology. This very diverse grouping of universities, spread out over nearly a dozen locations around the United States, is devoted to developing new technology, systems, and regulations in this new field. An in-depth R&D agenda has been developed for the center, with safety-related matters heading the list. The center is also working to create an active cooperative partnership with the IAASS, the ISSF, and the McGill University Institute for Aeronautical and Space Law. [9] COMSTAC, which represents the industry, continues to grow in size and at times seems quite zealous in seeking to keep the FAA/AST (the Office of Space Commercialization) in a low-key regulatory role. In particular, COMSTAC and the CSF have generally resisted the specific discussion of new regulations or innovations such as creating a process for space traffic management. Perhaps the most important change in terms of commercial space development is that NASA and the FAA in 2012 signed an agreement to define their respective roles in terms of addressing space safety issues and the regulation and control of commercial space vehicles and private space stations.

The CSF has continued to evolve and add a number of new Executive Members and Associate Members. These include many of the various spaceports in the United States as well as some abroad. Members include many of the new start-ups in the field, and they range from the well-established Virgin Galactic, XCOR, Space Adventures, Bigelow Aerospace, Scaled Composites, and Sierra Nevada to much newer and highly entrepreneurial ventures such as Moon Mission (space mining and commerce) and the Golden Spike Corporation (a company seeking to fly private entrepreneurs to the moon two at a time). The most significant shift that seems to be occurring is that some of the large conventional aerospace companies such as Raytheon, the United Space Alliance (Boeing and Lockheed Martin), AeroJet, and Pratt and Whitney-Aerodyne have let their membership lapse. The CSF leadership has also changed. Brett Alexander was followed by the former astronaut Michael Lopez-Alegria. But he has stepped down, and the new president is Eric W. Stallmer, formerly head of external relations for Analytical Graphics. Stallmer has assumed the helm and argued the case for the CSF quite competently before Congress and the ICAO in recent months. [10]

Is There a Clear Understanding of the Risks?

As the start date for a Virgin Galactic maiden flight, as well as for XCOR Lynx and others, finally seems to be on the near-term horizon, the subject of who will be the best oversight organization for these ventures becomes a topic of increasing interest. Will it simply be national agencies such as the FAA/AST or regional bodies like the EASA? Or do international organizations such as the ICAO and the United Nations Environmental Program eventually become involved? Others have suggested that the most urgent area for regulatory concern is the area just above commercial air space (21 kilometers) or below where satellites can operate (140 kilometers). This is an area sometimes called "protospace," the "protozone," or sub-space. The concern comes not only with regard to space tourism flights, but balloon flights, balloon communications systems, HAPS, robotic air freighters, and proposals for intercontinental hypersonic transport—all of which might one day fly into or occupy this "protozone" region, which is currently not internationally regulated. [11]

Questions of safety got the attention of the U.S. House of Representatives in March 2012 at a hearing before the House Sub-Committee on Space and Aeronautics. A Democrat from Illinois, Jerry Costello, put the issue as follows: "The public needs a clear understanding of the risks involved with commercial space transportation and will need to be convinced those risks are being effectively managed. The Office of Commercial Space Transportation will be at the center of establishing those expectations as it will have a critical role in ensuring the safety of would-be space tourists and potentially even of NASA astronauts or other spaceflight participants."Nield and Wilbur Trafton, chairman of COMSTAC, appeared before the committee to answer those very questions. As the government agency tasked with ensuring that the commercial industry maintains the highest levels of safety for those in the air and on the ground, the FAA has a demanding role to play in the growing private spaceflight sector. [12]

Perhaps in response to the hearing, the FAA announced that it would open talks with the industry from August 2012. "We're going to be setting up monthly public telephone calls to ask them about certain topics," said Pam Melroy, former NASA astronaut and senior technical adviser in the FAA's Office of Commercial Space Transportation, in a July 2012 interview. "We do plan on having these once a month for the foreseeable future. We really want maximum participation, and we want technical people to really help us understand what the thinking is out there." [13] These kinds of statements reflect a more conservative and less of an advocate role by FAA/AST than in the past.

Despite this subtle shift, the industry, at least in the U.S. context, retains a prime role in developing safety for spaceplanes, with the federal government playing more of an oversight and reactive role in response to specific safety issues or accidents that might arise. Since that time, the situation has not substantially changed. The next milestone will be what Congress decides to do about the licensing and regulation of space tourism flights, which will come before the end of 2015. The action at the end of 2009 was to create a three-year extension of the power of the FAA/AST to license, on a "case by case basis," experimental launches of space tourism and human crew spaceplane launches, as noted earlier. This was done again to cover the period from 2012 through 2015. The most likely outcome is that the FAA/AST will extend the experimental licenses for flights by such spaceplanes for another three years.

New Questions for Doctors

It is unlikely that only completely healthy people would be allowed to fly, as is the case with professional astronauts. The commercial spaceflight business is based on the idea of making space access more available in due course to people who can afford the price of the tickets. In terms of economic probabilities, these people may not be "spring chickens."

Therefore, doctors will need to be prepared to better understand how special conditions such as pacemakers or diseases like osteoporosis are adversely affected by microgravity and elevated gravitation weight experienced in the descent from space tourism flights. Even less severe symptoms such as motion sickness and loss of appetite, common among fully trained astronauts, could represent a risk for less experienced space travelers. According to former NASA astronaut Millie Hughes-Fulford, now a University of California-San Francisco (UCSF) scientist, guidelines have to be established to face every condition that the tourists may encounter as a result of the space trip. [14]

In the December 2012 issue of *Space Safety Magazine*, Hughes-Fulford said, "Most physicians will have traveled by plane and many will have attended to a passenger in need of medical assistance while on a commercial flight. They are, however, unlikely to have experience of space travel."

The magazine also reported that the FAA and the Aerospace Medical Association Commercial Spaceflight Working Group are proposing medical recommendations for regulating commercial space travel. To date, the FAA's Office of Commercial Space Transportation has not set specific medical requirements or disqualifying characteristics for space tourists, most probably so as not to hinder the development of the sector with over-regulation. The FAA does not propose to regulate medical aspects of space passengers beyond requiring informed consent. Therefore, doctors who clear patients for space will share the responsibility of issuing that clearance with the commercial travel company selling the ticket if something happens to a patient's health condition due to the flight.

And there is some empirical evidence that the standards for a two-hour space tourism flight with four minutes or so of weightlessness may not require very vigorous health standards. In 2012, the noted English astrophysicist Stephen Hawking, who is severely handicapped by ALS, also known as Lou Gehrig's disease, took a flight with Peter Diamandis of the ZERO-G Corporation with at least 18 parabolas to experience about 40 seconds of weightlessness on each high arcing parabola. Afterward, he announced that he wanted to go up on a Virgin Galactic flight. If Hawking can be medically qualified to go on a spaceplane flight, then it would seem that almost anyone could be medically approved to go up as well.

The Case for Health Testing

But on this point there is disagreement. In the United States, a leading medical establishment has suggested screening pilots and passengers for heart disease, balance disorders, and other medical challenges posed by out-of-this-world travel. The Mayo Clinic in Arizona has teamed with a Southern California training lab and the University of Texas to provide such comprehensive testing and training for passengers, pilots, and other commercial space travelers.

"Many people have the perception that this (commercial space travel) is something that is completely innocuous," said Dr. Jan Stepanek, director of the Mayo Clinic Aerospace Medicine Program in Arizona. "This is not the same situation such as the NASA astronauts, where you can select as many as you want and you end up with the fittest of the fit and the healthiest of the healthy. Many of these people (commercial spaceflight customers) may be in their 50s and they may have some health issues, cardiac issues."

Doctors at Mayo's Scottsdale campus have already performed

medical checkups on NASA astronauts. Stepanek and another Mayo Arizona physician, Robert Orford, are board certified in Aerospace Medicine and Internal Medicine. "The most important thing is to maintain safety and provide patients with correct information," Dr. Stepanek said. "That would include what would be safe, unsafe, and potential for injury." [15]

Health Advice for Virgin Galactic Passengers

The Virgin Galactic group turned to a U.S. company for specialist health and safety advice. Virgin Galactic has given Wyle Life Sciences Group a contract to provide its passengers with medical advice. This group is also providing a chief medical officer, management services, and analysis of medical data. As part of spaceflight preparations, many of the first 100 of SS2's passengers (including Branson himself) have already been through medical assessment and centrifuge training at the National AeroSpace Training and Research Center (NASTAR) facility in Philadelphia.

Wyle is no newcomer to the world of space travel. For 40 years the company has been working with NASA by helping with analysis, medical tests, training, and other support services needed to prepare astronauts for spaceflights. The position of chief medical officer falls to Wyle's Dr. James Vanderploeg, and his responsibilities will include setting up requirements for training, doing medical checks, and putting medical protocols in place. Dr. Vanderploeg has been enthusiastic about his appointment to this position, commenting, "Supporting one of the leading proponents of private human space travel is both exciting and professionally rewarding. Wyle has the right blend of experience and knowledge to successfully support this rapidly maturing industry."

Tai of Virgin Galactic was no less enthusiastic about Wyle's support: "We are delighted to bring Wyle on board the Virgin Galactic team. It became obvious that they bring invaluable experience and resources into this important element of the program." George Whitesides has now replace Tai, but he also strongly touts Wyle's support in their role of health adviser. Of course, when ticket prices are so very high, one does expect first-rate medical advice and support both for the training and post-flight checkout procedure.

Areas of Concern

The FAA regulations went through a lot of public scrutiny as the rule-making process took place over a protracted discovery and hearing process in 2006 and 2007. But until there is actual experience with commercial passengers riding to the edge of space, no one is sure whether the safety controls for the emerging new space tourism business are indeed adequate. Certainly, areas of concern remain. The procedures that operators must follow for the notification of passengers of potential dangers currently seem to be reasonably stringent. But spaceplane operators do not currently have to report on the performance of "prototype and experimental" versions of the craft they operate. The process by which operators can perhaps shift a craft from experimental to operational status by renaming it or perhaps making only superficial changes in the design is an area of some concern.

Even more significant is a review of the formal rule-making process that the FAA followed in creating the regulations now in place. On one hand it appears very clear from the record that the FAA considered very carefully the various interventions made by participants. In some cases, the FAA acknowledged the validity of industry interventions and made changes. In other cases the FAA rejected the arguments and adopted alternative language that it believed provided greater protection and safety. Those who provided comments in the rule-making process were almost all from those seeking to operate space tourist businesses and to develop spaceplane technology. There were no individuals and organizations with the knowledge of this new technology to address proposed safety regulations from the perspective of the customer or the safety of the general public. A key issue that has never been fully addressed concerns "range safety officers." In the case of conventional commercial launches of spacecraft, and even crewed launch vehicles, a range safety officer has the responsibility of "terminating" a flight that goes off course and heads from the Kennedy Space Center toward Miami. In this case "terminating" means blowing a rocket up before it does harm to perhaps thousands of people. Nowhere is there a clear understanding of whether space tourism flights would entail a flight termination system and who would exercise control over such a system. This is why most spaceports are currently licensed in rural and remote areas away from cities, not in busy flight corridors. The Midland, Texas, spaceport that has been licensed by the FAA/AST is quite near the city of Midland and is neither on a seacoast nor in a remote area as is Spaceport America. XCOR, as noted earlier, is planning to move its flight operations for its Lynx vehicle to flights in and out of this airport with horizontal takeoff and landing. It seems clear that this facility would not have been licensed by the FAA/AST if there were to be vertical rocket takeoffs or landings.

Lessons from the Red Bull-Stratos Project Parachute Jump

The Red Bull-Stratos Project, headed by Art Thompson, successfully allowed a parachute jump from the stratosphere down to earth and set a new world record. This carefully conducted experiment showed that it was possible to survive the extreme cold and the minimally thin atmosphere of extreme altitude. This project, however, raises serious questions as to whether passengers in spaceplanes on a space tourism jaunt should do so in a "shirt sleeve" environment or whether they should be in specially designed pressurized suits that would prevent their bodies from expanding and virtually blowing up if the hull of the spaceplane were breached and the plane de-pressurized to perhaps 1/220,000th of ground air pressure. Thompson contends that at least for early flights, pressurized suits should be a mandatory requirement. Others in the space tourism business are trying to develop emergency repressurization systems that allow space tourists to be more comfortable and to float around the cabin during weightlessness. One of the prime questions concerns the kind of space insurance that companies will require of space tourism companies in order to provide coverage. If the insurance companies say pressurized suits should be mandatory, then companies that do not comply would seem to be greatly exposed to liability claims in the event of accidents. [16]

The FAA Is Regulator and Cheerleader

As far as the FAA's regulation of commercial spaceflights, Congress has placed the FAA in a conflicted situation. On one hand the official Office of Space Transportation Report—"2007 U.S. Commercial Space Transportation Developments and Concepts: Vehicles,

Technologies and Spaceports"—indicates that its mission is to "license and regulate" U.S. commercial space activities and to "ensure public health and safety and the safety of property while protecting the national security and foreign policy interests of the United States during commercial launch and re-entry operation," However, the FAA is also directed "to encourage, facilitate and promote commercial space launches and re-entries." [17]

As noted earlier, current U.S. law has directed the FAA to play the role of referee and umpire as well as cheerleader for this emerging new market. In the case of aviation, it regulates and licenses, but the FAA, since 1982, is no longer chartered to aid in the promotion of the air transportation industry.

In short, the FAA Office of Commercial Space Transportation, unlike the rest of the FAA, has been directed by legislation and Executive Order to promote the development of this new industry while also developing and enforcing safety regulations. Further it has been directed to await experience with regard to safety issues before making modifications in its safety regulations.

Experience with regard to NASA's role as the builder, operator, and risk controller of the Space Shuttle, through its Office of Safety and Mission Assurance (OSMA) and Chief Engineer, suggests that conflicting priorities can be both real and deadly. In the case of NASA, however, this was not the fault of the OSMA but rather of OSMA concerns being overridden by those of other operational units responsible for Space Shuttle missions.

In many cases, OSMA opted to sign a statement of safety reservations; these individuals were then overruled at a higher level of NASA management. Nevertheless OSMA was involved in reviewing and approving an excessive number of waivers—in some cases, these waivers led directly back to factors that, in turn, led to deadly accidents. Some argue that rockets do not kill people but waivers do.

Comparing NASA to the FAA

The roles and responsibilities of NASA and the FAA are in many ways quite different. The current duties of the FAA are largely of a regulatory nature, and it is in no way operational in terms of spaceplane development or test flights. Its primary mission is to oversee the maintenance and improvement of safety in aviation. In the future it will also be oriented to oversee and improve safety in space tourism and other commercial space ventures. As such, it might be appropriate to rename the FAA the Federal Aerospace Regulatory Administration.

The FAA is now responsible for overseeing the safety and reliability of high-altitude platform systems, high-altitude commercial balloon flights, and private space transportation systems, and it appears possible that its domain will extend in time to protospace and to LEO. It is possible that its regulatory domain—or perhaps that of the ICAO—might eventually extend all the way to geosynchronous orbit. From the regulatory perspective, the FAA can monitor key statistical elements of safety performance and risk mitigation. Although the FAA operates tracking and traffic control networks and researches new and improved systems, it does not operate airlines nor does it develop new airplanes or space systems. These tasks are left to the operators of airlines or the manufacturers of aircraft and various types of airships.

NASA, on the other hand, has a wide range of duties and responsibilities. This includes R&D for aircraft, launch vehicles, satellites, and new space exploration systems; operation of various types of these systems; and operation of networks to communicate, command, and control these networks. It has a special challenge of not only developing and operating the most experimental and state-of-the-art systems but also maintaining the safety of their operation—as an operator as opposed to as a regulator.

The FAA has one great advantage when seeking to improve the safety of aircraft. When first licensed to operate, the aircraft has had extensive flight experience. Over time each certified vehicle will have accumulated millions of hours of operation. NASA and other space agencies, on the other hand, are responsible for developing systems that are really not comparable in terms of their testing because they are typically chemical-fuel rockets that "burn explosively" for just a few minutes. It is thus much easier to test and certify aircraft produced in volume to the same specifications than it is to test and certify short-term burns of chemically fueled expendable rockets and space vehicles. Even re-usable launchers have rather short-term periods of explosive propulsion that last only minutes at a time.

NASA's attempts to improve astronaut safety are very heavily focused on developing new technology and new launch systems that can perform more reliably. The FAA safety-related processes are much more focused on management, performance goals, and operating processes. These functions include such activities as setting safety measures and objectives for aircraft, airports, weather information networks, and operational personnel as well as upgrades to existing aircraft infrastructure. NASA is perhaps 80 percent focused on new technology and 20 percent on operating processes and performance goals for equipment and personnel, whereas the FAA might be said to be 20 percent focused on new technology and research and 80 percent on operating processes and performance goals. There may be logic in both agencies shifting toward a more balanced effort as between R&D/new technology on one hand and improved operating procedures and safety culture issues on the other. [18] However, when one looks to the future of safety in high-speed travel, and particularly for space systems, it seems likely that only totally new technology can produce better performance and lower fatalities. Thus, new space transportation technologies may be the only hope for improved safety as well as "greener" systems that produce less pollution. Looking 30 to 50 years downstream, technologies such as those that use tethers, space elevators, or new types of propulsion systems (such as electric and ion drive systems rather than chemical combustion) may have won out as the way out of earth's gravity well.

Certainly in 50 years, the idea that the only way to get into space is to put humans on top of a controlled bomb may seem to be a very dangerous and dirty way to undertake space transportation.

FAA and NASA in Partnership

Not surprisingly, the FAA and NASA have been collaborating on many of these issues since forming a joint Executive Committee in 1998 "to partner on aviation safety, airspace systems, efficiency, and environmental compatibility." Their deliberations resulted in a significant announcement in June 2012 to align commercial spaceflight requirements between the two agencies and to clarify which agency governs under what circumstances. [19] Lopez-Alegria, then president of the CSF, welcomed this news. At the time he said, "With proper implementation, the agreement … will prevent over-regulation and conflicting requirements and regulations that would result in costly delays for the growing American commercial spaceflight industry."*Space Safety Magazine* commented, "Until 2015, commercial spaceflight participants fly at their own risk and will be required by the FAA to sign off that they absolve the U.S. government—and presumably their flight provider—of any liability should something go wrong." After 2015, which is likely to be when commercial spaceflights are actually ready to carry crews on a more systematic basis, the regulatory picture is uncertain, although as noted earlier, there may be another three-year extension of the experimental licensing approach.

The magazine continued, "This at-your-own-risk approach does not work for NASA, however. While NASA is not a regulatory body, it does have jurisdiction over vessels that carry its astronauts. Since nearly all the commercial space companies in the United States are vying to carry astronauts or cargo to the ISS under contract to NASA, they must also comply with that agency's safety requirements. The latest announcement serves to clarify that duality." [20]

NASA, the U.S. Air Force, and the FAA have previously reached agreements to determine how these entities will work together and to clarify interfaces, but because their missions are quite different the level of cooperation has always remained a challenge. The prime focus of these agreements actually tends to address what to do if something goes wrong.

The International Perspective

By the time regular space tourist flights actually begin to spread to spaceports around the world, it also seems desirable to clarify international law; set minimal standards and registration procedures; and create common standards for ground operations, flight termination, and space traffic management (STM). Unfortunately this will not happen. The area of STM, for instance, seems to be a number of years away and is still strongly resisted by the CSF and most of those involved in commercial spaceflight.

There is no need for these international regulations and standards to be overly restrictive, and national governments (in the country where the commercial entity is registered) could and perhaps should continue to exercise most of the regulatory oversight. This would be largely to support this new and emerging space tourism market in terms of providing certification of minimal safety standards and to ensure that nationally regulated programs do not interfere with or endanger one another at the global level—especially in the context of the space transport systems creating hazards to aviation and vice versa. And then there is the additional concern of objects that fly in or through protospace, or near space, which is currently very much like a no man's land.

In addition to the various spaceplane and "access to orbit" projects, there are an even larger number of sites where spaceports are under development, with some already in operation. This number is continuing to grow, and some form of structured control and regulation appears fully justified and necessary. However, no public international organization has so far been chartered with the responsibility to regulate the safe operation of launch vehicles, international space stations, or space exploratory missions. Some legal scholars, however, note that the Chicago Convention chartered the ICAO to regulate "vehicles" that fly, and thus they interpret this as contending that ICAO could undertake this role on an evolutionary basis.

The IAASS—A New International Initiative

One significant initiative was the formation in 2005 of the IAASS. This organization is sponsored by the ESA and the national space agencies of the United States (NASA), Japan (JAXA), Russia (Roscom), Canada (CSA), France (CNES), Germany (DLR), China (CNSA), and Italy (ISA), plus the United States' FAA and Europe's EASA, among others. This entity is dedicated to the idea of creating international standards for space safety and is currently embarked on the development of new books, monographs, and articles that record in a single place a great deal of stored wisdom and scientific data concerning space safety. It also considers and helps develop space safety regulations and standards.

The IAASS has thus sought to strengthen professional training and academic study of space safety. It is also developing more effective and unified ways to manage safety on board the ISS. The IAASS-sponsored handbook "Safety Design for Space Systems" was published in 2009. This book represented the first attempt at creating a totally comprehensive text on space systems safety design as compiled by safety engineers and space systems practitioners from around the world. [21] This was followed in 2010 with "Space Safety Regulations and Standards" [22], in 2011 with "An ICAO for Space?" [23], and in 2013 "Space Safety Operations." A new book on human factors and space safety is currently underway. All of these books have explored and documented innovations in the space safety field that will come with new technology, standards and regulation, and improved operational techniques and improved safety practices. Perhaps most importantly the IAASS, together with the U.S.-based partner, the ISSF, has sponsored the creation and publication of the new *Space Safety Magazine* and its highly informative website. As noted earlier, these organizations are also seeking to work collaboratively with the FAA's Center of Excellence for Commercial Space Transportation.

The IAASS has also undertaken a study of how commercial or private space initiatives and private space stations might be regulated and licensed at the international level. This study is being carried out in cooperation with the . The reasoning is that many of the legal and regulatory practices and approaches to safety that have been developed for global aviation might be productively applied to private space initiatives as well. A workshop on this subject under the sponsorship of ICAO, the IAASS, the ISSF, the Secure World Foundation, and other entities was held in Montreal, , in May 2013. This was followed by the joint U.N. Office of Outer Space Affairs and ICAO Symposium, held in Montreal March 18-20, 2015; it was carefully titled "Emerging Space Activities and Civil Aviation—Challenges and Opportunities." There were many interesting presentations, but no conclusions were reached and there were no proposals that ICAO should seek to assume a regulatory role in this arena of spaceplane flights and other activities that are now occurring in protospace. [24]

One suggestion that has been put forth again at the ICAO Symposium is the creation of an Independent Space Safety Board that would provide international safety certification services to the space-tourism industry.

The advantage of this approach is that it would operate on a commercial (but non-profit) and presumably streamlined basis. Thus it would not involve the elaborate and often costly operations of a U.N. specialized agency. Further, this approach would presumably leave much of the regulatory responsibility with existing national regulatory agencies. Others have proposed that all these responsibilities simply be formally assigned to the ICAO. [25]

The Role of the International Telecommunication Union (ITU)

At present, the U.N. has no specialized agency directly charged with regulating or setting international safety standards for space activities. Nevertheless some degree of international regulation and standardization is already in place, and several "space treaties" that have been signed by many space-faring nations are currently in effect. These include the omnibus Outer Space Affairs Treaty of 1967 and the follow-on treaties that address the more specialized topics of the moon and other celestial bodies, liability provisions associated with space activity and space objects, and the international arrangements with regard to astronauts.

The ITU, headquartered in Geneva, Switzerland, in particular, has developed recommended standards for the launch, operation, and registration of applications satellites, space stations, and other types of space systems in terms of registration, allocation, and use of radio frequencies. These ITU regulatory processes also include "due diligence" activities carried out by national governments with regard to the safe launch of satellites. These procedures address, for instance, how satellites, space stations, and space vehicles are to be de-orbited from low, medium, or geosynchronous orbit.

These procedures also cover the review and inspection processes that governments are requested to carry out to ensure that the separation of various stages and components in a rocket launch is conducted so as to avoid the accumulation of space debris in orbit. The ITU, however, has no directly designated role in the regulation or control of space safety, except so far as it involves radio transmissions, although its recommended "due diligence" review and inspection procedures are generally observed around the world by current space-faring nations.

[26] Another U.N.-mandated agency, the Committee on the Peaceful Uses of Outer Space (COPUOS), headquartered in Vienna, Austria, considers the broader legal issues involving exploration and use of outer space. Its deliberations have helped to develop treaties and international conventions related to the exploration of outer space, including the international status of the moon and other celestial bodies. It adopted a form of voluntary procedures to minimize space debris, and these were largely based on the procedures previously developed by the members of the Inter-Agency Coordination Committee on Debris (IADC). Currently COPUOS plays no direct role related to space safety and the reduction of risk related to space launches and space operations, except through its creation of a working group to address the sustainability of space. This working group may help to develop improved procedures to reduce space debris. The new voluntary procedures agreed to by COPUOS require unanimous consent, and this process took almost two decades to develop. Many feel these new regulations add up to "too little, too late."

COPUOS is thus, in truth, simply a coordinating committee of the U.N. organization. Its current membership includes all space-faring nations as well as 65 other nations that have an interest in space applications. It essentially gathers and publishes information about global space activities. Its small Secretariat also conducts training sessions and symposia around the world. Despite its lack of regulatory powers, it can certainly be a useful sounding board for concerns with regard to commercial space practices. It could be a key forum if it is decided that a new regulatory vehicle or convention is needed to control dangerous or harmful practices involving launching vehicles into outer space or space operations—whether these involve launch ascent or descent, in-orbit operations, or extra-terrestrial issues. COPUOS might well have input into the issue of STM, which is now a subject of increased interest and discussion around the world, but COPUOS does not currently have the resources or staff to implement such procedures.

Will Europe Follow the United States?

The EASA is well positioned to play a key role with regard to the emerging spaceplane and space tourism industry in Europe. It is already a partner in the Enterprise project, led by EADS, which is currently underway to develop a reliable new spaceplane that will operate from a European spaceport, as well as other initiatives described earlier. The key regulatory question is whether Europe will follow the U.S. example of giving the EASA the responsibility for licensing spaceplanes, space tourism businesses, and spaceports and thus controlling safety issues and concerns with regard to European space tourism operations. Project Enterprise may therefore play a role in defining how the European space tourism business will be controlled, licensed, and overseen from a safety perspective. Recently EASA reduced staffing in the area related to spaceplane safety issues, and its top regulatory authority in this area has left EASA to go to work with the Swiss Space Systems. This is interpreted by some to mean that EASA intends to follow the U.S. regulatory lead in this area and not seek to develop its own rules unilaterally.

There will be further questions as to whether national space entities will play a shared regulatory role with EASA or whether EASA will have a clearly defined transnational responsibility for European spaceplane activities. A key issue to work out is the difference of regulatory approach between the FAA in the United States and EASA in Europe. However, discussions and interchanges on this subject took place at the Sixth Conference of the IAASS and at the follow-on Manfred Lacks Conference on the Regulation of Emerging Modes of Transportation, which both took place in May 2013. These discussions suggest that ultimately the FAA licensing approach and the EASA certification process will somehow evolve to be more and more consistent. [27]

There will also be the issue of how EASA will interface with the ICAO in this area if it does indeed become the responsible global entity within the U.N. system to regulate space tourism safety and to set international safety standards. [28]

ICAO—The Way Forward for International Regulation?

The bottom line is that few international treaties or international agreements are actually in effect to provide a solid legal framework for future space activities near earth and beyond. Efforts to create such rules and regulations in the past came afoul of Cold War differences between

East and West. Even after the Cold War tension had subsided, many have sought to keep a very open and unregulated approach to future space activities. Some would say that a clear legal framework that establishes the "rules of the road"—or the rules of the "space way"—and sets forth clear legal liabilities, standards, and regulatory guidelines would actually help the development of private enterprise in space. Others, especially in the United States, suggest taking a more laissez-faire approach. This would be to let private enterprise develop capabilities under national regulations and then coordinate international standards and practices as needed.

Some of those who have studied the issue have suggested that an international entity such as ICAO could and should help in setting safety standards for commercial space launch systems, international spaceports, and even ocean-based and high-altitude launch systems. ICAO might also be tasked with conducting "due diligence" in order to protect the public good and to avoid damage to life and property where international liability might be at issue. Also, ICAO might assist in establishing explicit guidelines such as for establishing legal liability or creating a global registration process that would cover all private launch operations with a "country of record" agreeing to provide some oversight or regulatory functions related to public and air safety, frequency assignment, etc. This type of responsibility would require coordination to ensure that the ICAO and ITU did not overlap or conflict in carrying out these functions.

Is International Regulation Premature?

Most spaceplane advocates in the United States disagree. They argue that private space missions, space exploration, and space development represent a new high frontier where innovation and experimentation should be actively encouraged and that national regulatory controls are all that are needed. In light of the literally hundreds of new ideas and concepts that are only beginning to realize their potential, such efforts to set the international rules for such effort at this time would be premature.

Despite these concerns about limiting ingenuity, there would seem to be some value in chartering the ICAO to at least minimally coordinate and structure international activity related to LEO and protospace. An "intermediate position" might be something like an International Space Safety Board, operating on a commercial basis, under the auspices of the IAASS. This approach would presumably accomplish the needed minimal international coordination and leave to national governments the bulk of regulatory oversight.

During the early days of spaceflight, and continuing on until today, the ITU has registered radio frequencies associated with the launch and operation of satellites or manned missions, and it continues to do so. Each national government is charged with the responsibility for the safe operation of its launch sites.

In the 21st century, however, we are seeing an increasingly new and different space and protospace environment, where international regulations, liability provisions, and safety controls seem more and more necessary. There are spaceport projects all over the world. There could be launches from balloons, ocean-based launch sites, and cargo planes at almost any location. We also are seeing projects to launch people to LEOs via lighter-than-air craft and by the use of ion engines that will spiral out to LEO over prolonged periods of time, with an increase in the possibility of being hit by space debris and micrometeorites. Further, there are no internationally recognized standards for radiation exposure and other hazards associated with a prolonged stay in space. A study of how ICAO might assume some or all of these international regulatory duties has been carried out by the McGill Air and Space Law Institute, Booz Allen and Hamilton, and members of the International Institute of Space Law. The study is available at http://www.iaass.org and has also been published as a short book. At this time, safety regulators at NASA and the FAA have shown little enthusiasm for bringing what are often seen as high-cost and bureaucratic ICAO regulators into the business of spaceplane regulation.

International Regulatory Controls and By Whom?

At this stage it is too early to conclude what international regulatory action needs to be taken with regard to commercial space ventures. Nevertheless, action at the international level probably needs to be taken, and sooner rather than later. Already unscrupulous business enterprises have undertaken to "sell" tiny parcels on the moon and Mars with so-called deeds of ownership. "Flag of convenience" spaceports might offer launch or landing facilities that might prove unsafe in terms of interference with private, commercial, governmental, or military aircraft. The new issue of protospace, or near space, also poses national and international safety issues.

For these and other reasons, action must be taken. There is a clear need to provide for the safety of both current and future "space tourists" with some form of international regulatory controls and rules of legal liability. To date, expanding the charter of ICAO to provide for the coordination of international activities in these areas may represent the most logical and most forthright way to proceed. Such action to widen ICAO's "aerospace role" would only be a preliminary step. The first step would leave open, for the time being, whether national commercial space transportation safety standards would continue to apply to the regulation and licensing of spaceplanes, or whether some sort of non-profit organization entity might in the future help establish global standards for safety regulation. Such an entity would operate independently from ICAO but nevertheless in close coordination with it.

The Issue of Environmental Protection

Another critical question would be whether an international entity might assume regulatory control with regard to protecting the ozone layer and other environmental concerns. So far, the ESA-backed A2 craft, which would use hydrogen fuel, is the only response to such environmental concerns that would seem to offer a reliable way to diminish greenhouse gases from a growing number of spaceplane vehicles. The option of using solar energy would, of course, be an even better solution because even water vapor is, to some extent, a greenhouse gas.

The efforts to create a sub-orbital flight space tourism business are generally seen as being the first phase of a wide range of new space businesses. These enterprises may include flights to LEO, space walks, stays in private space habitats, and dark sky stations maintained at altitude by lighter-than-air craft. Beyond these ventures, there could also be commercial flights to and from the ISS under contract to NASA as well as private missions to support space sciences, materials research, manufacturing for the pharmaceutical industry, and even support to a solar power satellite industry. In the longer term, commercial spaceplane technology might be replaced by other ways to access space, such as by using nuclear propulsion, tether cables that act as a "slingshot" to lift payloads to higher orbit, and even very advanced concepts such as the construction of a space elevator system.

This new and now surging development of a space commercialization industry on a global basis clearly implies the need for a regulatory process to provide for the safety of crew and passengers as well as to protect people who work at spaceports or those residents and businesspeople who live in proximity to the launch and landing facilities.

In the United States, the FAA's rulemaking and regulatory framework establishes health requirements for spaceplane crews and sets safety standards for the operation of spaceplanes as well as for the safe operation of spaceports; these standards protect not only the passengers but also those who might live or work near spaceports. In the U.S. regulatory framework, all passengers are required to sign waivers with regard to any U.S. government liability, and the onus is largely placed on the passengers to recognize that they have embarked on an activity that entails sizable risk.

Certainly, standards for spaceplane operations as well as for the training of pilots are important issues that are still evolving. There are clearly few international facilities available for this purpose since U.S. and Russian manned space programs have dominated human exploration programs to date. In Canada, the merged Canadian Arrow and PlanetSpace organization established a sophisticated training facility for space tourism pilots, but this unit has now been shut down with the end of this initiative in February 2013. [29]

Other Dangers for Space Tourists

Constraints have been imposed by international due diligence procedures in force through the ITU, the U.N. COPUOS-backed voluntary guidelines, and the IADC recommended procedures. Nevertheless, the increase in orbital debris has continued to escalate. Although the IADC has finally come up with "guidelines" to address the problem, these guidelines, based on international consensus, are vague in several ways. Certainly the U.N. COPUOS procedures that were eventually approved and that were based on the IADC guidelines contain a number of serious loopholes in terms of implementation. Perhaps more significantly, they are not backed by any explicit incentives or sanctions and are "voluntary" in nature.

The collision of the Iridium satellite and a defunct Russian Kosmos weather satellite in 2007 created a large amount of orbital debris (about 3000 trackable objects). Then the Anti Satellite (ASAT) missile firing by the People's Republic of China in 2009 destroyed an aging Chinese meteorological satellite and created another 3000 new debris elements in LEO. These two events, coupled with the breaking apart of several Russian satellites, have only accentuated the problem. Most recently, in March 2013, debris from a Chinese test in 2009 has collided with a Russian satellite and has raised orbital debris concerns anew.

International treaties, regulatory procedures, and technical solutions for removing such debris currently remain elusive. Some experts suggest that we are reaching a "critical point," where the breaking apart of a number of satellites will create a deadly avalanche effect that will lead to LEOs becoming increasingly deadly. In particular NASA scientist Donald Kessler warned as early as in the 1980s that we could, in time, reach this type of tipping point, where the cascading effect would only increase the amount of debris elements. The only good news is that this orbital debris problem is well above the altitudes where sub-orbital spaceplanes are now expected to fly, but the hazards to the ISS and to private space habitats are very real.

Another major concern with regard to the safety of space tourists and the further development of the space tourism industry is the deployment of weapons—either offensive or defensive—in outer space or on the moon. Strict international treaties and conventions that prohibit the deployment of weapons or anti-satellite devices are seen by some as key elements of a longer-term strategy for not only the development of space tourism but successful human travel to the moon and colonization of space assets.

The U.S. space policy announced by Executive Order in August 2006 suggested that the United States will not be bound by existing space-related treaties if it feels its national security is threatened by attacks on critical space assets. The latest statement of U.S. Space Policy from the Obama Administration, however, takes a less outspoken tone. These security-related issues are addressed in more detail later in the book.

The Next Safety-Related Steps

The FAA rules now appear to be a more or less acceptable basis for proceeding with the development of spaceplane systems and for the initial operation of spaceports and the space tourism business. For better or worse, these regulations will apply for the first few years because the first flights seem destined to come from U.S. spaceports and by U.S. developed and manufactured spaceplanes. Projects from Europe, Canada, Japan, Russia, and elsewhere are now scheduled to follow suit a short while after. As these international programs emerge, U.S. precedents will undoubtedly be closely observed by others and most likely followed unless major accidents suggest alternative courses of action.

Clearly, as experience is gained in the United States and abroad, these regulatory processes and licensing operations will need to be adjusted and perfected. Thus it is our considered opinion that the FAA, in cooperation with NASA—and perhaps other international partners such as ESA, EASA, and JAXA—should try to develop some broadly agreed international "rules of the road" and "best practices" related to commercial space travel and safety standards and procedures. In addition, another initiative within the United States might also have merit. This would be to ask the White House to appoint a Commission on Personal Spaceflight Travel Safety and Security. This activity would be undertaken as a direct parallel to the previously successful White House Commission on Aviation Safety and Security.

This commission, assisted by the expertise of researchers from the FAA-created Center of Excellence and the IAASS/ISSF, might be given a specific charge to carry out the following possible program of work: It should devise ways to reduce personal spaceflight accidents and enhance flight safety by developing targeted and realistic objectives.

- It should develop a charge for the FAA (and NASA, as appropriate) in terms of creating standards for the commercial industry for continuous safety improvement, and these goals should create targets

for its regulatory resources based on performance against those standards. It should develop improved and more vigorous standards for ultimate certification and/or licensing of spaceplanes as well as for their operations, spaceports, and training and simulation facilities.
- The FAA Rules for Private Spaceflight should be rewritten with statements in the form of performance-based regulations wherever possible.
- The FAA should develop better quantitative models and analytic techniques to assess spaceplane (and especially spaceplane subsystem) performance and to monitor safety enhancement processes. These should be based on best industry practices as new operational vehicles as well as safety and emergency escape systems come on line.
- The FAA Office of Commercial Space and the Department of Justice should work together to ensure that full protections are in place, including new legislation if required, to allow full "whistle-blower" protection for employees of the space tourism business. These include, but are not limited to, manufacturers and/or operators of spaceplanes, maintenance and other ground-based crew, owners and operators of spaceports, and owners and operators of personal spaceflight training and simulator facilities. These employees should be allowed to report safety infractions or risk factors of concern regarding safety violations or security infractions to government officials without fear of retaliation or loss of employment. This would include creating a safety hotline that is similar in operation and purpose as the current NASA Safety Reporting System.
- This effort should use the White House Commission on Aviation Safety and Security as a model for this activity. [30]

Some would argue that it might be too soon to undertake such an effort and that more experience needs to be gained. The best time and place to reform and upgrade standards are when they are not fixed in concrete and where the learning experience is fresh.

The above program for the presidential commission would deal with public and passenger safety requirements for commercial space activities and systems operated or manufactured in the United States. Yet, in order for the space tourism business to flourish safely on a worldwide basis, other space-faring nations will need to develop parallel measures—and a degree of international coordination and cooperation will also be an essential ingredient. For the time being, bilateral coordination among countries may be sufficient, but sometime in the near future, some of the issues regarding the role of ICAO in space will need to be definitively answered, including the issue of what organization will be responsible for STM as well as whether one U.N. agency needs to be assigned responsibility with regard to environmental concerns, frequency allocation issues, issues related to orbital debris, and even radiation safety standards.

REFERENCES
[1] "ICAO Air Traffic Flow Projections for 2020, 2030, and 2040," http://www.icao.int.
[2] "A380: The Best Solution for 21st Century Growth," http://www.airbus.com/aircraftfamilies/passengeraircraft/a380family/.
[3] Boyle, A., *Cosmic Log*, October 24, 2006.
[4] *Wall Street Journal*, "How Safe Is the Race to Send Tourists into Space?" April 19, 2007.
[5] Ibid.
[6] "Space Tourism Set to Lift Off," May 29, 2007, http://theage.com.au.
[7] Ambrose, K., "The 30 year anniversary of the crash of Air Florida Flight 90," January 13, 2012, http://www.washingtonpost.com/blogs/capital-weather-gang/post/the-30-year-anniversary-of-the-crash-of-air-florida-flight-90/2012/01/11/gIQAEVH4tP_blog.html.
[8] The Aerospace Corporation, George Washington University, and M.I.T., "Analysis of Human Space Flight Safety – Report to Congress," November 2008.
[9] DOT/FAA Centers of Excellence, http://www.faa.gov/about/office_org/headquarters_offices/ang/offices/management/coe/facts/images/currentCoEs.pdf.
[10] The Commercial Spaceflight Federation, http://www.commercialspaceflight.org/membership/member-organizations/.
[11] Pelton, J. N.,"A New Integrated Global Regulatory Regime for Air and Space: The Needs for Safety Standards for the Protozone," Manfred Lachs Conference, McGill University, March 2015.
[12] http://www.msnbc.msn.com/id/46811246/ns/technology_and_science-

space/t/house-panel-insists-safety-private-space-travel/.
[13] http://www.space.com/16710-commercial-space-safety-faa.html.
[14] http://www.spacesafetymagazine.com/2012/12/20/health-clearance-space-tourists/.
[15] Alltucker, K., *The Arizona Republic*, May 16, 2007.
[16] Presentation by Art Thompson on the Red Bull-Stratos Project at the Sixth International Conference of the International Association for the Advancement of Space Safety (IAASS), Montreal, Canada, May 21-23, 2013.
[17] FAA, "2006 Commercial Space Transportation Developments and Concepts: Vehicles, Technologies and Spaceports", Washington D.C., January 2006.
[18] Pelton, J., and Novotny, E., *NASA's Space Safety Program: A Comparative Assessment of Its Processes, Strengths and Weaknesses, and Overall Performance*, Washington D.C., September 2006.
[19] *Space Safety Magazine*, June 20, 2012.
[20] Ibid.
[21] Musgrave, G. E., Larsen, A., and Sgobba, T., *Safety Design for Space Systems*, Elsevier Press, Netherlands, 2008.
[22] Pelton, J. N., and Jakhu, R., *Space Safety Regulations and Standards*, Elsevier Press, Netherlands, 2010.
[23] Sgobba, T., Jakhu, R. et al, *An ICAO for Space?* 2011.
[24] U.N. OOSA and ICAO Joint Symposium, "Emergency Space Activities and Civil Aviation: Opportunities and Challenges," Montreal, Canada, March 18-20, 2015.
[25] International Association for the Advancement of Space Safety, http://www.iaass.org and IAASS Report, August 2006.
[26] International Telecommunications Union, http://www.itu.int.
[27] See the "Final Report of the Manfred Lachs Conference on Emerging Modes of Transportation" and various reports and documents of the Sixth Conference of the International Association for the Advancement of Space Safety. For further background, see Pelton, J. N., and Jakhu, R., *Space Safety Regulations and Standards*, Elsevier Press, Netherlands, 2010.
[28] European Aviation Safety Agency, https://www.easa.europa.eu.
[29] Corporations Canada, "Federal Corporation Information—6363253," Industry Canada, Ottawa: Government of Canada, archived from the original on September 30, 2014.
[30] "Final Report of the White House Commission on Aviation Safety and Security," http://www.sourcewatch.org/index.php/Final_Report_of_the_White_House_Commission_on_Aviation_Safety_and_Security.

CHAPTER 10

Recap of Other Key Issues and Potential Show Stoppers

Sir Richard Branson's high-flying parties on his private island and his beaming, ruggedly handsome face on the cover of *Time* have stimulated public interest and fired the imagination of potential space tourists around the globe. His well-tuned publicity machine has kept private spaceflight in the headlines, including the announcement in April 2008 of the first space wedding, scheduled to take place on one of the first Virgin Galactic flights.

Apparently imitation is the highest form of flattery. The latest entries into the space tourism business are those now offering commercial "space tourism" high-altitude balloon ascents—and their offerings include "space weddings." These new high-altitude balloon capsule ascent companies are known as "World View" (in the United States) and "Zero to Infinity" (in Spain), and they are offering hour-long ascents to 30 kilometers (98,000 feet). They are particularly promoting a package that includes a wedding with champagne and flowers. [1]

Branson reportedly was even going to partner with Google in a new space exploration enterprise called "Virgle," which captured some headlines before it was recognized as an elaborate April Fools hoax on April 1, 2008!

Charles Simony's 13-day "First Nerd in Space" trip to the ISS in April 2007 was a tremendous PR success—for both him and the space tourism industry. Simony's girlfriend, the American TV personality Martha Stewart, certainly helped extend the coverage. Her exotic lunch bags packed with duck Pt for her boyfriend in space gave the mission a romantic twist.

The enthusiasm for space tourism has certainly commanded a lot of coverage in the worldwide media. The latest burst of publicity has been stimulated by the Golden Spike's announced intention to fly space tourists to the moon like a mini Noah's Ark, and then that was trumped by Dennis Tito's plan to fly a couple to encircle Mars. Film impresario James Cameron, Simonyi, Peter Diamandis, and others have apparently tried to one up even these exploits by forming Planetary Resources to mine asteroids as the very latest in 21st-century space exploits by entrepreneurs seeking to push the envelope.

Clearly gossip columnists are poised for the day when a superstar like Madonna or Bill Gates joins the celebrity rush to zero G. Against this background almost any commercial space project begins to sound reasonable. Bigelow Aerospace is still pursuing the idea of relatively near-term trips to private space hotels. Mars One, amid much publicity, is continuing its process of "selecting" astronauts that it would send on a one-way colonization trip to Mars. Like Alice of *Alice in Wonderland* we sometimes feel as if we are being asked to believe six impossible things before breakfast.

The real-world truth is that a host of issues must be faced and resolved before commercial space transportation and so-called space tourism can become viable and succeed. This chapter explores a range of issues that still need to be explored and resolved before true industry breakthroughs are achieved.

Real-World Issues Still to Be Addressed

More than a half dozen real-world issues still need to be definitively addressed, and some very thorny problems still must be solved. Not least of these is that this industry is taking on one of the highest-risk enterprises that one might reasonably conceive of as a viable commercial venture. The Aerospace Corporation Study, in partnership with George Washington University and MIT, that we undertook in 2006 and 2007 indicated that the "churn rate" of corporations involved in space tourism will likely remain high. For every space tourism company that succeeds, perhaps two or more will fail. Against the background of this fast-changing industry, the team at the FAA Office of Commercial Space Transportation has striven hard to develop safety regulations that will protect the U.S. government and property adjacent to spaceports yet will allow the new and emerging companies plenty of maneuvering room. People like Patricia Grace Smith, George Nield, Pam McElroy, Ken Davidian, and Kenneth Wong, among a number of other talented FAA/AST officials and engineers, have clearly been devoted to developing a new U.S. spaceplane industry and making it safe—especially in terms of public safety. The regulations developed by the FAA Office of Space Commercialization also provide reasonable safeguards for crew and passengers and provide a regulatory framework that can allow this industry to survive and perhaps ultimately even thrive. But regulatory provisions are not the only things that the space tourism business needs to succeed. In fact, several key questions still need to be addressed. These include the following issues, which could end up constituting potential "show stoppers."

Environmental Concerns and Issues

As described earlier, the flights of the supersonic Concorde high into the stratosphere presented a serious concern in terms of their potential damage to the ozone layer. Many "greens" breathed easier when the SST was grounded. The prospect of potentially thousands of flights by spaceplanes into the stratosphere raises anew these environmental concerns. Likewise the near-term development of hypersonic commercial executive jets as a parallel industry raises similar questions of even greater concern. The truth is that damage to the ozone layer may be a more urgent concern than global warming. Genetic damage due to lethal ultraviolet and X-rays from the sun could kill off the human race much faster than rising temperatures. This may seem like a quibble to some, but survival of the species seems deserving of some serious thought.

The FAA currently has overall regulatory responsibility for the space tourism industry. Associate Administrator Patricia Smith, before leaving office at the end of 2007, explained at a public forum sponsored by the Center for Strategic and International Studies (CSIS) that it does have environmental engineers examining this issue. Yet before full-scale service begins it would seem that the Environmental Protection Agency (EPA), which has ultimate jurisdiction in this area under the National Environmental Protection Act (NEPA), should examine in some depth the potential environmental impact of a large number of flights into the stratosphere and near-space by spaceplanes and supersonic commercial executive jets. This should be an examination in terms of both the ozone layer and other greenhouse gas pollutants. Further, this is much more than an American concern.

Environmental scientists around the world should examine this issue and render an opinion on both the threat level and even more importantly possible solutions. What we do know is that most of the spaceplanes now under test and development spew out either greenhouse gases such as carbon oxides or, even worse (some 400 times worse, according to some ecologists), nitrous oxides. The particulates from the aluminum polyimide fuel now used in the Virgin Galactic SS2 are a new concern, and data on this fuel and its environmental impact are of key interest. We also know that new technology such as hydrogen-fueled engines would produce water vapor as its residue. This too can trap heat within the earth's atmosphere, but this, of all options, is currently the best.

Most aircraft emit a good deal of pollution, and, indeed, it has been estimated that 15 percent of the pollution that contributes to global warming may come from aircraft. One of the reasons why the Concorde SST generated so much concern was the fact that it flew very close to the ozone layer, where destructive chemical reactions are more pronounced.

The Concorde, created as a joint venture between the British and French nearly 50 years ago, was designed to travel at a speed near Mach 2 for transatlantic flights at the high end of its trajectory. The main concern for environmental groups at that time was the effect of the plane's engines on the ozone layer, which resides at the top of the stratosphere. Further the exhaust from the Concorde was seen as compounding the air pollution problem. The plane was designed to climb to an altitude of 19 kilometers (60,000 feet), nearly 7000 meters higher than "normal" airplanes. This high altitude for the SST was key to achieving very high velocity, but it also put the plane proximate to the ozone layer. The engine also required an exotic fuel that when burned emitted nitrogen oxides (NOW) directly into the ozone layer.

Another exhaust problem is the emission of water vapor and hydroxide radicals (OH). Each type of exhaust emission contributes considerably to the greenhouse effect. This means that even a hydrogen-fueled hypersonic jet may not be a magic solution to high-altitude pollution. This is because such high-performance jets would allow the upper level of the atmosphere (the stratosphere) to cool. Even minor changes in temperature in the atmosphere have a dramatic impact on weather patterns. And this is certainly more than just a national concern. The United Nations Environmental Programme (UNEP) and the World

Meteorological Organization (WMO) have given recent indications that they may undertake a joint investigation of high-altitude jet emissions and their impact on climate change. In addition, there are other concerns such as very nasty and poisonous hypergolic fuels that are still used for spacecraft maneuvering and only now are being replace by "greener fuels," such as energetic ionic liquids, which are much less toxic and dangerous.

Risk Management for a Space Tourism Industry

The first words in the waiver statements that "Citizen Astronauts" will be asked to sign indicate that such flights are of high risk and that the spacecraft are "years away" from being certified as "safe" by the FAA. It is not clear whether a reliable insurance or reinsurance market will be able to sustain the space tourism industry on an ongoing basis. There is an even bigger question as to whether the various start-up companies could possibly provide self-coverage. This might work on a short-term basis but certainly not for the longer term. So far, third-party liability coverage by insurance companies has been elusive. This is in part because insurance companies are leery of the fact that the majority of passengers signed up for flights are fat cats with lawyers at their command. Of the 600 or so passengers who have posted tens of millions of dollars in deposits with Virgin Galactic, virtually all are high-net-worth individuals whose families would command the resources needed to hire high-profile lawyers who might seek to overturn signed waivers of liability by the passengers if a fatal accident occurred. This would be especially true if oversight by the operator or the FAA seemed demonstrably at fault.

Thus the challenges to the space tourism industry clearly go well beyond coping with environmental concerns (even though coping with the environmental impact may be a significant part of the various risks the industry will have to try to manage). In short, the space tourism industry, even more than other space applications, needs a robust and well-informed risk management system that can help these new enterprises cope with the many challenges—and the potential huge losses—that it faces. Risk issues become even more vital as a new and incipient technology such as space tourism takes off. It is believed that as spaceplanes mature and are licensed by the FAA, the number of passengers choosing to fly will increase. Prices will drop; launch operations will become smoother, easier, and cleaner; and economies of scale will take over to help stabilize the industry.

The problem is how does the industry bridge the gap from high-risk start-up to ongoing reliable and safe operations? Courtney St add, the well-seasoned former official at both the U.S. Department of Transportation and NASA, who has some three decades of experience in commercial space efforts, has had this to say about the prospects for space tourism:

> "Space entrepreneurs still tend to be seduced by the tendency to mistake technical possibility for market opportunity. Nonetheless, I feel that, by and large, today's entrepreneurs represent a particularly sophisticated and seasoned group of business managers who stand a better chance of navigating the many daunting technical and market challenges associated with the new commercial space industry sector. Although it is still a work in progress, the good news is that the U.S. government is doing a better job of fostering a more stable and predictable regulatory and policy climate for space entrepreneurs." [2]

Despite Stand's optimism, the fact remains that many of today's space tourism businesses are busily designing and testing spaceplanes with a focus on the technology. The "gee whiz" technology generates the adrenaline rush that comes from opening up a new frontier. As we know from long experience, pioneers—both territorial and technological—can often end up dead on arrival from "bleeding edge" technologies.

Will the Insurance Industry Provide the Needed Solutions?

Most of the communications satellite ventures that today rake in billions of dollars in revenues and large corporate profits would never have made it through the early development years if they hadn't been able to purchase launch insurance against the sad event wherein their satellites and/or launchers failed during lift-off or initial deployment.

Lloyd of London and some innovative brokers of the 1960s and 1970s were able to convince large insurance companies and re-insurance alliances to spread the risk of such launches broadly across the global risk-management markets. Over time they would be able to make money from this new industry by playing the odds that a few launches might fail, but that by due diligence and proper calculation of the risk a business could be built on "managed risk."

The problem with the space tourism business is that many consider it too early in the development curve to provide reasonable levels of coverage at reasonable rates. In the 1960s, when the first coverage began for the satellite industry, it involved insurance against the potential loss of the satellite, the loss of the launcher, and even the very remote possibility that the rocket might go awry and land in Miami Beach or some other populated area before the destruct button could save the day. (Actually, in the early years, the wary insurance companies were only willing to provide Intelsat, then a monopoly global consortium, with insurance against the second launch failure in a row.)

The big difference for space tourism is that there are people on board. Liability claims for people and their lives are something else, even when compared with multimillion-dollar satellites. The FAA regulations have attempted to address that issue with a rule-making process whereby all passengers engaged in a space tourism flight would receive a detailed briefing on the craft in which they are planning to ride, plus a detailed review of the possible high level of risks that would be involved.

Then most space tourism companies would proceed to require their passengers to sign an iron-clad waiver of all legal rights in light of the disclosures made to them of the risks involved. Thus one waiver holds the U.S. government completely harmless in all regards against any loss of life or damages. Then the space tourism companies ask their customers to sign a parallel waiver. The first five "Citizen Astronauts," who flew with the Russian Soyuz under arrangements made by Space Adventures, signed a similar release for their flights. [3]

It is too early to know whether the global risk management companies, the insurance and re-insurance industries and others will be willing and able to provide the extra layers of coverage that the various space tourism companies need to become viable. If there are no accidents and several years of positive experience are achieved before the first space tourism mishap materializes—especially if there are perhaps injuries, but no fatalities—the viability of the overall enterprise will become clear. Some daring entrepreneurs from the global insurance industry have made possible the now vibrant space applications businesses, enabling the launch of communications and remote sensing satellites, which contributes over $100 billion a year to the world economy. It is still too early to know how today's men and women in gray suits and equipped with actuarial tables and computers will respond to the new challenge.

Will the United States extend its liability protection provisions to shield the new commercial spaceplane industry against truly catastrophic losses? This is just one of the issues to watch in terms of U.S. Congressional action in this area. Can the international commercial insurance or re-insurance markets fill the gap? Will state governments in the United States, or perhaps other countries wishing to attract this new business to their jurisdictions, step up to meet this large risk management need? Already, France, within its space legislation, has moved to offer greater liability protection for conventional space launches. Clearly this is an area to watch closely.

Other Regulatory and Taxation Concerns

Even after the waivers and insurance questions are resolved, a number of other regulatory concerns are awaiting the space tourism companies almost at every turn. These companies, at least those in the United States, need to make sure that even if they have jumped through the right hoops at the national level, state laws or local ordinances don't hold them responsible for large claims if there is an accident involving public safety on the ground or for claims by the families if there is an accident on a flight. In Europe, other issues seem rather prosaic or just details, but they nevertheless involve a large price tag.

In Europe, for instance, Value Added Taxes (the dreaded VAT) often run to 25 percent. Thus a $200,000 trip could end up costing the passenger $250,000. In Sweden the Esrange airport is a site that might be converted into a spaceport for Virgin Galactic passengers wanting to see the Aurora Borealis up close and personal. Thus appropriate tax relief legislation is being considered; this would allow space tourism flights to be taxed in the same way as hot air balloon flights, allowing passengers to pay far less than the current 25 percent VAT.

The EASA may ultimately be ceded much more control of spaceplanes in an attempt to give it control of all "aviation" flights in the region. This would lessen the ability of a country seeking to provide regulatory, tax, or liability relief to do so if the European Union and the EASA have assumed overall control. The space tourism industry will presumably have to address and resolve such issues in each legal and taxation regime around the world.

Finally, in the United States, those seeking to develop new spaceplane technology are today significantly constrained by ITAR, which limits the export of spaceplane-related technology that can be and will be increasingly available from other sources in Europe, Japan, Russia, and perhaps China and India, if not other parts of the world. In recent months, Congress has passed legislation that does transfer the ITAR review responsibility for application satellites from the U.S. State Department to the Commerce Department. This move will streamline ITAR approvals for these types of satellites, but launch technology remains under a stricter approval process. [4] U.S. firms may currently lead in the spaceplane field, but this lead might not last for long if ITAR restrictions are not further relaxed in the area of spaceplane technology. The bottom line is that regulation of safety for spaceplanes may today be the prime focus, but the truth is that a host of other regulatory and tax issues still need to be addressed as well.

Orbital Debris

One of the problems with orbital space debris is that there is no official definition as to what this means. All of the U.N. space treaties refer only to "space objects." In general, however, orbital space debris refers to material that is in orbit as the result of space missions but that is no longer serving any function. There are many sources of such debris. Collisions between debris elements or between debris elements and active satellites or missiles are today the largest sources of debris.

Fig. 10.1 Visualization of low earth orbital debris. *(Courtesy of NASA.)*

In addition, there are micrometeorites, so-called meteor showers (such as the Leonids), and other natural sources of hazards to satellites and spaceplanes. These hazards include cosmic radiation, the solar wind, and very powerful solar storms. These sources of natural hazards such as micrometeorites, cosmic radiation, and other phenomena that could harm spacecraft or endanger astronauts for the most part remain fairly constant, and the magnitude of their danger can be reasonably calibrated or special alerts can be issued. Debris from various space programs and the aftermath of ASAT attacks, however, continues to increase in number, and the average size of this debris has increased in recent years. Although some graphs show a leveling of debris due to increased due diligence against explosive bolts and controls on fuel tanks and de-orbiting, other studies suggest that breakup of larger debris could start an "avalanche effect" that could greatly increase orbital debris. At this point we have competing models as to what the future holds.

One source of debris is discarded hardware. For example, many launch vehicle upper stages have been left on orbit after they are spent. Many satellites are also abandoned at the end of their useful life. Another source of debris comes from parts of spacecraft that result from deployments and separations. These parts include everything from separation bolts to momentum flywheels, clamp bands, auxiliary motors, launch vehicle fairings, and adapter shrouds. Material degradation due to atomic oxygen, solar heating, and solar radiation has resulted in the production of particulates such as paint flakes and bits of multilayer insulation. Solid rocket motors used to boost satellite orbits have produced various debris items, including motor casings, aluminum oxide exhaust particles, nozzle slag, motor-liner residuals, solid-fuel fragments, and exhaust cone bits resulting from erosion during the burn.

A major contributor to the orbital debris background has been object breakup. More than 124 breakups have been verified, and more are believed to have occurred. Breakups are generally caused by explosions and collisions with other objects in space, but the majority of them have been caused by explosions that can occur when propellant and oxidizer inadvertently mix, residual propellant becomes over-pressurized due to heating, or batteries become over-pressurized. Some satellites have been deliberately detonated. Explosions can also be indirectly triggered by collisions with orbital debris.

Some 70,000 Objects Detected in Space

Three major collisions are known to have occurred since the beginning of the space age. The cause of approximately 22 percent of observed breakups is unknown. Approximately 70,000 objects estimated to be 2 centimeters in size have been observed in the 850 to 1000 kilometer altitude band. NASA has hypothesized that some of these objects are frozen bits of nuclear reactor coolant that are leaking from a number of Russian RORSATs, although this has not been conclusively determined.

At altitudes of 2000 kilometers and lower, it is generally accepted that the man-made orbital debris population now dominates the natural meteoroid population for object sizes 1 millimeter and larger. The issue of greatest concern is that the amount of man-made debris continues to increase and that an avalanche effect, once triggered, could make the problem much worse.

Various computer models with names like EVOLVE, CHAIN, CHAINEE, and Nazarenko have been developed by NASA, the ESA, a German university, the Russian Federation Space Agency (Roskosmos), and others. All of these programs seek to project future trends in debris buildup and the likely number of collisions that might be expected to occur in coming years. These models have used different assumptions and different experimental results, but despite variations, the conclusions are generally parallel. The major finding is that the growth of debris is expected to accelerate unless there is a halt to space launches and improved controls are placed on satellite operators to limit collisions that are the major source of debris. The problem is clearly most severe in LEO, below 2000 kilometers. [5]

The increase in orbital debris, particularly in LEO, continues to expand despite the due diligence requirements developed by the IADC and formally adopted under the auspices of the U.N. COPUOS.

Indeed, the growing problem with orbital debris is taking on scary proportions, particularly since the Chinese ASAT test in January 2007. This test destroyed the Fengyun 1C meteorological satellite and led to the largest single source of orbital debris since the start of the space era. This missile test generated well over 2000 debris elements exceeding 5 centimeters in size, which is more than large enough to destroy a space shuttle window or significantly damage the ISS. Recently a piece of this debris struck a Russian satellite, and the orbits of the debris from the missile strike could directly intersect with the ISS.

The solar wing on the Hubble Telescope was severely damaged by a piece of orbital debris. The number of significant pieces of space junk—the size of a baseball or larger, or about 10 centimeters or larger in diameter—has continued to increase, with some 22,000 objects of this size or larger currently being tracked. As long as orbital debris continues to increase, the effective use of space for commercial, experimental, governmental services, or strategic purposes can only be said to decrease.

Fig. 10.2 Heat-shield damage on the Hubble Telescope, which astronauts repaired in 2001. *(Courtesy of NASA.)*

In February 2007, there were four other new sources of major space debris. The breakup of two Chinese satellites and of two large Russian launch vehicle components added hundreds more debris elements. Corrective action could have prevented the spread of much of this debris. A year later, in February 2008, it was reported that the U.S. spy satellite known as satellite 193 was destroyed in orbit by a missile fired from a warship in the Pacific Ocean. Reports indicated that 99 percent of the debris from this satellite de-orbited within days of the missile firing. This happened, in part, because the satellite was already in de-orbiting mode and because the missile impacted the spy satellite in a downward arc rather than ascending upward.

Equipping all satellites with fuel to either de-orbit or move beyond the GEO orbit into a safe parking orbit is actually the solution that U.N. guidelines suggest, but there are no enforcement penalties and these are only voluntary. Venting fuel so that a satellite does not explode is another critical step. The bottom line is that orbital debris, despite efforts taken to date, is a mounting problem, especially for LEO.

Fortunately orbital debris is much less of a problem for sub-orbital flights, but for those planning trips to space stations, the risk is very real. Just a mere chip of paint a few millimeters in size is enough to crash through a space shuttle window or tear through a space suit and cause a fatal accident.

The U.N. COPUOS has had this problem "under study" since 1994, but the problem has only grown worse since then. The IADC committee did finally agree on seven "guidelines" to address the problem, but these guidelines, now largely adopted by COPUOS, are based on international consensus. They are vague, riddled with loopholes, and legally non-binding. The adoption of these guidelines on a unanimous basis within the U.N., in fact, represents only a baby step forward on orbital debris. [6]

Meanwhile hundreds of billions of dollars of space assets from remote sensing and telecommunications to spy satellites are increasingly at risk. Growing amounts of space junk in orbit—especially LEO—could easily damage GPS navigation satellites, among others, that are used not only to target missile firings but also to allow planes to take off and land safely all over the globe. From the Internet to global electronic funds transfer, from close satellite monitoring of hurricanes and tornadoes to our defense and civilian space assets, we live in a modern technological society that is more and more vulnerable if our satellites in the sky should fail. A new effort is clearly needed to address this growing menace to spaceflight, space exploration, and space applications —not just for space tourism, but for all space-related systems and missions.

The good news is that sub-orbital flights as now planned have a very, very low risk of being hit by orbital debris. However, Bigelow Aerospace and others have serious plans for LEO commercial space habitats. These private space stations and habitats could be very much at risk from orbital debris. There are many reasons why the issue of orbital debris must be taken on to preserve the use of outer space, but commercial space exploitation just adds to the sense of urgency that this mounting problem be addressed sooner rather than later.

In recent years, this problem has become of sufficient worry that the development of technology to engage in active debris removal has continued to move forward along with techniques for in-orbit repair, maintenance, or even construction. Some techniques to address orbital debris would involve using energy or laser systems on the ground to change debris orbits to avoid collisions. Other systems involve sending up missions to assist with the de-orbit of debris elements. Such efforts would address the largest debris elements first because collisions of the largest objects create the most new debris. [7]

The Weaponization of Space

The August 2006 National Directive on Space from the White House indicated support for private initiatives in space and space commercialization. The same directive also indicated that the United States would take all necessary actions to protect space assets and otherwise implied that further "weaponization" of space could occur. The Obama Administration has now issued a new U.S. Space Policy that did not renounce the earlier Bush Administration's National Space Directive, but it certainly softened the tone of and muted the possibility of potential military actions in space. In fact, the United States has now actually indicated large areas of agreement with the ESA's proposed international Code of Conduct for space, which calls for a more collaborative and international approach to space. But is even this enough of a context to move ahead with rapid commercial use and development of space? Clearly, it would be even more desirable to have an international convention or treaty that clearly ruled out the weaponization of space for all space-faring nations and that set the stage for more national and commercial collaboration in the space arena.

Today, the aerospace industry is a several hundred billion dollar enterprise worldwide, but the launch vehicle business is only a small percentage of this total. In fact, the commercial launch business experienced a state of decline in the early 2000s, with commercial liftoffs sinking to a level of only 12 to 15 launches per year during this period. More recently, there has been an increase in demand, and the number of commercial launches has since grown to an annual level of around 20 to 30. Further, it should be noted that a number of these launches are lifting multiple payloads, with rocket systems like Ariane 5 and Atlas V capable of putting over 10,000 kilograms into orbit. Nevertheless, the demand for military aerospace systems has outpaced all commercial space growth in recent years. Around 2000, many predicted that commercial space activities would outpace military space programs and would represent more than twice the economic level of expenditures by 2007. This quite simply has not occurred. Military space is still the largest gorilla in the global economic jungle of outer space affairs. Warfare in Iraq and Afghanistan may well have been part of the reason. Stiff competition from fiber optic cable systems has certainly limited communications satellite growth, whereas remote sensing by satellite has yet to achieve its potential. Space positioning, navigation, and timing satellite systems have certainly been among the most rapidly growing areas of space applications in the past decade.

The U.S. National Security Policy Directive of August 2006 concerning American space policy was announced rather unilaterally by the White House and proceeded to set forth at least two principles that put the utilization of outer space in both a security and military context:

> "The United States considers space capabilities—including ground and space segments and supporting links—vital to its national interests. Consistent with this policy, the United States will: preserve its rights, capabilities, and freedom of action in space; dissuade or deter others from either impeding those rights or developing capabilities intended to do so; take those actions necessary to protect its space capabilities; respond to interference; and deny, if necessary, adversaries the use of space capabilities hostile to U.S. national interests.

> "The United States will oppose the development of new legal regimes or other restrictions that seek to prohibit or limit U.S. access to or use of space. Proposed arms control agreements or restrictions must not impair the rights of the United States to conduct research, development, testing, and operations or other activities in space for U.S. national interests." [8]

In short, the 2006 Space Directive did not sound entirely promising for the civilian-oriented space tourism business. The sabers in space seemed to be rattling. New international agreements to control spaceplane traffic, under the auspices of the ICAO or other entities, would seem difficult to implement if the 2006 U.S. Space Directive were to be taken as meaning what it says.

This statement of principles, as set forth in the directive, would seem to suggest that the United States strongly views outer space first and foremost in a strategic and military context. Such a stance, if taken literally and compared to statements by the USSR during the Cold War, would seem to make it at least questionable for commercial organizations to invest in "space deployed infrastructure" or even orbital tourism enterprises if they feared that their high capital value investments were to be subject to military attack or perhaps appropriation for military or national defense purposes.

The new U.S. Space Directive from the Obama Administration does not entirely undo the Bush Administration's more bellicose statements, but it clearly softens the military and strategic elements and seems to set the stage for more international agreements. Now is very much a time of a crossroads in space. On one hand, rogue states like North Korea are seemingly threatening possible military uses of space and even suggesting the potential delivery of weapons of mass destruction against their enemies via missile systems. On the other hand, there is increasing discussion of an international code of conduct for space activities, exploration of an international system for space traffic management, and the prospect of many multilateral and cooperative space projects including international commercial undertakings. This clearly is a confusing set of circumstances, with winds blowing from two completely different directions. This is such a key area of concern that the next chapter will address this issue in much greater depth.

Clarification of Conflicting U.S. Governmental Roles in Space

The law under which the FAA now regulates space tourism indicates essentially two things. One is that the FAA is directly responsible for the public safety of the surrounding populace and proximate air space as well as that of those involved in spaceplane operations. This is the FAA's prime duty. Passengers and pilots and crew for spaceplanes fly at their own risk with the knowledge that what they do is very high risk. The other current responsibility of the FAA/AST is to promote the growth and development of this new industry. The FAA/AST has worked closely and effectively with the CSF to develop what appear to be reasonable safety regulations, and the rest of the world seems to be following similar approaches in response to the U.S. lead.

As highlighted in earlier chapters, putting agencies in the role of both "cheerleader" and "referee," however, is always a troublesome concept. NASA's role, serving as its own safety monitor for the Space Shuttle and the ISS, was clearly exposed as a problem by the Columbia Accident Investigation Board and the earlier *Challenger* investigations.

Fig. 10.3 The Columbia Accident Investigation Board. *(Courtesy of NASA.)*

The nuclear industry has its regulatory controls handled by a separate government agency, but this agency is not charged with developing nuclear power as well. For a number of years, the FAA was charged with promoting civil aviation, but legislation was enacted to remove this responsibility after a major airline crash over a decade ago. All these issues, such as whether the Department of Commerce rather than the FAA/AST should promote space tourism, should be reviewed after regularly licensed flights begin to take place (after the experimental period ended in 2012). However, in April 2012, the U.S. Congress extended this "learning period" by three years, and it seems likely that another three-year extension may move this period forward until 2018.

At the same time, consideration should also be given to the possibility that aerospace research might be transferred from NASA to another governmental agency. The progress being made by JAXA with its Mach 4 spaceplane and the ESA efforts to develop the A2 hypersonic craft—both of which are fueled by liquid hydrogen and oxidized by LOX—are important developments. It is certainly not clear that NASA is keeping up with these advances in Japan and Europe. Currently languishing NASA and FAA aeronautical research programs suggest that as the space tourism industry moves forward, Congress needs to consider serious policy issues. It can be argued, however, that commercial developments, instead of governmental research, will be the best way forward in terms of U.S. research efforts in this field.

Health Concerns: Radiation Effects and Genetic Mutation

Many believe that worries about emissions in the stratosphere by aircraft are just a matter of environmental concerns and all about global warming. Admittedly, this is a very serious problem. The greenhouse effect is leading to the warming of the lower atmosphere across the globe. But beyond the environmental concerns, there are specific health concerns for people here on earth today.

In particular, ozone depletion from jet emissions allows more ultraviolet radiation into the lower atmosphere. This UV radiation at high enough levels leads to genetic mutation in all animals, including humans. Trust us when we say genetic mutation is a real "no no" when it comes to preserving the human race. And some would thus say that this is a problem we should be looking out for in case it happens in coming decades. However, these people would be way wrong in their surmises. We have serious warning signs to indicate that this problem is happening already. [9]

Frogs in the extreme southern latitudes are already showing genetic mutations. People in places like New Zealand and Southern Australia are experiencing elevated incidence of skin cancer. No one really knows if this is due to problems with the ozone layer and the so-called ozone hole or other phenomena such as what NASA research satellites have recently detected: the appearance of what seem to be "cracks" in the earth's geomagnetic field, and these might be allowing more radiation and deadly gases to penetrate down to the ground surface. [10]

Environmentally, it was estimated in the 1970s that a large fleet of supersonic aircraft in constant operation would reduce the ozone layer faster than the total worldwide output of chlorofluorocarbons (CFCs). These CFCs, in terms of length of chemical interactions, are still considered to be among the largest threats to the ozone layer today. Interestingly enough, the debate in the 1970s centered exclusively on the Concorde, whereas the United States—and other countries—also had, and still have, a fleet of supersonic bombers and fighters that produced the same emissions at similar altitudes.

In the 1970s and 1980s, the issue of high-altitude pollution was of major concern, but for obvious reasons the fears about these problems, which are now linked to the health of humans and wildlife—plus global warming—are even greater today.

What Next?

The space tourism industry excites our imagination, and we believe that in time commercial spaceflight will represent a key step in the evolution of humankind. One may hope that commercial space ventures can move the human race forward to new vistas where the sky is no longer the limit. Space Adventures has already announced plans to send "Citizen Astronauts" around the moon, and commercial stays in space habitats and space walks may also be coming. Credible individuals have even been seriously discussing commercially designed and operated trips to the moon and a human orbit around Mars.

The best way for space tourism to succeed, however, is for the above potential "show stoppers" to be addressed and resolved now rather than later, when retrofitting and retooling systems can be much more expensive and difficult.

The concept of "Citizen Astronauts" has gone from science fiction to tomorrow's fact. However, much more needs to be done. We have suggested that attention needs to be given to key issues such the environment, radiation and health-related issues, adequate liability coverage and risk management arrangements, minimizing and ultimately eliminating orbital debris, and addressing the very real concerns about space weaponization.

These and other related issues will become increasing problems if we do not undertake to solve them now—and in coordination with space-faring nations and enterprises around the world. If we proceed to advance from spaceplanes and space tourism to stratospheric and hypersonic transportation systems, the scale and seriousness of these issues will escalate rapidly. The next steps we take will make many of these issues of greatest import. If we cannot address and resolve these types of issues in the incipient days of spaceplane travel for the purposes of "tourism," then trying to address them as we enter the age of hypersonic transport will be twice as impossible.

At this stage, a large increase in the number of jet airliners traveling in the stratosphere is a much greater worry than the increase in the number of spaceplanes. The scope of environmental and health-related problems will only escalate. We need to develop a space insurance capability to allow space tourism businesses to thrive, but the precedents could also allow new modes of air travel to evolve. Perhaps most importantly, we need to forthrightly address the dangers of the weaponization of space in an uncertain world where terrorism is a global concern. A misstep here could not only destroy the space tourism business but also lead to much, much larger concerns. If weapons are deployed in space, this could truly lead to World War III. Or without the right protections and controls, a rogue state or technologically sophisticated terrorist cell could create terrorist destruction on a scale previously unknown in the world.

The FAA intends to review its rule making based on experience. We have elsewhere recommended a White House Commission that will undertake work that closely parallels its effort with regard to civilian aviation and its safety enhancement. Further, a safety hotline, similar to

NASA's Safety Reporting System, is also recommended. But this is only for starters.

The issue that should be addressed in the near future is how to eventually separate the FAA's duties and responsibilities as "safety referee" from that of market "cheerleader." It is never a good idea to combine these roles, at least over the longer term. This is why Congress eventually removed the FAA's promotional role from aviation and why at some point NASA should either be in charge of either space transportation operations or space safety, but not both.

The haunting question remains: What might have been the outcome for the ultimate *Challenger* and *Columbia* Space Shuttle missions if we had just separated NASA's role as space system developer from safety standard enforcer? In this case, the role of Congress in setting the mission goals and then also controlling the purse strings and ultimate schedule could be questioned in terms of their culpability as well.

Anyway, NASA has at times seemingly deserted its role as developer of advanced aviation and aerospace technology. The military and aerospace companies have essentially inherited this role although NASA Glenn is still doing some work of significance to develop new jet engines that burn much more cleanly. This work, however, does not extend as far as seriously investing in developing hydrogen-fueled engines or more advanced R&D. We thus see Congress as "missing in action" when it comes to key questions about setting a clear course for U.S. R&D as well as safety oversight in a number of aspects of space, aerospace, and aviation. Lord knows we probably do not need another study commission, but perhaps Congress might task a qualified non-profit study group that reports directly to Capitol Hill staff to address such issues as the following:

- Should development of advanced aviation and aerospace technology and safety innovations be transferred from NASA and thus make it only the National Space Administration? (These tasks along with the Langley Research Center and parts of Glenn might, for instance, be transferred to the Department of Transportation along with a viable budget.)
- Should the FAA continue to be charged with regulating the safety of

space tourism while at the same time being asked to promote its development? If not, should these functions be moved to the Department of Commerce or elsewhere?

- Is NASA responsible just for developing new rocket technology? Exciting new possibilities might provide lower-cost, safer, and more environmentally sound technology to lift humans and payloads into space. These include tethers, space elevators based on nano-tube systems, solar sails, nuclear propulsion, advanced ion engines, and even lighter-than-air craft. It seems that in many ways NASA's strategic plans are flawed and unbalanced in terms of space versus advanced aviation, and in terms of rocket systems versus everything else. NASA's goal, at times at least, seems skewed to the interests of powerful political and corporate forces, and NASA has an unimaginative approach to developing truly breakthrough technologies. NASA's only attempt to think outside of the box and envision truly new technology was known as the NASA Institute for Advanced Concepts (NIAC). But this was shut down due to the all-too-familiar budgetary constraints. Such efforts probably need to reside within the National Science Foundation (NSF) or a Civilian Advanced Research Projects Agency, if it were to be resurrected in future years.

- Further NASA's sprawling and decentralized research centers sometimes end up overlapping in their missions and remain unavailable to support other parts of the U.S. government. Reforms could start with making these centers into Federally Financed Research and Development Centers (FFRDCs), as recommended by several previous study commissions, including most recently the Aldridge Commission.

- To state the obvious, the ITAR regulations need to be further revised and streamlined. The process today is cumbersome, expensive, and ineffectual. The system clearly has shot U.S. aerospace industries in the foot by making it impossible to compete for many international contracts for communications satellites, remote sensing satellites, and other aerospace research awards. McGeorge Bundy said that if you have a process that protects diamonds and toothbrushes with equal vigor, then you will probably lose fewer toothbrushes and more diamonds. Certainly, ITAR needs to be revised to protect the diamonds and to worry a lot less about the strategic value of toothbrushes.

This is just the start of what could be a much longer list of issues. We hope the focus can remain on a new U.S. vision in space that can look a couple of decades ahead.

Congress should recognize that all is not well or right in the world of NASA, the Department of Transportation, or the FAA. This is not to say that these institutions do not do a great deal of good, nor is it to say that they haven't made key technical strides forward in many areas. However, it is to say that much greater progress is still possible, and that commercial innovation should be part of the future solution. U.S. legislators should attempt to work out some of the key "wrinkles" that have emerged in the fabric of aerospace research and development. U.S. agencies that address space and advanced aviation and aerospace technologies should be given new direction and new goals, and in some cases, hands need to be untied. Corporate innovation related to spaceplanes, new space and aviation transport systems, and so on should be strongly encouraged, and serious environmental, public safety, and regulatory issues need to be addressed.

NASA and the Department of Transportation need to be gone over with a "policy iron" that smooths out the "wrinkles" and restores leadership to U.S. space and aviation programs. In short, Congress and the White House need to work together to create a national space agenda that is a whole cloth and not a raggedy patchwork kludge sewn together by special interests and often quite short-term thinking.

REFERENCES

[1] "$75K 'Edge Of Space' Balloon Ride Gets FAA Approval," Space.com, http://www.space.com/23289-75k-edge-of-space-balloon-ride-gets-faa-approval-animation.html. Also see "Zero to Infinity," http://www.bloostar.com/#!company/c1s5y.

[2] Pelton, J. N., *Space Planes and Space Tourism*, George Washington University, Washington D.C., 2007, p. 5.

[3] Pelton, J. N., *Space Planes and Space Tourism*, George Washington University, Washington D.C., Appendix 3, 2007.

[4] Listner, M., "President Signs ITAR Reform Into Law But Not Everyone Is

Happy," January 16, 2013, http://www.examiner.com/article/president-signs-itar-reform-into-law-but-not-everyone-is-happy.

[5] Technical Report on Space Debris, United Nations Committee on the Peaceful Uses of Outer Space, 1999, http://orbitaldebris.jsc.nasa.gov/library/UN_Report_on_Space_Debris99.pdf.

[6] Pelton, J. N., "Orbital Debris and Space Hazards from Space," Springer Press, New York, 2013.

[7] Pelton, J. N., "New Solutions for the Orbital Debris Problem, Springer Press, New York, 2015.

[8] National Security Policy Directive of August 31, 2006, on U.S. Space Policy, http://www.fas.org/irp/offdocs/nspd/space.html.

[9] Pelton, J. N., "Orbital Debris and Space Hazards from Space," Springer Press, New York, 2013.

[10] Ibid.

CHAPTER 11

Spaceplane Systems and Their Strategic Application

Why spaceplanes? There are many reasons why they might be developed. These include space tourism, supersonic or hypersonic transportation, space science experiments, testing of new materials and systems, or even low-cost launch of small satellites. Then there is national security. The private sector can and probably will develop spaceplanes to serve all of these potential "markets," but governmental funding related to spaceplane programs, not too surprisingly, will focus largely on national security and defense-related applications.

NASA's various ill-fated spaceplane projects have now all been canceled. The U.S. civilian space agency abandoned its development of spaceplanes as a prime objective in 2004. This was when NASA was asked to refocus on missions to the moon and Mars. But it well could have been seen as somewhat of a rebuke for its series of failures to develop a successful spaceplane after many years of futile attempts in this direction.

The X-37 spaceplane (or Orbital Test Vehicle [OTV]) program was transferred to the U.S. Air Force and the DoD, although NASA continues to play a minor role in this program. In addition, the X-43, after achieving a record speed of Mach 10, has also been discontinued by NASA and moved to DoD. In short, NASA has redirected its efforts away from spaceplane projects. In light of two decades of failed attempts in this direction, many would say, "Hurray."

In truth NASA's efforts to develop one version or another of spaceplane technology have been an embarrassment. Anyway, NASA finally gave up. Failures, or at least discontinued efforts, included the HL-20, X-33, X-34, X-35, X-37, X-38, and X-43. Ironically, NASA had defined the development of spaceplane systems as an essential part of its mission up until 2003, but it then turned on a dime to say, "Oh no. This is a mission for someone else."

Fig. 11.1 The X-43 Unmanned Scram Jet Technology Demonstrator, whose development has now been turned over to the U.S. Air Force.
(Courtesy of NASA.)

Today, NASA is concentrating on working with companies such as OSC, SpaceX, Sierra Nevada, Boeing, and Lockheed Martin to develop commercial lift capability to the ISS. But in the latest awards, the concentration is exclusively on a capsule return rather than on a spaceplane.

Only the Sierra Nevada Dream Chaser, which NASA originally supported under a commercial development model, would have fit the spaceplane profile. However, in the last round of the COTS development program awards, nearly $7 billion in awards went to Boeing and SpaceX. Sierra Nevada and the Dream Chaser did not make the cut for the final two corporate developers.

The U.S. military, long a strong proponent of a spaceplane to orbit and back as one of its prime objectives, has clearly moved in to take over the main governmental role in this type of development. But there are clear differences between the NASA and DoD approaches. Defense applications are not driven by flying crews into space, but rather to acquire intelligence or to deliver weapons to targeted objectives. This means that unmanned and remotely controlled spaceplanes, much like UAVs, are of significant interest to defense organizations. Thus DoD spaceplane development appears to be increasingly moving in the direction of remotely controlled spaceplanes. These systems would not endanger the lives of pilots, they could operate in space for sustained periods of time, and they can be designed and implemented at lower cost because life support systems are not required.

Strategic Use of Spaceplanes

During the George W. Bush Administration, the strategic aspects of space systems and technology mushroomed in importance. On one hand, NASA was asked in 2004 to pursue the vision of exploring outer space and mounting a mission to the moon and Mars. On the other, the White House asked the DoD to take over developing spaceplane systems and near-earth military space applications. The nature, purpose, and scope of U.S. spaceplane development programs in the defense arena, however, are not entirely clear. In fact, because some of the programs are classified, no one is entirely sure what capabilities have been or are now being developed. For instance we do not know which spaceplane systems are to be robotically controlled and which can be manned but perhaps will also be capable of remote control. What is known is that many advocates in the advance defense systems planning arena advocate a prime role for robotically controlled spaceplane systems.

There are ongoing efforts around the world to develop spaceplane technology that may lead the way towards space tourism goals in the private sector, but parallel governmental programs seem to have strategic defense-related objectives as well. In short, all space-faring nations, at least the United States, Russia, China, Japan, and France, probably have as one of their objectives the improvement of their strategic space capabilities and related spaceplane technology. The potential strategic objectives for spaceplane projects certainly could include many different applications. These might, for instance, include better space reconnaissance systems, better communications and networking capabilities, and rapid deployment of certain strategic capabilities to space. Military spaceplanes might be deployed for other tasks such as use as an ASAT system, use in deploying small micro-satellites, active orbital debris removal, satellite servicing, on-orbit repairs, and even retrofit or manufacture.

There is, of course, even the possibility of future deployment of military weapons systems in space, the use of spaceplanes to disable space-based hostile laser systems, or carrying out other defensive actions against offensive space weapons. A spaceplane might also be able to deliver weapons of various types and forms. Unlike a ballistic missile, a spaceplane could change its orbital trajectory and be able to disguise its entry path for the delivery of a weapon system such as a bomb. It could also be configured to deliver weapons at extreme velocity that could, on impact, destroy a targeted underground and fortified bunker by employing the so-called "Rods from God" method of attack.

Deadly Start of Space Weaponization—In 1944

There is an interesting and deadly precedent here, and it dates back to World War II and the German rocket program developed by Wernher von Braun. Some 3000 V2 rocket bombs were launched against targets in the U.K. and Belgium. These weapons were Hitler's final attempt to reverse the outcome of the war by halting the advance against German forces through Europe and to terrorize the population of Great Britain.

These rocket bombs, fired from launch sites in Germany, Belgium, and Holland, each carried a warhead of 2150 pounds of Amatol, reached an altitude of 55 miles (88.9 kilometers), and had a range of some 200 miles. Over 6000 of them were built by Germany's forced labor workforce, and over 50 percent were actually launched—1402 of them targeting the U.K. Accuracy was a considerable problem, but the records show that these bombs, rained to earth from above, caused the deaths of 2700 civilians and more than 10,000 injuries. Prime Minister Winston Churchill's war memoirs make it clear that he was seriously concerned about this threat after learning the details of this startling new development in Nazi Germany from an intelligence source in 1943. This was when von Braun and his team gave a secret demonstration to Hitler at the Peenemunde research establishment in northern Germany. Hitler was so impressed that he gave the order for 10,000 V2 rockets to be built—a way, perhaps, for him to change the course of the war.

In London, Churchill's war cabinet recognized that this was a new weapon for which there was no defense, and that the potential for these new rocket-delivered bombs was to damage morale. The only defense was to attack the factories producing the V2's and supplying the rocket fuel—and to destroy the launch sites then being constructed. The

RAF began a series of risky low-level bombing raids targeting the main factory at Peenemunde, at a heavy cost to aircraft and their crews. Following the D-Day landings in June 1944, the V2 threat influenced the war strategy so that the direction of advance went northwards from Normandy to neutralize the operational launch pads. Hitler responded by moving the production and launch facilities back into Germany, and regular V2 attacks continued on southeast England and then on the Belgian port city of Antwerp after it had fallen to the Allies. Some V2 rockets also hit Paris after it had been liberated. [1]

V2 Rockets Continued to Strike the U.K. until March 1945

The U.K. had already experienced the pilotless V1 bombs—known colloquially as "doodlebugs" or "buzz-bombs." Londoners became accustomed to listening for their unmistakable engine noise, and then taking cover when the aircraft went silent. It was not until September 1944 that the first V2s began to fall on London, but Churchill waited until November before making a low-key statement to Parliament about "the German rocket attacks." By then, the Allied advances in France and the Low Countries towards the Rhine and Ruhr were making headline news, but rockets continued to fall on Britain until March 1945, when the last recorded V2 bomb hit Croydon, south of London, just four months before VE Day.

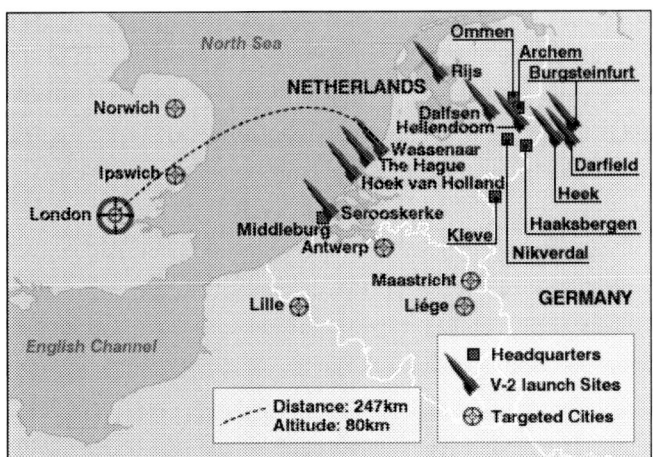

Fig. 11.2 The V-2 Rocket threat to Britain in 1944 and 1945. *(Courtesy of BBC.)*

As the war moved towards its close, the progress made by the German rocket industry was of great strategic interest to the Russians and Americans, who both sought to retrieve equipment and personnel to assist in their own space research programs. In fact, von Braun and his team surrendered to the U.S. forces rather than fall into Russian hands. And von Braun himself subsequently played a major role in the development of America's manned space program.

In the Cold War environment that ensued after World War II, the German rocketeers "rescued" by Soviet forces became a key component in the Soviet team that developed rockets and atomic weapon missile systems for the U.S.S.R. This was mirrored by U.S. efforts led by von Braun to develop ICBMs to deliver American atomic weapons. Thus, space systems of all types have, over the years, tended to become or evolve from weapons systems. Missiles were developed to deliver explosives and even nuclear devices. Communications systems were used to aid strategic communications. Spy satellites were developed for espionage. Geodetic and space navigation systems were used to target bombs and for strategic mapping and instant location determination for blue and red forces. Now spaceplanes could well become the latest capability to be deployed for strategic military purposes.

Cold War Proliferation of Weaponry

As the Cold War heated up, or should we say became more frigid, there was an ever-increasing proliferation of space weaponry to support the incredible strategy of Mutual Assured Destruction (MAD). For decades, ever more sophisticated missile systems kept the American and Russian populaces in a constant state of apprehension. New technologies such as Multiple Intercontinental Re-entry Vehicles (MIRVs) and anti-ballistic missiles (ABMs) evolved. Then came attempts to create a "shield" over an entire country against nuclear missiles. This was officially known as the Strategic Defense Initiatives (SDI), but the public called these attempts to create a nuclear shield by the more popular name used by the press—Star Wars.

With the fall of the Berlin Wall and the demise of the U.S.S.R., space-based MAD-driven nuclear-tipped missiles began to be decommissioned and from the 1990s the world began to breathe a bit easier. Despite these changes, strategic uses of space continue to be a part of the modern condition. Although there have been calls to create a treaty that would "outlaw" space weaponization, military experts claim that such a ban on weapons is not verifiable. Even though there is an Outer Space Treaty that obligates signatories not to put nuclear weapons in orbit, there is no ban on sub-orbital flights. And there is no clear definition of what constitutes "militarization" of space.

There is certainly a gray line as to when a communications satellite, a remote sensing satellite, a space navigation network, a laser system, or a steerable space vehicle "is or is not" a space weapon. Some have suggested that even the now-grounded Space Shuttle could have been used as a space weapon. Certainly a spaceplane designed for sub-orbital flights for space tourism could also be used for strategic applications, as both an offensive or defensive space weapon. For these reasons governments are likely to regulate the use of spaceplanes, and the U.S. ITAR trade restrictions will likely apply to the sale of spaceplanes or their components. As noted previously, the capabilities and technologies that are currently being developed around the world suggest that these ITAR controls will really have little effect if spaceplane technology can be purchased from other suppliers in the global aerospace market.

Ironically, Sir Richard Branson could not obtain technical information about SS2, which he was funding and purchasing for his Virgin Galactic fleet, due to ITAR restrictions.

What is clear from recent history is that space systems are a vital element in modern warfare in both a defensive and offensive context and that spaceplanes figure into the military planning of a number of space-faring nations.

Onwards to the 21st Century and "Project Constellation"

Space-based systems, whether commercial or defense based, are becoming more and more sophisticated and diverse. A growing number of potential defensive or offensive applications for space systems can today go beyond the delivery of explosives, nuclear devices, or bio-chemical weapons. There are indeed defense-based programs that are a combination of space and strategic ground systems. These include the so-called "C-2 Constellation" project of the U.S. Air Force. This project aims to integrate space assets, linking them together instantly or at least in "near real time" with various sensor, navigation, and communications systems. Such a project would interconnect space, air, and ground sensors via so-called "netcentric" communications networks that extend right out to the edge of military operations where actual hostilities may be occurring. Thus Project Constellation would hook up and integrate incoming and outgoing data via high-speed computers. Countless sensors of all types and at all locations around the world would be available for military operations, regardless of whether the sensor was located on land, at sea, on board aircraft, or even aboard UAVs or satellites.

The various elements of Project Constellation would, over time, grow to include the Mobile User Operational Satellite (MUOS) system and the Wideband Global Satellites (WGS). At one time these various capabilities were to have been linked by the super-fast laser-based satellite communications orbital ring known as the Transformational Satellite System (TSAT). This super-fast network, which would have encircled the globe, has now been canceled due to budgetary restraints. All of the existing and new satellite capabilities are currently planned for orbital insertion using conventional rockets, but spaceplanes might allow future elements of low earth orbiting systems, such as small space-based sensors, to be deployed and serviced at lower cost.

The deployment of these "netcentric" warfare systems by the United States and its allies depends on space sensors, space reconnaissance, space navigation, and space communications networks. Robotic spaceplanes could be another element in this overall net-centric information and action network.

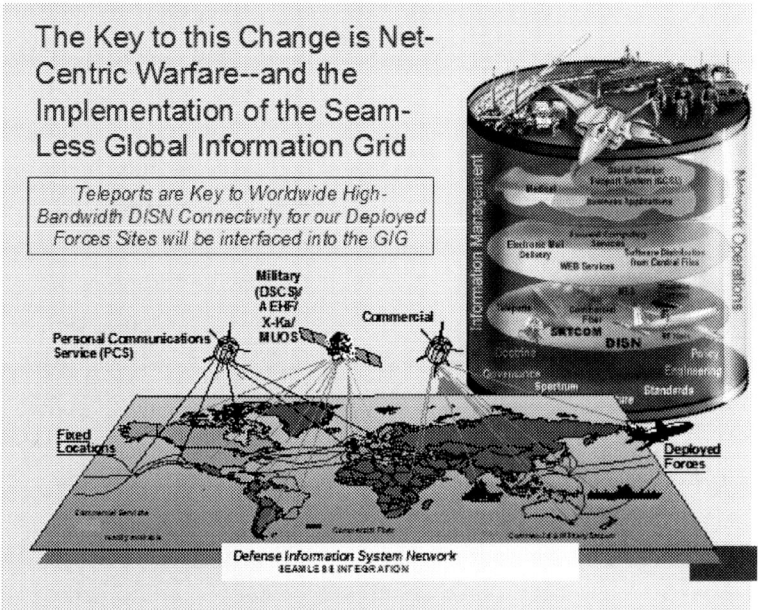

Fig. 11.3 The many dimensions of a Global Information Grid (GIG).

Clearly spaceplanes, once developed, will be utilized for a number of functions and will increasingly become a part of the strategic planning process. The public had visibility into the X-43 and X-37 spaceplane programs, but only because NASA initiated these programs. Now, however, these are classified projects. While they were still unclassified projects, these vehicles were successfully tested up to speeds on the order of Mach 10.

Space-Based Hardware Is Part of Today's Reality

The use of space-based communications relays, space sensors, and anti-satellite weapons (ASATs) is part of today's reality. Perhaps, in time, we may see spaceplanes or very high altitude UAVs with many types of capabilities that include advanced communications, reconnaissance, and surveillance. Others could be used to instantaneously deploy "theater communications systems" associated with a particular geographic hostility. Most significantly, we might see spaceplanes armed with what might be called space weapons. Crewed and highly maneuverable spaceplanes armed with various types of weapons would likely change the strategic dimension of space systems, and they would certainly create problems and complications for commercial space enterprises. It is clear that many of the U.S. commercial enterprises pursuing the development of sub-orbital spaceplanes and/or those seeking access to LEO are funded by or are otherwise being encouraged in their efforts by NASA, the DoD, or both. Essentially three types of enterprises are pursuing what might be generally characterized as the "space tourism" or the "commercial access to space" market. These are the major aerospace corporations, which have traditionally been heavily funded by defense as well as NASA civil contracts; the mid-tier aerospace corporations, which often support the major aerospace corporations in many defense-related contracts; and the true start-up companies, which may also find that their future success depends on governmental contracts and support from the defense sector. The bottom line is that perhaps half of the money for the spaceplane industry may come from defense budgets—not only in the United States but in other countries as well.

U.S. Spaceplanes from the Military Sector

Currently the DoD has at least two active spaceplane development programs. One is the X-37 system, which was jointly developed by NASA and the USAF but turned over to the DoD in late 2004, when NASA abandoned this project to focus its efforts on the moon/Mars project.

The Air Force is currently working on an unmanned spaceplane that is directly based on NASA's X-37 program. This craft was once planned to be the direct follow-on to the Space Shuttle. If successful, this spacecraft would be the first since the shuttle to be able to return experiments back to earth for analysis. What is unique about this particular development is that it has been redirected to operate robotically for sustained periods of time, with missions that last up to nine months in orbit. This reusable OTV, as now designed, would be about one-fourth the size of the Space Shuttle and would deliver objects into LEO in its experimental bay, much like the shuttle's payload bay except smaller. The OTV could then continue orbiting for months before bringing the objects back to earth. This would allow it to test how satellite components react to long stays in space, among other tasks. The OTV can and will help test technologies for more advanced reusable space vehicles.

The OTV's first flight occurred in April 2010, launching on an Atlas 5 rocket from Cape Canaveral Air Force Station in Florida. The first one or two flights did not carry any experiments. Instead, these flights tested the craft's ability to autonomously re-enter the atmosphere and land, and to probe new landing gear and lightweight structures that can withstand the heat of re-entry. The OTV is designed to land in California, either at the Vandenberg Air Force Base, which hosts expendable rocket launches, or Edwards Air Force Base, which acted as an alternate landing strip for the Space Shuttle and is a proving ground for experimental jets and rocket planes. [2]

Boeing's Wide-Ranging Role in Space

The Air Force Rapid Capabilities Office manages OTV. The Air Force Research Laboratory, the Defense Advanced Research Projects Agency (DARPA), and NASA also collaborate on the program. NASA's role is, however, largely defined by handing off its institutional knowledge from the preceding X-37 program.

Boeing, not too surprisingly, serves as prime contractor for the OTV program. This selection of course stems from the fact that Boeing was the commercial lead for the NASA X-37 technology demonstrator. There is accordingly a significant amount of continuity from the X-37 to the OTV program. The OTV is nevertheless a new program, and new technology that represents much more than simply a name change is involved with this development.

According to a statement from the Secretary of the Air Force, the OTV program focuses on "risk reduction, experimentation, and operational concept development for reusable space vehicle technologies, in support of long-term developmental space objectives." This is the rather vague type of statement that is often associated with a classified military program, and thus the actual use of the OTV could be for almost any purpose.

In just a few more years, a range of U.S. space vehicles could return experiments to earth. The OTV is unmanned, whereas the craft that Boeing and SpaceX are developing under the NASA crewed COTS program should be able to take astronauts to and from the ISS within the next two years or so. The OSC Antares launcher and its Cygnus capsule is just a one-way cargo system and burns up when returning to the earth's atmosphere, and thus it is not being developed as a human-crewed system.

When the X-37 was a NASA civilian space program in the late 1990s, it was to be the first of a planned series of flight demonstrators that NASA called Future X. At the time, the X-37 was described as a robotically controlled, autonomously operated vehicle designed to conduct on-orbit operations and to collect test data. Its special thermal blanketing system and shape were designed to withstand the super-hot temperatures associated with re-entry from earth orbit by deceleration from speeds as high as Mach 25. This is a velocity of around 26,500 kilometers per hour (about 18,000 miles per hour), sufficient speed to reach LEO.

NASA's Aborted X Projects

The initial plans for the X-37 envisioned that it might be lifted to orbit by either a space shuttle or a launch vehicle like the Atlas or Titan and would then be deployed in earth orbit. The vehicle would fly through space for some period of time and perform a variety of experiments before re-entering the atmosphere and landing on a conventional runway. NASA transferred its X-37 technology demonstration program to DARPA in late 2004 without completing this development program. This was but one of the Future "X" programs that NASA interrupted and canceled, in part due to delays and cost overruns, and in part due to the demands of the Space Shuttle program and the drain on expenses associated with the ISS. [3]

NASA had earlier planned on two X-37 unmanned vehicles. One of them, which was actually built, was known as the Approach and Landing Test Vehicle (ALTV). The other craft, which was designed but not built, was known as the Orbital Vehicle. The purpose of the ALTV was to test re-entry trajectories from within the atmosphere for possible later use on the more advanced Orbital Vehicle. This vehicle demonstrator, which DARPA "inherited" from NASA in 2004, was actually tested in April 2006 with less than ideal results. In the test, the vehicle was released from Burt Rutan's White

Knight. The White Knight, a very high altitude carrier aircraft, is, of course, best known as the aircraft that carried the SpaceShipOne to high altitude, from where it blasted its rocket engine to win the $10 million Ansari XPRIZE. The ALTV version of the X-37 ALTV, however, was not able to maintain course in landing and ran off the runway. The X-37 is not the only governmental program to develop spaceplane technology that NASA passed on to the military. The other program, known as the X-43, was designed to prove the viability of so-called scram jet technology.

X-43—Another Future X Program Discontinued by NASA

This is the first truly successful "air breathing" rocket engine to provide sufficient oxidizer via intake airflow to fuel a continuously firing rocket engine at multi-Mach speeds. This program is now proceeding on a classified basis.

NASA initiated this hypersonic craft in partnership with DARPA. NASA undertook to develop at least three X-43A craft for testing. DARPA also agreed to design in parallel its own version of a scram jet hypersonic vehicle, known as the X-43B and X-43C, to various scales. After successful tests by NASA, the X-43A reached nominal test speeds of Mach 10 in 2004. NASA turned over the subsequent development to DARPA after these tests were completed.

In addition to the X-43, the DoD is carrying out tests of other spaceplane systems, including prototypes developed by Lockheed Martin, Boeing, Northrop Grumman, OSC, and others. Further, in 2006, there were disclosures in *Aviation Week and Space Technology* (sometimes aptly nicknamed the "Aviation Leak") that there might be a now discontinued two-stage-to-orbit spaceplane operated by the U.S. Air Force. This unverified vehicle was supposedly designed by Boeing so that the first craft would fly to very high altitude and the second stage could then be "rocketed" into LEO. This elusive and perhaps entirely mythical craft called the Blackstar (also known as the SR-3/XOV) is, in theory, a now moot proposition, because the *Aviation Week* story claims that this secret program (if it ever existed) has been shut down, either for cost or performance reasons. [4]

Certainly, experts have considered the existence of such a plane as "unlikely" for a variety of cost and technical performance reasons. The so-called X-47A, however, is apparently quite real, and there could well be other classified high-altitude jet, scram jet, and rocket systems. These various defense-related developments have also led to potential spin-offs such as Lockheed Martin's new Quiet Super Sonic (QSS) prototype for a hypersonic personal executive jet for transoceanic service. [5]

Efforts to Develop Less Ambitious Hypersonic Aircraft

Although NASA has abandoned true spaceplane development, it has continued efforts to work with industry and military interests in the Navy and Air Force to develop much less ambitious new aircraft. In particular, NASA is working on three different prototypes known, respectively, as the N+1, the N+2, and the N+3 vehicles. The N+1 vehicle prototype is seeking nearer-term sonic boom reduction, the N+2 is seeking a supersonic vehicle ready for flight in the 2020-2025 time frame, and the N+3 is aimed to develop a vehicle for the 2030-2035 time frame. These developments are targeted first and foremost at sonic boom reduction, and secondarily to high altitude emission reduction and development of vehicles with speeds in the range of Mach 2. The joint development with the Navy and the Air Force has focused on such issues as sonic boom mitigation and more efficient input systems. [6] JAXA in Japan has recently reported about its development of hypersonic vehicles, which can fly at speeds up to Mach 4; it would appear that these will be available well before the N+3.

The problem is that NASA's SST development program essentially runs from the NASA Dryden Center, but this, at least, appears to operate independently of the work going on elsewhere in NASA and in the emerging new commercial spaceplane industry, which has far more ambitious goals and objectives for hypersonic travel.

The technical difficulties to be overcome are significant: (i) sonic boom mitigation; (ii) emission reduction; and (iii) input design and fuel efficiencies. There are, however, regulatory issues to be overcome as well. There is actually what might be called a "Catch 22." Aerospace developers such as Boeing, Gulfstream, and QSST look to the FAA to establish standards for acceptable noise levels for SST vehicles. Yet the FAA appears likely to await more technical development before adopting new standards. [7]

In the United Kingdom and Europe, similar types of research are also being carried out under the so-called Horizon 2020 program, the Clean Sky Joint Technology Initiative, and the longer-term Flightpath 2050 initiative. Although there are similarities with the U.S. initiatives, the European developments are far more centered on aircraft fuel efficiency, the reduction of carbon and other greenhouse gas emissions, and even electric vehicle propulsion. [8]

Similarly, European aerospace developments in aviation seem separated from the ESA's much more ambitious initiatives to develop projects such as the recent Bristol SpacePlane company's design for a 300-passenger spaceplane that could travel at speeds up to Mach 6 and complete trips between London and Australia in about three hours.

The future sometimes evolves as a series of small steps and refinements. Conversely, some spurts of development—what Larry Page, the CEO of Google, describes as moonshots—represent breakthroughs that are not in 10 percent increments but in giant 10X leaps forward. [9] We will discuss these topics in Chapter 12.

As far as these various R&D programs, it is difficult to sort out what exactly are commercial programs, what are strategic and military development programs, and what are so-called dual-use programs. Clearly the X-37 and X-43 spaceplane developments started out as civil space programs and then converted to military programs in the United States. The various hypersonic development programs that are being undertaken by NASA, the ESA, and JAXA are currently civil programs, but in future years they could end up being strategic/military programs. Currently the strongest element of separation seems to be that crewed vehicles are essentially civil development programs, whereas robotically controlled and piloted vehicles seem to be strategic/military programs. In the next chapter, we will discuss this key area of emerging technology in hypersonic transport in the commercial sector.

Inter-Agency and Governmental Tensions

For a number of years, there has been budgetary tension between NASA (as the civilian space agency), the U.S. Air Force, the U.S. Congress, and the U.S. Executive Branch (especially the Office of

Management and Budget). One of the issues clearly has been about which agency should develop spaceplane technology. It seemed that the solution to this problem emerged in 2004 with the so-called moon/Mars Vision Statement. This presidential directive assigned the top priority to the U.S. space agency to develop space systems to go to the moon and Mars. The bottom line of this moon/Mars initiative is that NASA has turned over the spaceplane initiatives (to wit, the X-37 and the X-43) to the U.S. Air Force and DARPA.

Many see NASA's abandoning of spaceplane development as both a logical and desirable decision. They ask, "Why should NASA fund development of spaceplane technology when the DoD has the funds and is willing to take over this development?" A follow-on comment from the Space Exploration Foundation says, rather cynically, that the Air Force and DARPA can certainly do no worse than NASA, which has unsuccessfully undertaken half a dozen spaceplane projects and terminated all of them when they were behind schedule and well over budget. NASA applications for a spaceplane are limited, especially since it has now awarded its latest COTS research contracts to SpaceX and Boeing. These contractors, however, are also receiving funds from the DoD. SpaceX is funded by the Air Force to develop the Falcon launcher to launch defense satellites at lower cost. Likewise, XCOR, in announcing its new Lynx spaceplane in 2008, released information that it has a contract with the U.S. Air Force to share information with regard to the testing and performance of this new vehicle. Boeing has significant funding from the U.S. Military, and Sierra Nevada has also won some U.S. military awards. In Europe, EADS, Thales, Alenia, and BAe are likewise receiving large defense-related contracts and support from military and strategic funding sources.

Human Control versus Robotics

The potential applications of spaceplanes for strategic purposes have been variously listed as: (i) reconnaissance vehicles that provide a fast response to emergency conditions (including natural and manmade disasters as well as dynamic and critical war conditions); (ii) rapid response and elusive systems for delivering military strikes and bunker-penetrating projectiles ("Rods from God"); (iii) quick deployment of small satellites on demand; and (iv) other applications that include research, testing of materials, and re-supply of various different application satellites. In many, if not all, of these applications, robotically controlled, remotely piloted vehicles would seem to be most well suited to these tasks and presumably could do so for the least cost. Several key debates are currently going on within the strategic defense community. One concerns whether it is possible to use space for military purposes under current space treaties. Some argue that you can use space for offensive military strikes if you believe you are under threat of attack. Others argue that you can employ defensive space military weapons such as anti-missile systems only to create a protective shield. Yet others argue that no military use of space is appropriate or even "allowable" under space treaties. Finally, the legal aspects of military applications in sub-space or the so-called "protozone" have yet to be worked out as well. There are questions as to whether there is a difference in the legal status of intercontinental ballistic missiles with limited steering control, unmanned but robotically controlled vehicles, and crewed vehicles with pilots aboard in terms of military uses of this region between altitudes of 21 and 140 kilometers. The protozone is the area above which commercial aviation vehicles do not generally fly (21 kilometers) and the altitude under which space objects cannot fly (140 kilometers). [10]

Remote-Controlled versus Human-Controlled Space Systems

Growing debate within military circles is emerging, not over the existence of space and protospace weaponry but rather over how it is controlled. This debate is between those who believe that military spaceplanes and like technology should be designed to operate under human control only and those who argue for robotically controlled vehicles. The advocates of robotic systems maintain that such space weapons systems are more efficient, less costly, and even more accurate in achieving their various potential objectives. Many would argue that having pilots in the cockpits of spaceplanes is an overrated concept. It might look exciting or even heroic in a "Star Wars" movie, but from the standpoint of cost, performance, duration, and effectiveness, robotic operations could make more sense (and could cost much less).

An article by Major David M. Tobin of the U.S. Air Force presents in great detail the reasons why the cost of providing life support systems for humans in space is enormous and explains that computers, avionics, and robotic control systems can often perform strategic missions better and at lower cost. It is a little known fact that the Space Shuttle could have been designed to operate without a crew and that a proposal to replace the *Challenger* with a robotically controlled shuttle without a life support system was seriously considered. [11] This would have made the *Endeavor* a much lower-cost vehicle, and had other shuttles been retrofitted for robotic operations, the tragic loss of life with the *Columbia* might never have happened. Over the last 50 years, however, NASA has consistently promoted "human-piloted" space programs because top U.S. space officials know from detailed public opinion polling that astronauts hold human interest, whereas "expendable" machines, in the form of robots, do not. Scientists from the Planetary Society have argued for years that we could find out much more about the solar system for much less by using remotely controlled machines that do not have to breathe oxygen or require expensive life support systems.

In the case of military spaceplanes, clearly arguments can be made on both sides of the "crew versus automated controls" debate for such vehicles. It is significant that at this point, both the OTV and the X-43 scram jet projects are currently unmanned programs. In the carefully analyzed and argued paper by Major Tobin, the conclusion is as follows:

"A first-generation Military Spaceplane (MSP) could function without a man on board—but whether it operates autonomously or under the close supervision of ground controllers remains to be seen. This first-generation MSP could execute at least a portion of all [defined] space-mission areas. It could overfly any point on the planet to deliver a strike payload or conduct a reconnaissance mission. On a counter-space mission, it could destroy hostile satellites using kinetic-energy projectiles or directed-energy beams. As a reusable launch vehicle, it could perform a simple yet critical space support mission—satellite deployment." [12]

Many factors support the development of a first-generation MSP without people on board. First, it could satisfy the near-term mission requirements as currently defined. As the less expensive alternative, it stands a greater chance of being funded. Finally, the absence of a crew, their life-support equipment, and a dedicated cockpit help reduce the vehicle's operating weight and allow sustained space operations. Given the technical challenges involved with single-stage-to-orbit flight, any opportunity to reduce the vehicle's mass is advantageous. [13]

Will International Agreement Limit Development for Military Use?

There are two pathways forward with regard to military spaceplanes. One involves space policy and international law, and the other involves technology. There is no doubt that new technology will bring forward new capabilities with regard to military spaceplanes, both the crewed and non-crewed kind. The key, however, is whether new international agreements will limit the use of space for military purposes, and whether in an offensive or defensive mode. There have been no new international space treaties since the 1970s. Therefore, the United Nations Office of Disarmament Affairs in Geneva, Switzerland, has lately tended to focus on approaches such as transparency and confidence building measures (TCBMs) and codes of conduct with regard to strategic space activities rather than seeking new space treaties.

The potential is certainly there for nations to agree to "de-militarize space," but in the new era since 9/11, a question that remains is "What about terrorist groups?" Who knows when techno-terrorists might attain access to space capabilities.

The strategic future of space, however, involves more than just military use of spaceplanes. Today there is also rising concern about space debris and its potential to have a very destructive impact on all types of private space commerce. In time, the alarming rise in space debris could be a threat to telecommunications, remote sensing, navigation, and meteorological satellites as well as to military satellites for surveillance and weapons monitoring.

We also need new ways to address the strategic issues raised by the increasingly sophisticated space program. This may mean creating a new entity to address such issues as the international regulation of space tourism and related travel through the air space of various countries. Likewise, we need more effective international control of orbital debris. Certainly, we need improved global capabilities to monitor near-earth objects as well as comets and asteroids that might pose a hazard to humanity. A step further seems to be the December 2013 action by the U.N. General Assembly to create a new International Asteroid Warning Network (IAWN) and the Space Mission Planning Advisory Group (SMPAG). Further, it would be a good idea to figure out a way to implement the de-militarization of space and effective regulatory control of space systems of all types. So far, the U.N. has had limited success at addressing such space issues effectively. [14]

The Outer Space Treaty of 1967, which has been signed by 100 nations and all space-faring countries, seems to prohibit the weaponization of space. Article III reads: "States Parties to the Treaty shall carry on activities in the exploration and use of outer space, including the moon and other celestial bodies, in accordance with international law, including the Charter of the United Nations, in the interest of maintaining international peace and security and promoting international co-operation and understanding."

Article IV further indicates: "States Parties to the Treaty undertake not to place in orbit around the earth any objects carrying nuclear weapons or any other kinds of weapons of mass destruction, install such weapons on celestial bodies, or station such weapons in outer space in any other manner. The moon and other celestial bodies shall be used by all States Parties to the Treaty exclusively for peaceful purposes. The establishment of military bases, installations and fortifications, the testing of any type of weapons and the conduct of military maneuvers on celestial bodies shall be forbidden. The use of military personnel for scientific research or for any other peaceful purposes shall not be prohibited. The use of any equipment or facility necessary for peaceful exploration of the moon and other celestial bodies shall also not be prohibited." [15]

This treaty, however, needs to be read in terms of what it does not say as well as what it does say. It does not prohibit the launching of sub-orbital weapons of mass destruction; it does not ban the deployment of military communications, navigation, targeting, or surveillance satellites nor of laser, particle beam weapons, or other non-specified military or defense-related space systems.

Albert Einstein once said that the hard part of any issue is finding the right question. At this point, the right question seems to be how can we organize the world's affairs so that space-related issues and concerns can be better managed and controlled—especially to avoid star wars and to make it safe for civilian passengers.

Other Space Programs—For Warfare or Prosperity?

It is likely that military spaceplane programs are currently active in other countries around the world. These could well be underway in China, Russia, and Europe, plus possibly under classified programs in other countries. At this time, no specific unclassified information is available on such programs. The European or Russian aerospace firms that would be involved in developing such a capability would, in any event, be the same companies that are involved in developing new spaceplane capabilities for civil governments or private space tourism. There is also widespread interest in the concept of developing reusable spaceplanes that could operate robotically. The design of vehicles that could be operated via ground control commands would appear to be an attractive safety feature if the pilot (or co-pilot) of a spaceplane became incapacitated. The tragic March 2015 Lufthansa crash in the Alps, which was caused by the co-pilot's deliberate suicidal actions, vividly demonstrated there could be other emergency situations as well.

The escalation of warfare into outer space—whether manned or unmanned—should be a major concern to all. As new technology evolves, it will be more and more difficult to put the genie of sophisticated space weapons back into the magic lamp of world peace. Today, the Aladdins that can rub the magical lamps of scientific research and produce increasingly menacing space weapons are most likely to be allies. But in another decade or two, these weapons can find their way into the hands of terrorists and rogue nations, and they could hold global society hostage.

In the past, some have foolishly suggested that we can restrict the "nuclear club" to a handful of responsible nations. This has proven to be wishful thinking. Seeking to develop space vehicles that can deploy a plethora of weapons of mass destruction is equally misguided. The best way forward for our troubled world is not more and more weapon systems—particularly space-based weapons. No, the best path to stability and reasonable tranquility will involve better global education systems, economic development, and ecologically responsible practices. In all of these areas, space systems offer economic efficiency, effectiveness, and equity. A space race to create better and better weapons in the stratosphere and beyond will ultimately end in despair.

It is ironic that as "billions" are being spent to pursue destructive wars around the world, it is difficult to get appropriations for "millions" to support such projects as the Global Legal Information Network (GLIN). This is an electronic system accessible via a browser, and it was designed to allow scores of nations plus the U.N., the EU, the OAS, and the Arab League to share their legal system of laws and thus to develop a worldwide rule of law. It is likewise true that a few million dollars could also help start a global television university modeled after the Chinese TV University, which was started in 1986 under Intelsat's Project Satellites for Health and Rural Education (SHARE). This began in Beijing with the support of the Ministry of Education and Central China Television as part of the 20th-anniversary celebration of Intelsat. This system is now bringing education, training, and health to over ten million students and citizens in remote parts of rural China. India has now deployed its Edusat satellite system, which, as in China, is supporting over a million students. Indonesia, Mexico, Malaysia Nigeria, and other countries have done the same. A similar satellite tele-education system deployed globally could ultimately show that, over the longer run, space tele-education can do more for world stability than space weapons.

In conclusion, then, the question here is what will be the future of spaceplanes? Will new space systems evolve from the space tourism/space adventure activities to also provide commercial transportation systems at hypersonic velocities, or possibly support new strategic weapon delivery systems? Or will all these applications and yet others arise—the good with the bad?

Commercial space systems could come sooner and at lower cost, if the right objectives and synergies were set into motion now. Currently, environmental and health concerns, future transportation needs, defense-related requirements, and even anti-terrorist concerns come into the picture. Space can be used to fight star wars, support future transportation needs, and develop new space-based businesses. Contemporary trends suggest that all these forces will be at work. In the next chapter, we explore in greater detail how aerospace transportation at hypersonic speeds may fit into this future picture.

REFERENCES
[1] Churchill, W., *The Second World War, Vol. 6*, Cassell & Co, London, 1956.
[2] David, L., "U.S. Air Force Pushes for Orbital Test Vehicle," Space.com, November 17, 2006, http://www.space.com/news/061117_x27b_otv.html.
[3] Pelton, J., et al, "Space Safety Report: Vulnerabilities and Risk Reduction in U.S. Human Space Flight Programs," George Washington University, Washington D.C., 2005.
[4] "Two-Stage-to-Orbit 'Blackstar' System Shelved at Groom Lake?" http://www.collectspace.com/ubb/Forum30/HTML/000303.html.
[5] "All Sonic, No Boom" *Popular Mechanics*, Vol. 270, No. 3, pp. 64-67, March 2007.
[6] Wilson, J. R., "SST Research: Breaking New Barriers," *Aerospace America*, pp. 26-30, January 2013.
[7] Ibid.
[8] "Europe's New Plans for Research and Funding," *Aerospace America*, pp. 4-5, January 2013.
[9] Levy, S., "Big Thinker: Seven Massive Ideas That Could Change the World," *Wired*, pp. 66-71, February 2013.
[10] Pelton, J.N., "A New Integrated Global Regulatory Regime for Air and Space: The Needs for Safety Standards for the Protozone," Manfred Lachs Conference, McGill University, May 2014. Also see Pelton, J.N., "Regulatory Issues in Space Communications and Space Traffic Management—And Especially the Importance of the Protozone," ABA Conference on Space Law, June 6, 2013.
[11] Workshop at George Washington University of Space Shuttle Safety, 2007; and interview with Lewis Peach, former NASA long-range space systems planner.
[12] Major Tobin, J., U.S. Air Force, "Man's Place in Space-Plane Flight Operations Cockpit, Cargo Bay, or Control Room?"
[13] Major Tobin, J., U.S. Air Force, Air Command Air University, Report Number AU/ACSC/285, 1984-04, Maxwell Air Force Base, Alabama.
[14] Pelton, J.N., Sgobba, T., Bahr, N., and Jakhu, R., "Orbital Debris and Space Security," *Space News* editorial, June 2007.
[15] Treaty on Principles Governing the Activities of States in the Exploration and Use of Outer Space, Including the Moon and Other Celestial Bodies, United Nations Office of Outer Space Affairs, http://www.unoosa.org/oosa/SpaceLaw/outerspt.html.

CHAPTER 12

Hypersonic Transport: The Golden Goose of Commercial Space?

When Virgin Galactic signed up its 500th passenger to ride into outer space on SS2, people started to think maybe this space adventures business might prove to be not so crazy after all. Perhaps this "Citizen Astronaut" boondoggle might become a viable business. After all, $200,000 a pop times 500 does calculate add up to $100 million. Surely, that is serious money.

But although serious revenues have come in, even more serious money has flowed out. Angel investors have poured billions into this incipient industry. As noted in Chapter 2, the estimated total investment in the private space enterprises is now on the order of $3 billion (U.S.), and total revenues (with generous accounting) might be estimated to be as high as $800 million. This is not the sort of balance sheet an executive would want to take to stockholders.

Paul Allen admits to putting several tens of millions of dollars into developing SpaceShipOne just to collect the $10 million XPRIZE. We also know that the SS2 and WK2 carrier aircraft have now been in development for a decade, and Virgin Galactic has put serious money into this development. The crash of the first flight model of SS2 on October 31, 2014, with one fatality was a significant setback. The earlier accident, in which three people were killed at the Mojave Air and Spaceport facilities of Scaled Composites, was tragic, but this was essentially an on-the-ground industrial accident involving a fuel tank explosion and not a flight accident. Each accident, especially with a loss of life, involves huge financial liabilities.

The truth is that the development costs for these vehicles, spaceports, and other infrastructure have now run into very big bucks indeed. It sounds as if there will be a lot of red ink for a number of years to come for the "space adventures" industry, and there will be many more expenses. If there were a spaceplane accident with paying passengers on board, the costs associated with legal fees and efforts required to recover from a fatal crash could soar very much higher.

We know that building the New Mexico Spaceport had a price tag of over $300 million because the voters in that state approved a huge bond issue to cover the cost of this space age enterprise. And that is just one of many spaceports either under construction or planned. And then what about the cost of all the people who are now employed to undertake marketing, to cope with regulatory forms and the signing of waivers, and to arrange for passenger training, pilot training, and so on? You need some very deep pockets to invest in this new enterprise, unless you are totally a service company like Space Adventures, or unless you have an established revenue stream and well contained capital investment costs like ZERO-G Corporation. No one knows for sure what will be the true profitable market in the field of commercial space transportation, but one will likely need deep pockets indeed—plus a lot of patience—to succeed.

Unique Personalities

The truth is the so-called space tourism business has gone from a zero million dollar business to a zero billion dollar enterprise in the last few years in terms of media hype. The space billionaires like Branson, Bigelow, Bezos, Carmack, Allen, and Musk are in a "bleeding edge" enterprise at this stage because of the excitement factor and not the bottom line profits. Branson can probably justify his Virgin Galactic's currently low revenue enterprise (at this stage) because it also enhances Virgin's global brand identity. All of these daring space billionaires are super Type A personalities who actually make money off their celebrity-hood as unique personalities willing to take on a gamble against the odds. Consider the evidence. Branson has taken death defying balloon rides and bought a Caribbean island for hijinks vacationing. Bigelow has offered a $50 million America's Challenge, perhaps not so much to develop new rocketry systems but to get his name in the news. Allen has sufficient money to found a Science Fiction museum and purchase the Seahawks pro football team in Seattle, Washington. And so he can truly afford to also take a flyer on developing rocketplanes. A billion here and a billion there, and yet he still has quite a large pile of cash from his Microsoft founder's stock. Carmack, the creator of computer games Doom and so on, is not your typical cautious businessperson. Musk is no shrinking violet, either. In fact, he is the classic multiple venture capitalist who has taken the profits from PayPal and used the capital not only to found SpaceX but also to back the Tesla electric car company, which is poised to become the world's largest battery producer.

These guys, with their billions in cash, can afford to be on the "bleeding edge" of this new technology. To them $100 million may not be pocket change, but it is money that they can risk—and afford to lose. And based on their track record to date, they are perhaps more likely to turn a profit than lose their investment—at least in the longer term.

But if "space tourism" is perhaps a decade away from anything but a sea of red ink, then the more profound question is whether there can be larger longer-term profits to be made somewhere in this new commercial space business. Can these risk-takers (and other businesspeople out there) also see a way to embrace this new technology and shape it into longer-term business models that not only make sense but that can also actually produce billions of dollars in new revenues and show true profitability?

The answer to this question seems to be yes. New markets could be highly profitable within a decade. A new space transportation industry of the future would challenge satellite communications as the most important and profitable space business.

What is not clear is whether this new market is "supersonic transport" or "hypersonic transport" or low-cost launch of small satellites, or all three. It is unclear whether the tortoise or the hare will win the race to faster air transport.

Two types of development are now underway for innovations in commercial aeronautical transportation vehicles. "Official" development in the United States, Japan, and Europe is currently largely focused on new vehicles that present incremental advances for new types of advanced jet transport. The focus of this development includes sonic boom suppression systems and new smart materials to make airframes lighter and stronger. These airframes would fly on "greener fuels," would use electrical propulsion systems, and/or would fly at speeds of no more than around Mach 2, although longer-term efforts aspire to around Mach 4 or 5. The highest-profile programs in the United States are those of NASA, Boeing, Gulfstream, Lockheed Martin, and others. JAXA has prototypes now flying at Mach 2 and geared to velocities of Mach 4 within a five-year window. European Union programs involve cooperation with EADS/Astrium/Airbus, Dassault, BAE, Fokker, and others. The European Commission has just agreed to invest over eight billion euros in its new Horizon 2020 program. There are over half a million skilled workers in the European aeronautics industry, and some 60 percent of its production is exported. Therefore, the European Union is keen to support this industry and its ongoing advancement. A clear objective is to maintain at least 40 percent of the world aeronautical market, and that market—now dominated by Airbus—is foreseen to remain largely subsonic for some time to come. [1]

However, the European Horizon 2020 program (largely governmentally funded) and the Clean Sky Joint Technical Initiative (largely industry supported) are typically centered on incremental improvements. In short, these research and technology funds concentrate on new aircraft innovations that are a decade or even less away from market. This type of new aircraft is likely to be greener, more fuel efficient, and constructed of smarter materials, but it will involve speeds that range from subsonic to no higher than Mach 2. Considerable research activity is focused on subsonic electrically powered aircraft as well. The overall goal is to keep intra-European flying times (door to door) down to four hours. This objective clearly does not involve the need to develop even supersonic jets. [2]

And in Japan, JAXA has similar programs under development. We will discuss these programs, which have been quite successful in the last few years, in greater detail later in this chapter.

The NASA N+1 to N+3 Programs

If one looks at NASA's aeronautical developments, which are largely operated from the NASA Dryden Center, the type of advances as seen from America are again quite similar to those in Europe and Japan in terms of pursuing incremental goals. Although it would not be fair to describe vehicles that fly up to Mach 2 as tortoise like, it is certainly true that these speeds are certainly far below the Mach 5 to Mach 10 or even higher speeds that are associated with truly fast and rocket-propelled spaceplanes. NASA's efforts to work with U.S. industry to develop new higher-speed and higher-performance jet aircraft are currently devoted to three different levels of performance with three different timescales for arrival in the marketplace.

The N+1 program is currently underway, and its prime focus is on sonic boom mitigation. It includes programs such as Garfield Investigation of No boom Threshold (Fa Int). The N+2 program is seeking to develop new aircraft designed to fly at speeds such as up to Mach 1.8, with prototypes flying for certification in the 2020 to 2025 time period. The seating capacity for the N+2 is a modest 80 passengers. Finally, the NASA N+3 program is looking to longer-term capabilities for the 2030 to 2035 time period. Here, the objective is for a high-capacity spaceplane that could fly at even higher speeds. [3]

Fig. 12.1 Icon II Design concept for NASA N+3 scram jet airliner for the 2030 time frame. *(Courtesy of NASA and Boeing.)*

The question that thus arises is whether in this case the hare can overtake the tortoise. In short, we wonder if we should skip over the Mach 2 type of jet aircraft that reprise the Concorde Supersonic Transport (SST) and the Letup 144, which were developed in the 1960s. Instead, in the coming decade, do we see true spaceplanes that fly at speeds in the region from Mach 4 to Mach 10? [4] This complicated question involves a host of problems and issues such as developing new propulsion, noise abatement, and environmental technologies. Other issues, relating to safety, include researching the environmental effects of high-altitude emissions as well as the setting of standards in the United States, Europe, Asia, and elsewhere as to safety certification, allowable noise levels, and environmental emissions limits. There are even such concerns as the availability of capital financing; insurance coverage; and even the market viability of various aircraft designs in terms of seating capacity, fuel efficiency, range, affordability of fares, and especially passenger safety and safety certification.

The Potential of Hypersonic Aircraft

The Anglo-French Concorde SST was a technological success in the 1970s, but ultimately it proved to be a disastrous attempt to create a new type of air travel option for business executives and jet setters with unlimited amounts of cash to spend. Only 20 Concorde planes were ever built, and just 16 entered service. The Concorde's design, which looked and felt like a supersonic military jet that expanded to serve as an aviation carrier, had almost everything imaginable going against it. It had very limited range, the seating arrangements were rather cramped, the overall seating capacity was limited to fewer than 100, it used an enormous amount of fuel per passenger, and its tremendous sonic booms grounded it from flying within the United States.

The aircraft entered service with British Airways and Air France in 1976 and operated on their commercial routes for 24 years. But in July 2000, a tragic accident occurred after takeoff from Charles DE Gaul airport in Paris, shaking confidence in the aircraft's safety. First, British Air grounded its fleet, and then Air France reluctantly followed suit. In the fast-paced world of global commerce and multinational enterprises, the promise of rapid travel across the oceans has remained a dream of the aviation business for many decades. The experience with the Concorde SST at least helped to establish a list of design objectives for such a new aircraft in the future:

- Higher seating capacity—NASA's N+1 experimental vehicle is sized at only 80 passengers. In contrast, design concepts recently developed by Bristol SpacePlanes for ESA are aiming for at least 250 to 300 passengers and at much higher speeds in order to make a much more compelling business case. Of course, others are looking at hypersonic vehicles for executives, but passenger-based vehicles would likely need higher seating capacity than the Concorde to create a profitable business case.

- Fuel efficiency—Greater fuel efficiency is needed to cope with the rising cost of jet fuel. If hypersonic spaceplanes could use bio fuels, so much the better.
- Improved materials—Advanced materials can make the craft lighter in weight, more fuel efficient, and safer.
- Transoceanic range capacity (especially Trans Pacific).
- Ultra-rapid transit capacity—This would be something like three hours from New York or London to Sydney, Australia, or from Washington D.C. to Tokyo or Beijing. The slower speed for SST vehicles is currently the problem with many of the U.S. and European design studies for incremental improvements. The Mach 1.8 speeds envisioned by the NASA N+1 vehicle would cut total transit times in only a modest way, while also entailing major fare increases. This suggests that incremental improvements in plane speeds with limited seating capacity and perhaps limited range would result in a failed business model. (Note the total transit time will need to include a cool-down period for hypersonic transport.)
- Suppression of powerful sonic booms (on landing or takeoff)— These would need to *meet all national and international standards at higher and lower altitudes.* (The FAA and EASA, among others, do not have clearly established standards with regard to sonic boom suppression, and so this represents at least a partial barrier to new vehicle development.)
- *Clear emission standards*—No clear standards exist for emissions associated with high-altitude jets or spaceplanes. The U.N. Environmental Programmer and the World Meteorological Organization have indicated plans to consider this issue, and various national legislative initiatives are under way. Europe has been a leader in trying to develop cleaner jet emissions—especially for high-altitude craft. This includes the European Clean Sky Joint

Technology Initiative. The bottom line is that without international agreement on emissions standards, it will be difficult to develop a commercially viable new hypersonic transport capability.
- Proven safety and full certification of airworthiness.
- Airfares that are not significantly greater than the cost of first-class tickets for similar flights today.

Various feasibility and engineering design studies carried out in the United States, Japan, Russia, and Europe suggest that the new technology—along with viable capital and operating costs, as well as other business and safety objectives—is certainly within reach, perhaps within 10 years. Nevertheless, threshold questions remain.

One of the key questions is the fundamental issue of whether hypersonic spaceplane transportation is even the way to go. It has been proposed that transcontinental tunnels to support mag-lev transport could achieve hypersonic speeds—especially if these tunnels were subjected to a high-quality vacuum. These proposals suggest that conventional air transport could then support world traffic if one earth-circling tunnel were constructed with a route such as from New York to Los Angeles to Tokyo to Shanghai to Mumbai to Frankfurt and finally back to New York. The argument is that this would curtail upper air emissions and would be more environmentally friendly than hypersonic air travel—and perhaps it would be safer. Perhaps capsules could be dispatched every 10 minutes along such mag-lev tunnels and carry a much higher volume of traffic at lower cost. Of course there remain key safety and economic issues for this mode of transport as well. Just one safety-related issue is the increase in earthquakes and human-triggered seismic events, and the economic trade-off studies have not been carried out in a systematic way. In short, no serious techno-economic-environmental-safety assessment of proposed options for future intercontinental transportation has yet been undertaken in any definitive way. There is, of course, significant momentum behind supersonic and hypersonic aircraft initiatives. In contrast, the tunneled mag-lev options have no parallel industrial lobby behind this approach.

Certainly, mounting concerns about global climate change suggest that we need to take quite seriously the issue of high-altitude emissions. There are clearly rising concerns about upper atmosphere air pollution and possible negative impact on the ozone layer. The challenge to hypersonic spaceplane operations is to find propulsion systems that would not add major new pollution to the stratosphere. New reports from the United Nations Environmental Programme (UNEP) and the United Nations World Meteorological Organization (WMO) have suggested that there is reason for concern about jet contrails that already exist as the result of flights by high-performance military aircraft. Even beyond the substance of concerns about upper-altitude emissions, there is also the concern about having globally-agreed-upon standards that could apply to this issue so that each country or region would have consistent rules to follow. This would apply to both sonic boom mitigation as well as upper-altitude emissions.

Finally, there is simply the issue of whether a viable business case can be devised for hypersonic transportation. Here, there are concerns about the availability of sufficient supplies of low-cost and "green" fuels and whether the aircraft would be safe enough, fast enough, of sufficient seating capacity, and with sufficient consumer traffic levels to create a viable business. In this case, the "build it and they will come" strategy could result in a multi-billion-dollar blunder. When one looks to this new aspect of the space business and its feasibility, it is good to recall a seemingly pertinent parallel in the history of space industry development. This is the case of the premature development of the satellite broadcasting industry. In the 1970s, the Comsat General Corporation undertook development of a broadcasting satellite service in the United States, purchased communications satellites for this purpose, and sought partners within the entertainment industry and among those that would install consumer TVRO terminals. This multi-billion-dollar project went bankrupt, and the satellites were abandoned at the manufacturer's plant. The problem was that the satellite industry had yet to develop sufficiently high-powered technology that would make the service cost effective and viable. Some 20 years later, satellite broadcasting technology had evolved so that satellites could provide signals that were 100 times more powerful, and the industry took off. Today, direct broadcast satellite service is the most profitable space application business because satellite technology and the user terminal industry have soared ahead and transformed a new service from economically unfeasible to a highly profitable business. This happened because technological and regulatory systems were finally suitably matched with consumer demand and needed cost efficiencies.

Technical Feasibility and Engineering Studies

Research and feasibility studies being carried out in the United States have, for good reason, given much more attention to the issue and regulatory problem of sonic booms. The Concorde was banned from U.S. airports because of the sonic boom issue. Thus, companies such as Lockheed Martin, Boeing, Northrop Grumman, and Quiet Supersonic Transport (QSST) have sought to perfect technology that, instead of producing one large sonic boom, would actually generate a number of "micro-booms" that would be much less of a problem. Extensive design studies and tests have been conducted using supersonic planes with extendable on-demand, long needle-like noses, and these tests have created a series of these much softer "mini-booms."

One successor to the Concorde could be the work of the Aerion Corporation of Reno, Nevada. This company is currently developing an 8- to 14-passenger business jet that will be capable of speeds up to Mach 1.6 and will have a range of more than 4000 nautical miles. However, Aerion is not without competition. Michael Paulson, son of the founder of Gulfstream, has founded his own firm, Supersonic Aerospace International (SAI), in Las Vegas, Nevada. Paulson is actively pursuing the same market niche. Teamed with the fabled aircraft design house Lockheed Martin's Skunk Works (Palmdale, California) SAI has developed the QSST design, featuring a radical "inverted V-tail."

Fig. 12.2 The Mach 1.6 business jet envisioned by the Aerion Corp.
(Courtesy of Aerion Corp.)

As of December 2010, NASA had awarded three R&D study contracts known as N+2 studies to Northrop Grumman, Lockheed Martin, and Boeing. These study efforts were valued at $2.65 to $5.3 million and were all geared to study new supersonic aircraft that could enter service by 2025. In addition, all would allow the design of new aircraft that would be quieter, faster, and more fuel efficient. Meanwhile, in Tokyo, the Japan Aerospace Exploration Agency is developing its Silent Supersonic Technology Demonstrator in support of future SST aircraft design and propulsion. Prototype designs for this unmanned test aircraft are undergoing wind tunnel testing for use on transcontinental vehicles capable of carrying 100 to 300 passengers at Mach 1.6 and faster.

The reality of such developments can be seen in the reports that NASA and the U.S. Air Force have actually demonstrated a hypersonic craft known as the X-43A, which has now been turned over entirely to the U.S. Air Force. This craft, developed by OSC, flew in public trials while under NASA development at speeds exceeding Mach 10. In fact, the X-43D is still at a conceptual stage but would in theory be able to achieve speeds up to Mach 15.

Boeing and Pratt Whitney have also developed the unmanned X-51 scram jet system (also known as the Waverider) under the Air Force's

Hy-TECH program. This vehicle has also achieved velocities of up to 6400 kilometers per hour (4000 miles per hour) in initial flights. The X-51 Waverider, however, is an unmanned vehicle that is essentially and clearly meant for military applications rather than commercial transport of people. [5]

In Europe, birthplace of the Concorde project, work is going on to develop the A2 hypersonic aircraft, capable of flying 300 passengers at a speed of over Mach 5 (or 4000 miles per hour). It is being led by the LAPCAT consortium (a convenient abbreviation for Long-term Advanced Propulsion Concepts and Technologies), sponsored by the European Union and ESA. The project received some $12 million in funding from the European agencies and private investors, and new funding is anticipated.

The potential routes for the A2 are between Europe and Australia and between North America and Asia. It would not be efficient to operate a passenger service between London and New York (the world's busiest business-class route) because the distance is too short for the aircraft to reach the necessary altitude. [6]

As we have seen, upper altitude emissions and sonic booms are issues, yet another issue deserves careful attention: the aero-braking process. This process will occur with the deceleration of the spaceplane as it re-enters the earth's atmosphere and slows to landing speed. Several issues here need to be addressed: the very high-temperature gradients involved in terms of the materials of the craft's outer structure, the need for cooling the craft on its landing, and the safe egress of passengers. These are yet other aspects of the safety evaluation and certification for hypersonic flight.

Reaction Engines Ltd.

A leading participant in the LAPCAT consortium is the U.K.-based company, Reaction Engines Ltd. It has also been working for several years on propulsion systems for its Skylon project. This is an unmanned Horizontal Take Off and Landing (HOTOL) spaceplane that can be launched into LEO after taking off from a conventional runway. It is described as "a self-contained, single stage, all-in-one reusable space vehicle with no expensive booster rockets or external fuel tanks."

The vehicle's hybrid SABRE engines use liquid hydrogen combined with oxygen from the atmosphere at altitudes of up to 26 kilometers and speeds of up to Mach 5, before switching over to on-board fuel for the final rocket-powered stage of ascent into LEO. The Skylon is intended to cut the costs involved with commercial activity in space, delivering payloads of up to 15 tons including satellites, equipment, and even people into orbit at costs much lower than those that use expensive conventional rockets. [7]

Fig. 12.3 The Skylon spaceplane as designed by Reaction Engines.
(Courtesy of Reaction Engines Ltd.)

This development is now linked to the A2 hypersonic spaceplane, and Alan Bond, managing director of Research Engines, told the BBC in April 2012 that engine tests were underway. He added, "We can reduce the world to four hours—the maximum time it will take to go anywhere. It will also give us an aircraft that can go into space, replacing all the expendable rockets we use today." [8] In an earlier interview with the U.K.'s *Guardian* newspaper, Bond said, "The A2 is designed to leave Brussels international airport, fly quietly and subsonically out into the north Atlantic at Mach 0.9 before reaching Mach 5 across the North Pole and heading over the Pacific to Australia. The flight time from Brussels to Australia, allowing for air traffic control, would be four hours, 40 minutes. It sounds incredible by today's standards but I don't see why future generations can't make day trips to Australasia." [9] One of the most significant parts of the design is the hydrogen-fueled engine, which Bond and his team have developed over several years. It would avoid either nitrogen oxides or carbon oxides being spewed into the fragile upper atmosphere. There remains the issue of water vapor exhaust, which is also a greenhouse gas, but nevertheless it is much less of a problem than carbon or nitrous oxide emissions. The key element here, therefore, is the increasing interest in developing hypersonic transport craft to ultimately move people around the globe rather than just taking a quick jaunt up to peek into space. However, the design is still at the conceptual stage, and it is estimated that it is still many years away from the commercial market. [10]

Swiss Space Systems

One of the newer entries into the spaceplane race is Swiss Space Systems (SSS). This project also involves HOTOL. The spaceplane would, at takeoff, be carried aloft via a carrier vehicle. The spaceplane would then be released at very high altitude and then fly LEO satellites weighing up to 250 kilograms (550 pounds) for insertion into orbit. These could be for commercial purposes or scientific research. SSS has indicated that it would charge on the order of $10 million (U.S.) for such a launch, using its unmanned sub-orbital spaceplanes. These LEO launch costs were expected to be significantly reduced in comparison to those for conventional expendable rocket launches for three reasons: the reusable nature of the spaceplane, the lack of launch facilities (since launches will take place from the air), and the somewhat lower fuel consumption than conventional systems. SSS has indicated plans to start such flights by 2018. [11]

Fig. 12.4 The Swiss Space Systems spaceplane with carrier vehicle.
(Courtesy of S-3 Corp.)

SSS has announced three areas of intended operations. It intends to use unmanned spaceplanes to launch smaller satellites into LEO. It also has announced planes under "Clean One" to capture and remove a cube satellite from orbit to demonstrate the ability to remove space debris from orbit. Finally, it intends to fly passengers for space tourism on sub-orbital flights. More significantly, however, SSS intends to create a hypersonic aviation transportation business. The flights for small satellites to LEO are to take place from the Canary Islands, and the passenger flights are intended to take place from the Colorado Spaceport in the United States.

According to SSS founder and CEO, Pascal Jaussi, the prime longer-term objective is the hypersonic transportation flights. At the "launching" of SSS in 2013, he said, "Far from wishing to launch into the space tourism market, we want rather to establish a new mode of air travel based on our satellite launch model that will allow spaceports on different continents to be reached in an hour. [12]

The Bristol SpacePlanes Project

Another U.K. group working on hypersonic aircraft is Bristol SpacePlanes, which was commissioned by the ESA to study what type of hypersonic aircraft design might be feasible within a decade and commercially viable. The result of this study was a 250-passenger hypersonic scram jet aircraft that could fly from London to Sydney in a little over three hours.

As described in Chapter 7, the concept is based on a carrier aircraft that could fly subsonically to above 40,000 feet (about 11 kilometers), where it would then release a rocket plane scram jet that could fly as an air-fueled rocket plane in a sub-orbital arc at speeds in the range of Mach 7. When this rocket plane was in descent, it would decelerate at an appropriate point over the Pacific Ocean, where it would break out of a suborbital arc and brake to subsonic speeds and then land in Sydney, Australia, as would a conventional jet. One of the key elements of the design would be a landing process in which the thermal gradients would be minimized so that the plane would require only a modest time for cooling off before passengers deplaned. Bristol SpacePlanes envisions that this type of hypersonic transport will develop in a series of stages. The first phase would be an Ascender spaceplane, which could make sub-orbital flights with a limited passenger capability.

The Ascender Rocket Plane is also envisioned as being able to fly small satellites to LEO or to engage in various forms of space tourism/space adventures flights with a much smaller number of passengers.

Fig. 12.5 The Bristol SpacePlanes Ascender—Phase One System.
(Courtesy of Bristol SpacePlanes.)

The Ascender phase would be followed by a second phase, the "Spacecab," which would employ scram jet technology. This vehicle would be twice the size of the Concorde SST, would be designed to carry 50 passengers on a hypersonic flight, and could land at conventional airports.

The third phase would be a rocket plane called the Space Bus, which could carry 250 passengers and could complete an entire intercontinental hypersonic flight for what seems to be a very ambitiously estimated cost of perhaps a quarter of a million dollars.

Fig. 12.6 The Space Bus concept. *(Courtesy of Bristol SpacePlanes.)*

Sierra Nevada Dream Chaser

Recently, NASA made its contractor award to develop an astronaut crew transportation system that can deliver crew and cargo to the ISS in coming years. This three-way competition for NASA's Commercial Crew Transportation Capability (CCTC) included Boeing, SpaceX, and Sierra Nevada's Dream Chaser system. The Boeing and SpaceX systems involve the use of human-rated expendable rockets and a capsule return system.

Fig. 12.7 Sierra Nevada Corporation's Dream Chaser spaceplane.
(Courtesy of the Sierra Nevada Corp.)

Only the Sierra Nevada involves a re-useable spaceplane, yet the company lost out for these billion-dollar NASA awards. Sierra Nevada, who felt its Dream Chaser system was best for ferrying NASA astronauts to and from the ISS, formally appealed the NASA decision, but to no avail. The GAO rejected the appeal as of January 5, 2015. Despite the setback, Sierra Nevada is continuing to develop its spaceplane in the hope it will find commercial customers. [13]

JAXA Mach 4 Spaceplane

Many see JAXA as Japan's equivalent to NASA, which it is, but they also incorrectly interpret this to mean it is Japan's "space" research and development agency. However, JAXA was formed by merging three units—two were involved in space-related R&D, but the third was the

National Aeronautical Laboratories, which was devoted to aerospace development. Today, both NASA and JAXA continue to have a mandate to develop new aviation technology in terms of performance, safety, and environmental issues.

Fig. 12.8 Visualization of the JAXA Mach 4 spaceplane. *(Courtesy of JAXA.)*

Accordingly, JAXA is actively undertaking the development of a hypersonic spaceplane. In the past five years, JAXA has moved ahead with the objective of having a crewed flying spaceplane that would fly at least Mach 4. JAXA has developed a series of increasingly larger and higher-performance prototypes and has recently successfully tested the performance of a liquid hydrogen fueled spaceplane motor. By 2018, it is seeking to successfully fly a Mach 2 prototype and then a Mach 4 full-scale spaceplane suitable for transoceanic hypersonic commercial flights across oceans. [14]

Other Initiatives

Currently, other supersonic aircraft are also under development to support travel for business executives, but many of these are essentially extensions of military jet aircraft rather than true spaceplane developments. Among those known to be researching designs and technologies primarily for such business executive supersonic flights are Gulfstream Aerospace Corp (Savanna, Georgia), Cessna Aircraft Co. (Wichita, Kansas), Dassault Aviation (Paris, France), and Tupolev PSC (Moscow, Russia).

The Dassault project is known as High-Speed Supersonic Aircraft (HiSAC), and the company is leading a European group—the S3 Corporation—in developing three Supersonic Business Jet (SSBJ) designs. Most of these concepts are based on developing business supersonic jets capable of carrying 8 to 19 passengers. Dassault's latest announced timetable is to begin flying demonstration aircraft for safety certification within a decade. [15]

Commercial Ventures to Fly to the Moon ...

In December 2012, on the 40th anniversary of the last Apollo flight to the moon, the Golden Spike Company boldly announced a new commercial space transportation mission to resume lunar missions.

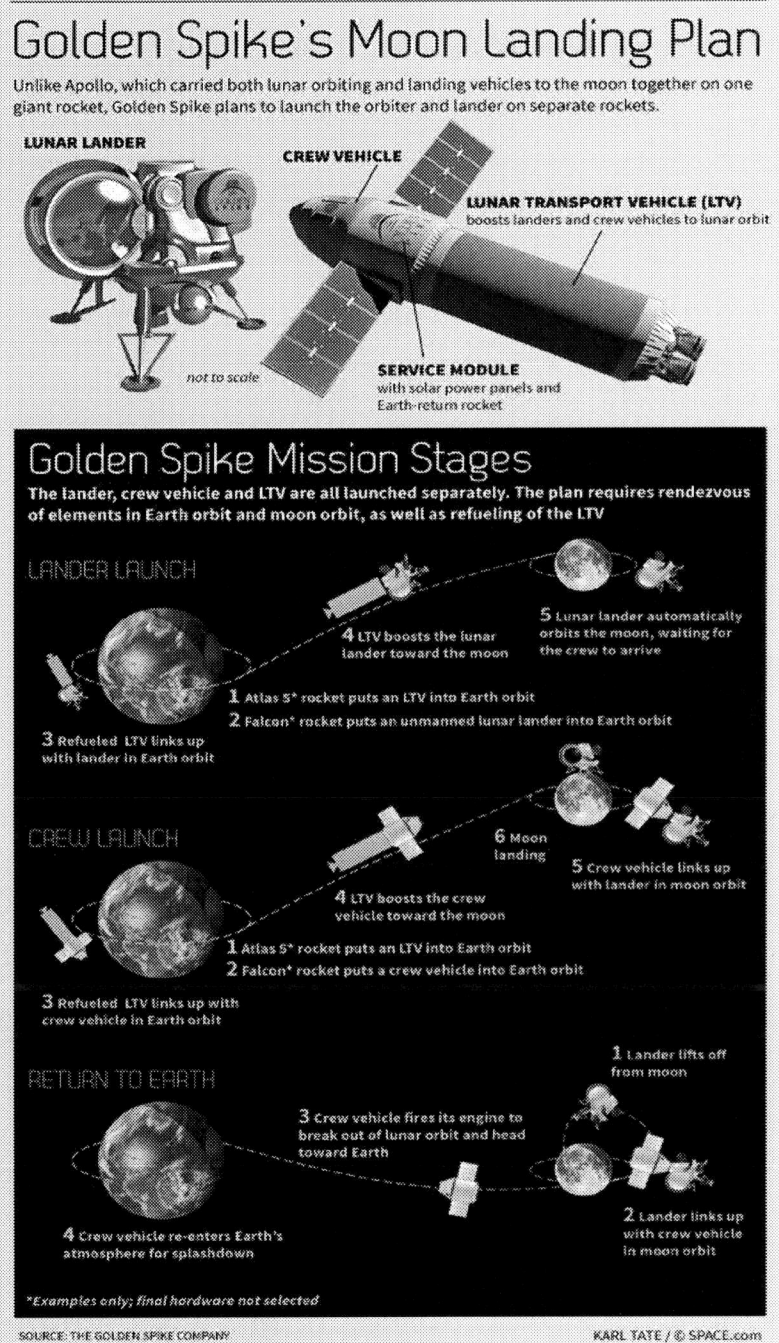

Fig. 12.9 Proposed transport system for the Golden Spike mission to the moon.

(Courtesy of the Golden Spike Company.)

Golden Spike, based in Colorado, outlined its plans to get humans back to the moon by 2020, using existing or underdeveloped technologies, to begin "a commercial lunar campaign." [16] The name comes from the railroad spike that heralded a new era of transportation with the completion of America's transcontinental railroad. Alan Stern, Golden Spike CEO and former NASA associate administrator, has confidently proclaimed, "We know how to do this."

The company plans to use existing vehicles or those already under development. It will use a two-step architecture, first launching an unmanned lander to lunar orbit and then sending a crewed vessel to rendezvous with the lander. Golden Spike will commission its own lander and its own space suits. The names of those involved with Stern include several ex-NASA experts and a Board of Advisors including the lunar base advocate and politician Newt Gingrich; Mike Okuda, the creative director behind "Star Trek"; and Jonathan Clark, a former Space Shuttle flight surgeon who was also married to astronaut Laurel Clark, who was killed in the Space Shuttle *Columbia* accident.

... And Then Mars!

When one thinks of hypersonic transportation, the normal reaction is to contemplate rocket planes from London to Sydney or from New York to Beijing, but there are always people who are willing to push the envelope to the limit. Our friend Peter Diamandis is fond of saying, "The meek shall inherit the earth. The rest of us will go to Mars."

And there are indeed entrepreneurial spirits who conceive of the commercial transportation systems of the future as much grander than spaceplane hops around the world.

In February 2013, the first "Citizen Astronaut," Dennis Tito, announced his "Inspiration Mars" project to send a married couple on a "fly-by mission" to the Red Planet in 2018. Tito claims this can be accomplished using conventional rocket technology. Indeed, in terms of propulsive force, it does not require much higher impulse to fly to Mars, to orbit, and to come back than it does to do the same for the moon. The differences are that it takes a lot longer, and one must fly much farther. And thus, Tito has said, "Why not?" He is now seriously embarked on such a mission.

Another initiative that has received even more press attention is the Mars One initiative, which has envisioned the possibility of a one-way trip to Mars as an initial space colony. In short, the astronauts would go to establish a colony on Mars and would not return. As of mid-March 2015, Bas Lansdorp, CEO of this initiative, responded to a number of questions about the feasibility of the concept and the current time timetable. He noted that the selection process had narrowed 202,000 candidates down to 100 possible astronauts for the mission and confirmed that Mars One envisions that it can accomplish its mission for $8 billion (U.S.), although NASA's lowest estimate for a Mars mission is $36 billion (U.S.). Lansdorp also indicated that the current schedule is to send a spacecraft to Mars as an unmanned vehicle in 2020 and that the first mission for colonization would be in 2027. The idea is very exciting, but the skepticism about this project is very high.

At the opening of the March 18 Conference on Space, hosted by the U.N. Office of Outer Space Affairs (OOSA) and the ICAO, Canadian astronaut Julia Payette was sharply critical of the project in her opening address, saying it has unrealistic objectives.

No doubt, in the months and years ahead, more ventures will, if nothing else, maintain the interest and enthusiasm of not only the aerospace community but also the incredulous wider public.

Conclusions

At this stage, the future of commercial spaceflight is clearly filled with a lot of unknowns. In some ways, the wide range of efforts seems like a race between a Model T, a Prius hybrid, and a Lamborghini. Some want to create a reliable Mach 2 jet that can make global hops in about eight hours. Others want to create stratospheric spaceplanes and scram jets that can complete a four-hour flight from London to Sydney. And then there are even commercial ventures that claim they can fly to the surface of the moon and back, or even orbit Mars and return or even create a Mars colony. The technology to accomplish these things is currently under development, and all of these diverse programs actually may be achievable in coming decades. Sorting out the projects that can be achieved in a single decade from those that are unrealistic and merely aspirational is something only time will tell. New commercial ventures and accomplishments, however, depend on more than technology. There must also be regulatory standards, viable business models, and even customers who are willing to pay the fare to make these new commercial ventures actually succeed. At this point, it is unclear who will win these far-from-parallel races into commercial space. What is clear is that there is no lack of ambition and innovative thought as to what challenges commercial space might hold.

REFERENCES
[1] "Europe's New Plans for Research and Funding," *Aerospace America*, pp. 4-5, January 2013.
[2] Ibid.
[3] Wilson, J.R., "Breaking New Barriers," *Aerospace America*, pp. 26-31, January 2013.
[4] http://www.universetoday.com/84554/what-will-airplanes-of-the-future-look-like/.
[5] Johnson, R., "The Boeing Mach-5 Waverider Is Getting One More Shot at Unlocking Scramjet Flight," October 12, 2013, http://www.businessinsider.com/boeings-x-51-waverider-is-getting-one-more-shot-before-being-scrapped-2012-10.
[6] http://www.bbc.co.uk/news/magazine-16090841.
[7] http://news.discovery.com/space/skylon-spaceplane-budget-approval-110526.html.
[8] http://www.reactionengines.co.uk/sabre.html.
[9] *The Guardian*, February 7, 2008.
[10] http://travel.aol.co.uk/2012/04/29/Skylon-spaceplane-can-fly-anywhere-in-four-hou/.
[11] Swiss Space Systems, http://www.s-3.ch/en/home. Also see SSS presentation at the ICAO and U.N. Office of Outer Space Activities Aerospace Conference, April 18-21, 2015.
[12] Messier, D., "Swiss Space Systems Announces Plans for Crewed Suborbital Spacecraft," *Parabolic Arc*, June 17, 2013.
[13] Howell, E., "Private Dream Chaser Space Plane Keeps Marching Toward Flight," Space.com, http://www.space.com/28203-dream-chaser-space-plane-propulsion-milestone.html.
[14] Coopinger, R., "Japan's JAXA Wants to Send up a Space Plane or Capsule by 2022," http://www.nbcnews.com/id/49557677/ns/technology_and_science-space/t/japans-jaxa-wants-send-space-plane-or-capsule/#.VSsDzfnF_OE.
[15] http://www.compositesworld.com/articles/beyond-the-concorde-next-

generation-ssts.

[16] Dorminey, B., "Golden Spike Still Aims For Human Lunar Surface Expeditions By Decade's End," *Forbes*, January 30, 2014, forbes.com/sites/brucedorminey/2014/01/30/golden-spike-still-aims-for-human-lunar-surface-expeditions-by-decades-end/.

CHAPTER 13

The Top Ten Things to Know about the Future of Commercial Space

Within the next two years or so, more and more regular people—not super astronauts—will be flying "beyond the skies" on commercial vehicles. What does this mean for humanity and the future of the human race?

Most of you will have a ready first response: big deal; this is really not for "regular people" at all! So what if a handful of celebs can pay big bucks to fly for a few minutes into space. Well, actually, it is not even truly "outer space"—just so-called "suborbital space." The billionaires, movie idols, and rock stars do at times seem to live up to their reputation as the idle rich by finding odd ways to throw their money around. With greenhouse gases rising, wars in the Middle East, and people starving in Darfur, a $200,000 ride that lasts a couple of hours or so likely seems just a bit decadent to some.

Certainly, this new 100-kilometer-high club may get some high-g kicks and prove the earth is indeed not flat. But won't this trip to nowhere just burn up a lot of extra fuel? Might not all this galactic joy riding into space actually lead to faster global warming? Won't all those rocket planes burn a bigger hole in the ozone layer, to boot? In short, so what? Why should anyone who is serious about global warming, famines, better health care services, or other truly "serious issues" care if the suborbital space tourism business takes off or not? Serious space exploration on the moon and Mars may be of true scientific interest, but why should serious people give a hoot about space tourism?

Irrelevant? Maybe NOT

The man and woman on the street—let's say Mary in her condo or Ralph in his three-bedroom rambler—are today most likely indifferent to this outer space business. Mary and Ralph are likely to assume that these new commercial space programs are pretty irrelevant to themselves and their lives. They probably believe that suborbital space tourism has nothing to do with serious space exploration, saving the planet, or the long-term destiny of humankind. Right? We actually think, wrong!

First of all, the space tourism industry does need to be re-thought and its technology improved. But all of this will come in time. Likewise NASA and possibly some of the other space agencies around the world need to be re-invented to become more relevant, more cost conscious, and better focused. We actually believe that the most effective and innovative space agency in the world—based on its track record of accomplishing a lot on a little—may well be the Indian Space Research Organization (ISRO). ISRO has had some remarkable achievements over the years. It has developed important telecommunications; meteorological, remote sensing, and satellite navigation capabilities; and reliable rocket technology. In addition, it is now embarked on lunar and Mars exploration programs. ISRO has accomplished this within a much lower budget than the big space agencies. NASA, in contrast, has been the most "success challenged" over the past decade, particularly when judged in terms of "bang for the buck."

President Barack Obama, who campaigned in the 2008 U.S. election race on a platform of "Change," said he grew up on "Star Trek" and is awed by the potential of space, but he feels that NASA in recent years has "mostly lost its way."

Certainly, the FAA Office of Space Commercialization, which oversees commercial space and space tourism ventures, needs to keep evolving and adjusting its role to new realities if it is to successfully regulate the new commercial space frontier. At the top of the list for space and aviation regulators are some new and creative ideas to keep spaceplanes and high-altitude supersonic business jets from doing damage to the earth's atmosphere. It is critical to make sure that future flights of all types (conventional, supersonic, or hypersonic) do not end up depleting the ozone layer or overloading the atmosphere with greenhouse gases. This fragile layer now has a sizable hole in it at the poles. It is important to note that the Van Allen Belts and the ozone layer are what protect humans from rampant genetic mutation and species extermination.

Personally we feel that preserving the human race is sort of a high priority—up there in importance and newsworthiness with the shenanigans of a Paris Hilton or Britney Spears. The following is why we believe that space commercialization is truly important to the longer-term future of our great great grandchildren.

Give Entrepreneurs a Chance

We believe that commercial space entrepreneurs are the people who can and will find new solutions and who will help us develop clean and cheap ways to fly across the world and even into the heavens. The conventional space agencies are typically too large and bureaucratic to find the BIG SOLUTIONS rapidly and cost effectively. We need an innovation process that can quickly find the new technology and new solutions needed to get humanity through the 21st century. In short, we need to give a new team of start-ups and entrepreneurial wizards a chance to find totally new answers, new systems, and new environmentally smart technologies. So what are we talking about?

We need to find smarter ways to get people into space than just putting controlled bombs underneath them. To launch a rocket today we are, in effect, lighting the fuse on a controlled explosion. We need something like clean hydrogen-fueled vehicles that can fly people around the world farther, faster, and cleaner. In time, we need totally new systems like space tethers, space elevators, or advanced ion engines that can get people and "smart machines" into space in ways that are a lot cheaper and cleaner.

As the Space Tourism Society has recently said, "Exciting new ventures, funded by billionaire founders and backers, are dramatically changing the aerospace and new space enterprise industry. Markets are opening up internationally and the competition is heating up." [1] For many decades, the general pattern has been that we have moved from expensive, new technology backed by the government; then on to pioneering technology; and finally, as governmental backing has been phased out, into technology that has been commercially developed by the mass market. To date, human space travel has resisted this typical transition, which often happens over 10 to 20 years. Now that the space age is almost 60 years old, we hope we have reached that pivotal point of transition.

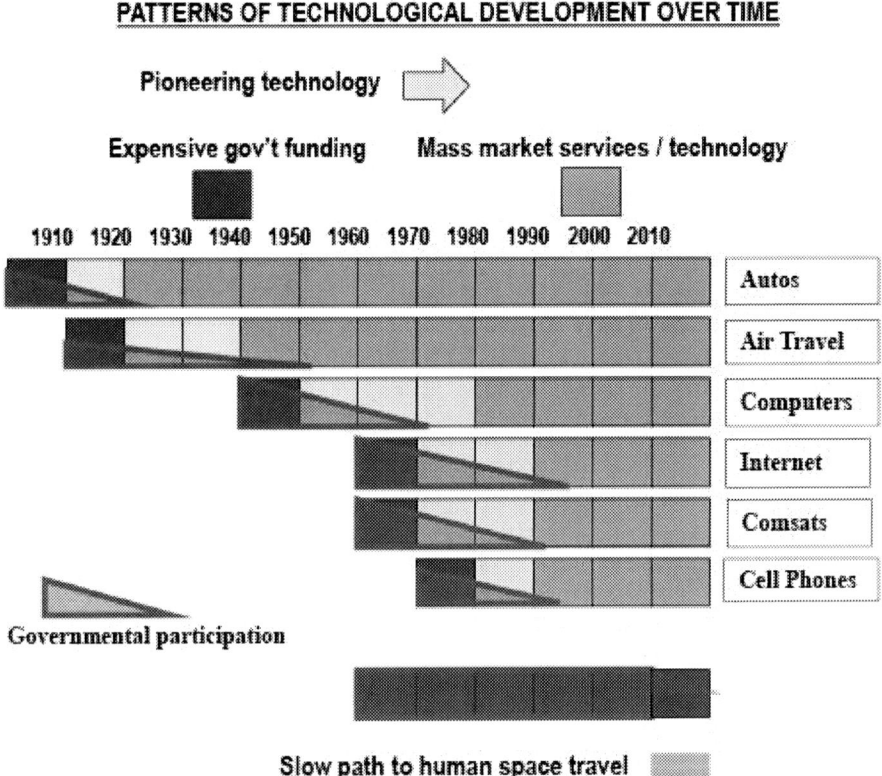

Fig. 13.1 Human space travel has not followed traditional patterns of transition from government-backed, expensive technology to mass-market systems.
(Courtesy of Dr. Eric Dahlstrom of International Space Consultants.)

We need to do a lot of things smarter and better, but do them we must. Cheap and reliable access to space is our destiny—or our destiny is to become yet another failed species. Over the longer term, access to space that is safe and lower in cost is essential to maintaining a viable human civilization on earth. What might be called "space tools" are key to combating global warming; detecting and taming hurricanes; warding off threats of comets, meteorites, and near earth objects (NEOs); developing better ways to cope with disasters; finding vital resources; improving agriculture; monitoring our oceans and coastlines; sustaining global communications networks; and supporting vital infrastructure. [2]

Meteorites and Other Risks from Space

And we should not forget that we need space technologies to protect ourselves from threats from outer space. A recent *New York Times* editorial noted that there are probably at least 1100 meteorites and asteroids out there in earth's neighborhood—NEOs—that could either destroy the planet or do a great deal of damage like wipe out an entire city such as New York, Beijing, or Rio de Janeiro. The asteroid that whizzed by earth inside of the Clarke Orbit in early February 2013 is a reminder that such risks are indeed very real. [3]

The current U.S. space program strategy, as defined by Congress and NASA, can be called the "Let's ignore it and maybe nothing will happen to us Program." [4] Apollo 9 astronaut Rusty Schweickart is sufficiently worried about the destruction of modern global society by a huge meteorite crashing into earth that he has formed "the B612 Society" in cooperation with the Planetary Society to create a systematic response to these dangers. In case this all seems way too theoretical, just be aware that Apopis, a quite solid chunk of real estate, is out there and is projected to come perilously near to earth in 2036.

The real lesson that is still to be learned in space is how to pool our resources more efficiently so that we can maximize global space cooperation—both at the level of space agencies and in commercial space. The space agencies should have a common focus on coping with threats from space, including asteroids, meteorites, comets, solar flares, and coronal mass ejections, and these agencies need the highest level of cooperation because this is clearly the highest level of national and planetary defense. This can and should be supplemented by efforts of non-governmental organizations such as the Planetary Society and B612, but commercial space organizations may also be able to provide assistance in time.

Commercial Space Can Innovate in Other Ways As Well

However, we are concerned about more than scary meteorites. We humans are a rather scary race. We could wipe ourselves out by nuclear accident, by a global pandemic, by rampant mutation of the species if the protective ozone layer dissipates, or by a slow roasting that comes with excess buildup of greenhouse gases. We need a backup Plan B if our biosphere here on earth somehow fails to sustain future generations. This may seem heavy stuff, and way too pessimistic, but a backup plan is part of what commercial space is all about. Let's build a new type of basket in which to keep and protect some of our eggs.

Cheap and effective access to space is also needed for many other good reasons. Seeking to "fly beyond the skies" is also required to discover the mysteries of our universe, tap into clean solar energy, monitor our oceans and atmosphere, and much more. We may not succeed in a quest to preserve the species in two different cosmic locations in the next decade, or the next century, or even the next millennium, but the challenge remains. The Leif Ericksons, the Columbuses, the Magellans, the Galileos, the Henry Cabots, and the Lewis and Clarks of the future just want a chance. We must not say no to the future. We should not say to explorers of the 21st and 22nd centuries—to our future progeny—that we lost our human curiosity, that we no longer dare to believe there could be a better and more profound future for our great great grandchildren.

We believe that it is part of the genetic code of *Homo sapiens* to say that we cannot know enough. We are destined to explore—to go where no one has gone before. We can never abandon our own sense of imagination, which has defined the human experience for millions of years. The human species will never survive if it suddenly says we are now content to stay home and abandon future voyages of discovery. We must not give up on the future.

We thus see commercial space as the potential for innovation and new paths of global cooperation. In short, commercial space travel may start as "space tourism," but it will not end there. The rest of the story must take us much farther, to the following:

- Hypersonic transport
- Cleaner propulsion systems to reduce emissions in the atmosphere

and to mitigate sonic booms
- Solar power satellites
- Support for global cooperative efforts to combat threats from outer space such as solar storms, a weakened geomagnetosphere, asteroids, and space debris
- Support for human colonies in space
- Human dreams yet to be dreamed

A New Stairway to the Stars

We humans badly need a new stairway to the stars. If we don't find this new way to get into space within the next few generations, we may be in deep trouble. Commercial space business, if nothing else, is rich in innovation. This is why space tourism must be recognized as just the start of something big. Let's just hope we get it right—and in the not too distant future.

Private spaceflight could be the beginning of a fundamental revolution in the feasibility, affordability, and accessibility of space for a growing number of humans across the planet. The vision is that in time, millions of humans will go into space or will at least take a flight into protospace on a hypersonic vehicle for transcontinental flights—reaching altitudes of 80 to 100 kilometers. In short, commercial space travel must become much more than a skylark. Instead of a flight to nowhere, it will become a flight to somewhere.

There will be missteps. There have already been mistakes, such as the tragic explosion at Burt Rutan's Mojave Desert facility, where the U.S. Office of Safety and Hazards Agency (OSHA) found lapses in risk mitigation; and the heart-wrenching October 31, 2014, accident with the SS2 flight over the Mojave Desert, caused by pilot error. Within the confines of best safety practices, new technology and systems concepts must be developed. However, in the long run, the history of humankind will ultimately be divided into two distinct parts. Phase one is when humans lived, breathed, bred, and incubated within the confines of the earth's gravity. Phase two will be when humans expanded the scope of their activities into the broader reaches of the solar system and reached beyond the skies. Affordable access to space will ultimately be critical to the survival of the human species.

Neil Armstrong had it right when he said, "One small step for a man. One giant leap for Mankind." Well, almost right. Perhaps he should have said, "One giant leap for Humankind." Any new celestial civilization won't get very far without some women around to sustain the species and to provide some much needed insight and common sense as well.

Overcoming Some Over-Optimistic Forecasts

In 1969, at the time of the first moon landing, our hopes were high and our expectations even higher. Futurists predicted that low-cost trips into space would be commonplace in a decade—20 years at the most. Cal Tech scientists actually projected that astronauts would fly to Mars—in the 1980s! In the early 1970s, Allen Drury wrote a popular novel about a trip to Mars, which he believed was only a short time into the future. Remarkable achievements in terms of projecting rapid and economical access to space via the Space Shuttle turned out to be wrong by orders of magnitude. Ever since Apollo, NASA's space program has had more than its share of troubles. The addition of a solid fuel rocket to the shuttle was only one of many misjudgments. On the occasion of Apollo 17, the last moon mission, in 1972, Wernher von Braun predicted that there would be a "permanent Lunar Colony" and that it would be achieved by 2009 or 2010. When news reporters asked why his prediction was so precisely timed, he responded that it took us 37 years from the time we first reached the South Pole until the time we established a permanent colony in Antarctica. He calculated that the challenge of setting up permanent shop on the moon, in light of ever-increasing technical know-how, should allow us to create a realistic colony on the moon's surface in 37 years.

Well, we all know he was wrong, and off by at least a decade, if not by two or three. One reason is that von Braun envisioned a space station as a stepping-stone to the moon. This was a BIG mistake. It turns out that a large-scale space station was an unnecessary and burdensome step to the moon and Mars. Clearly, NASA and its partners made a series of errors in the design, construction, and operation of the ISS. Likewise, NASA's mistakes involving the Space Shuttle design and its multi-billion-dollar refurbishment program are almost legendary in terms of compounded errors. The *Challenger* and *Columbia* disasters made all the headlines, but this was only the tip of the iceberg. The mistakes and errors made with the Space Shuttle could fill a number of books. (We should know, because a number of these have already been written, including one by ourselves.) [5]

The Strengths and Weaknesses of National Space Agencies

But the ISS and the Space Shuttle are now the "back story." These past misjudgments are today, at least in terms of galactic history, largely space flotsam and jetsam over the cosmic dam. The point is that national space agencies are pretty good at developing totally new and state-of-the-art technology at the conceptual level. The scientists and engineers at space agencies are very proficient at solving totally new and arcane problems. But NASA and its counterparts around the world are generally poor at implementing major new infrastructure projects against a budget and achieving a drop-dead schedule.

Rather astonishingly, when Ronald Reagan first announced the "Space Station" in the early 1980s, it was supposed to be finished by 1994, and it was to cost a tiny fraction of the $140 billion that the ISS eventually cost. This works out to well over a billion dollars a ton.

No commercial contractor could undertake a large-scale project and then finish it some 15 years late and with a price tag four or five times over budget. If NASA were a business, it would have been bankrupt and out of business years ago. Nor is the ISS the only example of some galactic errors of judgment and management. NASA has started and then abandoned at least six spaceplane projects at a total cost exceeding billions of dollars. The time has arrived for private enterprise to lend its hand to make space projects affordable and responsive to schedule. Some would say this is judgment in hindsight about incredibly difficult technology-development programs. The after-the-fact insights of "Monday morning quarterbacks" are legendary. But let's review the history. The entrepreneurial External Tanks Corporation, which was chaired by Dr. John McLucas, a former Secretary of the U.S. Air Force, and the very clever Randall "Stick" Ware showed how one could simply lift the shuttle's external tanks into orbit and put six to eight of these together to create a very cost-effective space station. The entrepreneurial

Robert Bigelow has shown how, with inflatable systems, one can create a space station with more usable volume than the ISS, again at a small percentage of the cost of the ISS and with very rapid deployment. One needs only to review the various proposals of the entrepreneurial talent that have been put forward as proposals to win the $30 million dollar Google Lunar XPRIZE to understand how different and innovative the thinking can be if freed of conventional space agency bureaucratic constraints.

We are indeed seeing a new type of "private space venture." These initiatives—often backed by cagey and creative billionaires—are seeking to develop commercial spaceplane technology as well as operate a profitable space tourism business. In earlier chapters, we have described the business organization and commercial plans of these new commercial ventures as the "anti-NASA," and the differences are starkly apparent. A well-focused 300-person organization up against a 15,000-person organization can often win in terms of schedule, price, and goal achievement.

Avoiding the Bureaucratic Processes

These new, agile, and entrepreneurial organizations totally reject the space agency model for the most part. They certainly do not have the thousands of employees found within space agencies or large aerospace behemoths. These totally new types of private space organizations certainly seek to avoid operating under bureaucratic processes that can squelch creativity, prevent innovation, and frown on flexibility of operation. The creative "Lone Ranger" has a difficult time making it to a leadership position within NASA. One-time "rebel" Pete Worden, who ran NASA Ames in unconventional ways for several years, was the exception. But not surprisingly, the entrepreneurial Worden, who ran the shoestring Clementine mission to the moon while in the Air Force, has now opted out of NASA's bureaucratic constraints.

The processes found within space agencies for bureaucratic procedures, group think, quality assurance, and independent verification and validation are so cumbersome that a concern for safety can get lost in the process. Then, after the space agency creates extremely tight standards, it allows thousands of "waivers" to safety standards. As described earlier, both the *Columbia* and the *Challenger* shuttle disasters can be traced back to waivers that were granted to explicit safety and performance standards. At times, it does seem a lot like Alice in Wonderland. One always needs to believe quite a number of impossible things before breakfast.

This fresh new private sector approach to space projects can, we hope, bring new economies, new applications and products, new opportunities, new technologies, and new ways to address a variety of problems both in space and on earth.

At this stage, no one can say what our future will look like. Pelton's law of forecasting explains why one should never make a 5-, 10-, or 25-year prediction. Always go for a 50-year prediction when, you assume, no one will be around to check up on your accuracy.

Current concepts for a spaceport waiting room may look amazingly retro in a few decades. The point is that most people imagine it as a Buck Rogers fantasy rather than as a true doorway into tomorrow. If we manage these spaceports right, they can be gateways in both a literal and a conceptual sense. The best of commercial space technologies could help us find new sources of clean energy, combat carbon pollutants, and even re-invent the global economy, where the cost of planetary sustainability is priced into the cost of goods and services.

Peter Diamandis' XPRIZE enterprise, which started the whole spaceplane industry, has already morphed into the XPRIZE franchise. The idea is to create a systematic "challenge" process to develop better and safer ground transportation, better and cleaner energy systems, and much more. One of the more interesting new XPRIZE challenges is the 100-mile-per-gallon automobile. For many years, Diamandis has said that the pioneering individuals who are willing to think outside the box and challenge today's conventions are the ones who change the world. Committees and groups of people "defend the status quo."

Larry Page, co-founder of Google, has summed up the difference that entrepreneurial thinking can make in contrast to the conventional approach taken by large established organizations. Page advocates not looking for the 0.1X improvement that provides modest gains, but rather for the "home run" 10X gain in performance. This type of innovative thought sets Google apart from the conventional large corporation. [6]

The Big Questions

The previous chapters have set out the story to date about the commercialization of space and the coming new era of "Citizen Astronauts." A number of billionaires, a gaggle of rocket scientists, and many dozens of innovative design engineers are hard at work. These designers are working in facilities that range from sophisticated labs to back-lot garages. All of these efforts are aimed at creating viable space tourism businesses, mostly in the United States, although others in several different countries are also hard at work.

However, many big, as-yet-unanswered questions about this new commercial space industry remain. When will safe, reliable, and economical spaceplanes and spaceports become operational? Will these efforts closely parallel the development of high-altitude supersonic executive jets and will these two industries ultimately be closely interlinked or not? How will these various efforts be regulated at the national and international level? Will the ICAO, the U.N. specialized agency for aviation safety and air traffic management, assume responsibility for space traffic management in coming years?

When will the cost for fare-paying passengers reduce to a level so that a significant number of people can be attracted to the experience of a flight in space? Will a competitive, commercial market lead to more opportunities to expand into a wide range of new space-related businesses—and how safe will these activities be? Can private space ventures lead to affordable stays in space in "private space hotels"? Can the private space industry lead us to clean and economically viable solar-powered satellite systems, or allow cost-effective in-orbit systems that can manufacture better and cheaper pharmaceuticals, building materials, or entirely new products? Can space commercialization lead to new clean hydrogen rocket engines or other "green" ways to approach both air and space travel? Will the long-time-coming promise of scram jet technology eventually be realized in a cost-efficient and safe way? One of our key challenges over the next two or three decades is to find ways that move people in much more environmentally friendly ways, not only from earth to Mars but from Cleveland to Milwaukee or from Singapore to Sydney. And, one may hope this is just the start.

And There Are More Questions …

We must hope for second-generation breakthroughs as well. In time, we could and should look to new ways to access space, such as space elevators, space tethers, advanced ion thrusters, nuclear-powered systems, or solar sail systems. What will be the relationship between private space ventures and national space agencies as traditional roles change due to changing economics and opportunities? In short, where are all these private spaceflight ventures headed within the next decade and over the longer term? Will this changing role be different in the United States and other countries? Will NASA and the FAA need to be restructured to adapt to the presence of serious commercial space activities? Where will ventures in Australia, Canada, China, France, Germany, Israel, Romania, Russia, Singapore, the United Arab Emirates, the U.K., and other countries fit into the new space tourism industry? Who will succeed and who will fail?

There have been some extravagant claims about dramatic changes that will come as the result of private space initiatives. However, it is simply too early to know if this is the hype one expects as a part of the marketing and PR strategy for these new companies—or not.

What will these companies be saying in 5 or 10 years? If they fail to get an acceptable return on their huge investments, it will likely be a much different story. Can space agencies adapt successfully to a new era in which private companies provide an increasing percentage of the new capital, technological innovations, and management skills? What does it mean to the global economy if there is indeed a fundamental shift in what happens in space? Can we truly look forward to what might be called an "orbital space industry," which involves solar electric power generation or new types of materials and drug production grown in a low-gravity environment? Can we eventually see true space tourism with extended stays in commercial space hotels with new types of space-based environmental protection programs? Are the people developing this new private space industry truly visionaries or are they simply a group of individuals out to make money from a seemingly "unique and sexy" new enterprise? Are they just pursuing "the exotic new high-risk experience market" with a business model that resembles that of an entertainment park, or are they on a quest for fundamentally new technology that will unlock a new era of human experience and discovery?

The last of these questions probably cannot be answered for another decade, but so far we find "all of the above" now at work in this crazy new enterprise. Dedicated engineers, space visionaries, travel consultants, crass businesspeople, and scammers can all be found in the personal space industry today.

And More Questions about Weaponization

There are also serious questions about the current trend toward the "weaponization of space." It is not clear how private spaceflight might relate to national defense, homeland security, and improved air and land transportation and navigation systems. The Bush Administration, over its eight years, resisted any attempt toward limiting options in the commercial space world as long as they did not impinge on military options. Then, it was a different story.

George "W" Bush, Condi Rice over at the State Department, and the entire neo-con community vigorously resisted any suggestions, especially from Russia, that there might be a need for a new treaty to ban weapons in space. The Bush White House claimed that such a ban could never be verified. The key question for the new administration is what to do about space—particularly military uses of space. ESA's Code of Conduct for Space, the U.N. Committee on the Peaceful Uses of Outer Space, and non-governmental organizations such as the Secure World Foundation, the IAASS, and the ISSF all seem to be focused on developing new "rules of the road" for space. These initiatives, among other things, appear to be seeking to stop the deployment and operation of space weapons whether they are deployed in orbit or from the earth's surface. Today there are also many serious questions about how the new spaceplane industry can be made safe from an environmental standpoint. How should we address the danger of debris in LEO? What about rocket and jet flights in the stratosphere in terms of both the ozone layer problem and greenhouse gases? In short, some really key issues need to be considered, and the space tourism industry and national air and space regulators only have about two years to resolve these problems before commercial flights are initiated for real—not just a few experimental flights that seem relatively easy to regulate today.

Governments Provided a Launch Pad for Private Enterprise

It is difficult to find clear precedents for this almost unique industry. It is certainly hard to find a previous start-up industry in which governments around the world have invested, upfront, such a significant amount of taxpayers' money. So far billions of dollars, euros, and other currencies have gone into direct and indirect incentives. NASA's COTS program represents over a half billion dollars of public monies, and the people of New Mexico alone have voted to put in over $100 million to finance the quarter-billion-dollar Spaceport America project. Seldom does government serve so completely as the "launch pad" for private enterprise. It is unusual to offer promotional support, set up an "encouraging" regulatory framework, and lend access to major governmental infrastructure. Nevertheless, this is the case in the United States, Europe, and elsewhere—sometimes on a rather grand scale. National space agencies, aviation regulatory bodies, national defense systems, missile and satellite defense systems, and the new private space initiatives are, for better or worse, closely linked together in the United States and abroad.

There is no doubt that privatization and private space ventures represent today's "big story" in the aerospace field. The "outer space banner headlines" were supposed to be NASA's efforts to develop new vehicles that could in time go to the moon and Mars. However, NASA has been upstaged. The billionaires and the other private space entrepreneurs who are rolling out systems that promised the public suborbital flights "from 2013 or 2014" on have definitely replaced NASA in terms of public zeal—and interest. But, of course, just one more tragic accident could quickly dampen the enthusiasm for private space flight.

The last five years have seen the rapid formation of corporations to develop spaceplanes, to market space tourism flights, to develop new spaceports, and to create an interest in the possibility of "space travel." As we have seen in earlier chapters, Space Adventures led the way by arranging for Russian Soyuz vehicles to take billionaires to the ISS.

Then, the XPRIZE incentives prompted a phalanx of innovative corporations to follow, not only in the United States, but also around the world. Literally billions of dollars have now been invested in developing and upgrading spaceports; developing, manufacturing, and testing spaceplanes; and developing other needed infrastructure including hangars, launch pads, roads, testing facilities, and elaborate training facilities.

This activity has quickly changed from science fiction fantasy to real worldwide business. Government space agencies around the globe have played a part in the development of this new industry. Nevertheless, many of these enterprises have eschewed government help and intervention. This is largely because the entrepreneurs felt they could move farther and faster without "governmental interference and involvement."

How Congress Helped

Congress has passed legislation that places the regulatory control of the space tourism industry with the FAA. At least through 2015, the FAA is charged with granting experimental licenses for such spaceflights case by case.

In short, the FAA/AST, rather than NASA, is in charge. As a result, FAA officials have undertaken an extensive "rule-making" process to establish rules for the pilots and co-pilots and rules for their health and safety. As a result of this process, the FAA/AST has adopted regulatory controls for spaceplane passengers, crew, and pilots. These regulations govern the "experimental stages" of the spaceplane and space tourism business. This process requires all passengers to sign waivers indicating that they hold the U.S. government entirely harmless against any type of accident and that they fully acknowledge that spaceplane flights are high-risk undertakings. There are clear-cut safety controls and training standards for crews, passengers, and ground control systems, but FAA actions, in general, are geared toward encouraging the growth and development of this new industry. Passengers who "insist on going" are required by organizations like XCOR and Virgin Galactic to sign away their rights or those of their estate to make a claim against any accident.

The motivation provided by cash prize competitions and

challenges has helped to stir the imagination of innovative aerospace engineers the world over. The success of the Rutan and Allen team in winning the Ansari XPRIZE in 2004 was undoubtedly the single most important event in the development of the space tourism business—at least until the maiden fare-paying flight by Virgin Galactic.

Prizes continued to lead the charge forward. Bigelow announced his own $50 million cash award under the name of "America's Challenge." This challenge sought the development of reliable and low-cost transportation to Bigelow's planned space habitats. Bigelow made sure that the challenge and the performance period were difficult enough that he had a low probability of having to pay out this substantial prize. Indeed, the prize expired without any attempts ever recorded.

The Google Lunar XPRIZE, at $30 million, actually achieved greater media and Internet exposure, and, indeed, this has captured the imagination of the commercial space community around the world. This new prize is for the first commercial robotic mission to the moon that sends back video pictures from the lunar surface and that completes a number of other tasks. About a dozen teams have officially registered for the Google Lunar XPRIZE to date (they were listed in Chapter 1). Others would be seeking the prize if Google were willing to share more flexibly in the release of the videos and news coverage that is anticipated from these private lunar missions.

In addition, various NASA "Centennial Prize" competitions are on offer. These are for far less money and for less awe-inspiring goals; they have not captured the same amount of media coverage. Even so, this program, so energetically led by Dr. Kenneth Davidian, could be critical to advances in space tethers, space elevators, and extra-terrestrial activities.

So what will the private space industry accomplish in coming months, years, and decades? What are the most important things to know about this evolving new industry? Certainly, we all have a number of questions about the future of private spaceflight. Well, here are some possible answers, distilled from earlier chapters. These, in our view, are *the top 10 things* you should know about this curious new industry.

1. How Will the Private Space Industry Be Regulated and Controlled in the United States and Internationally?

The FAA/AST regulates this new industry under U.S. legislation passed in late 2004 (Title 49 USC, Subtitle IX, Chapter 701). In addition, the Commercial Space Transportation Amendments Act of 2008 and the latest FAA Appropriations Act at the end of 2011 extended the FAA's authority to grant experimental licenses for spaceplane flights through 2015. In December 2006, the FAA adopted detailed regulatory controls that were designed to ensure safety; to develop appropriate licensing, training, and inspection standards; and to harmonize national safety and reliability standards with emerging international regulatory processes to provide oversight of a new global space tourism industry. It has also released regulatory controls for pilots and crews. Additional responsibilities and authority are assigned to the FAA under Executive Order 12465. Although European regulators have suggested that spaceplanes might be certified in the not too distant future, FAA experts have described the drawbacks of a full certification approach and concluded that licensing procedures may be the only realistic way to support private human spaceflight enterprises. [7] One of the regulatory issues, at least in the United States, is the respective roles of the FAA and NASA when it comes to space-related development and regulation. The FAA is involved with spaceplanes and sub-orbital flights as well as licensing of commercial space launches. NASA is now seeking to develop commercial launch capabilities, at least to LEO, as well as to develop the planes of the future. Clearly there are issues of overlap.

To address this issue, in June 2012, the FAA and NASA signed a joint memorandum that sets forth procedures for these two federal agencies to cooperate with each other. The purposes of the memorandum are to avoid conflicting or overlapping standards, to jointly advance both public and crew safety, and to facilitate cooperation. [8] In addition, the U.S. Air Force, the FAA, and NASA must coordinate with regard to spaceport operations.

For the future, the even larger questions concern regulating the "interface" between aviation and spacecraft of various types, and ensuring safety for all concerned. This has led to increasing discussion of the issue of "space traffic management"—as opposed to air traffic management—which is now well defined at the national and international levels. Increasing efforts to address this issue and the ICAO hosted a conference on this subject but carefully avoided so naming it. Instead it cautiously called the May 2013 sessions "Regulation of Emergency Modes of Aerospace Transportation." Although the ICAO itself has remained cautious, others are pursuing the idea that the ICAO be formally designated as the entity to regulate international space operations and spaceplanes. This idea, however, is still far from popular among regulatory officials in the United States, and EASA has not taken a position. At this point, the most popular idea is for national regulation to oversee such activities and then coordinate with other countries as necessary and appropriate.

In March 2015, ICAO held a news conference on space traffic management, and a number of entities from the United States , Europe, and Japan made interesting presentations about spaceplane flights and stratospheric balloon systems. There were also presentations by those that are actively pursuing hypersonic transcontinental flights as early as 0 years hence. The most interesting presentation, from the perspective of the author, was JAXA's presentation, which explained Japan's continuing prototype development of liquid hydrogen and oxygen fueled spaceplanes that are moving up from Mach 2 to Mach 4. Even more unique, JAXA's presentation indicated specific consideration of environmental effects of such hypersonic flights and efforts to address these concerns. In contrast, the presentation by Virgin Galactic avoided the serious question of the particulates that are emitted from the SS2 engine and the aluminum polyimide fuel that it now uses.

Clearly, the issue of space traffic management and the interface with air traffic management represents the next big step for regulating the spaceplane industry and possibly for commercial space launches.

The current regulations hold the U.S. government harmless against accidents that may occur by requiring all passengers and crew to sign waivers against all liability and by forcing everyone who seeks to fly into space to recognize that significant risks are involved in spaceplane flights. The risks are clearly much greater than those of riding on a commercial airplane, and the regulations and waiver statements are crystal clear in this regard. The CSF claims that if the government will just let it be free to innovate, it will come up with new vehicles and systems that are far, far safer—at least for suborbital flight—than anything NASA or the U.S. Air Force has come up with to date. What is clear is that the regulations currently in force in the United States, Europe, and elsewhere have placed the prime safety concern on protecting those who are on the ground or flying on commercial aviation, much more than on protecting the crew and the passengers on experimental spaceplane flights.

Currently, the FAA licenses governmental and commercial spaceports only after thorough examinations that are conducted every five years. The FAA also licenses and authorizes, case by case, every flight of a commercial spaceplane or rocket launch, and each instance is considered an "experiment." Licensing an experimental launch is nothing like certifying an airplane to fly multiple missions, and we are many years away from such a process.

As noted above, Congressional legislation first extended this "experimental period" through 2011 and then as of December 2011 extended this case-by-case experimental licensing through 2015. This case-by case experimental regulatory process is something that can work quite well today given the small number of spaceports and the limited number of lift-offs now occurring. In the future, however, things will change. When the number of spaceport facilities doubles or triples and the number of flights increases by a factor of 10 or more, significant problems could arise. These problems could concern adequate staffing and competence by FAA/AST regulators to proceed on the basis of current case-by-case licensing processes. There are currently no particular regulations with regard to space habitats, such as those Bigelow has proposed to operate. Here, issues of both public safety and national security could arise.

A final issue is that under U.S. law, each space launch must have an accompanying environmental impact statement associated with that launch. Again, the currently limited number of spaceplane launches does not create a large regulatory barrier to space tourism-related flights, but when the number of flights expands exponentially, this will be another issue to address, and new processes will need to be found.

As described in earlier chapters, the FAA is currently placed in a somewhat awkward role by Congressional regulation and the White House Executive Order. The FAA is supposed to be regulating the industry and providing for public safety, and yet it is also charged with encouraging the growth of this new enterprise. Things often end badly in an enterprise where the roles of referee and cheerleader are combined. Patricia Grace Smith, the former Associate Administrator for Commercial Space Transportation for the FAA, who left office in late 2007, made it clear that the FAA feels it will be a number of years before it will "license and certify a vehicle" as flight worthy, such as is currently done for a Boeing 777 or an Airbus 380. As Smith has said, "The proven safety is simply not there." [9]

Dr. George Nield, who has replaced Smith at the helm of FAA/AST, has echoed the same message—safety first. Yet he knows Congress has charged him with promoting this new industry at the same time. The FAA is thus placed in this somewhat conflicted role.

So far, the United States currently leads the world in this new industry, and other countries are more or less following the American regulatory formula. This formula has seemed to work, but once there is a real worldwide service, the problem will be more clearly defined, particularly if an accident claiming human lives occurs.

It would be prudent, at least at some point in the future, therefore, for the FAA's safety and security oversight role to be divided from any responsibility to promote the industry's growth and development. Once the industry develops, it may be appropriate for its regulatory role to be restricted to only safety and security issues. The promotional role was taken away from the FAA after it played the roles of both industry backer and regulator for many decades. This occurred after an airplane crash in Washington D.C. in the 1980s and the subsequent Congressional investigation. It is possible, therefore, that the FAA's promotional role with respect to spaceplanes could remain until there is a high-profile accident and loss of life.

Space tourism, of course, is not just a U.S. affair. As detailed in earlier chapters, many enterprises around the world are seeking to develop spaceplane technology and systems and to promote spaceports in such locations as Australia, France, Singapore, the United Arab

Emirates, and the U.K. The EASA has been closely involved in two spaceplane development programs that ESA is supporting. At the international level, several other entities may ultimately play a role. The ICAO and the ITU, in particular, are likely to be involved in regulating flight safety and the frequencies used for communications. Further, the IAASS and the U.S.-based ISSF are also pursuing space safety issues at all levels. The IAASS and the ISSF are seeking a process to create an MOU among space agencies and spaceplane regulatory organizations for agreement on common space standards. The Secure World Foundation is trying to develop what it calls international "Rules of Road" for space activities, whereas the ESA is backing an International Code of Conduct.

Finally, if the development of private spaceplanes for space tourism branches into the development of executive supersonic jets and their safe operation, it is possible that regulation and safety controls for both enterprises can and will overlap, at least in several key areas. This would be not only in terms of FAA oversight but also in terms of spaceport inspection and licensing, liability waivers, pilot and crew standards, safety certification and training requirements, and the nature of environmental oversight. In addition, some ambiguities related to insurance and risk management might be involved. Both the Private Spaceflight Federation and the FAA Office of Commercial Space Transportation, at this point anyway, would like to keep these two activities as separated as possible.

2. How Could Spaceplane and Space Tourism Safety Be Enhanced?

Congress should keep a close eye on FAA regulation of the space tourism industry and at some future time, it might consider whether the safety regulatory function and the licensing function should be separated by a solid wall from efforts related to the promotion of the industry. Currently, the office that promotes the industry is clearly separated from the safety regulatory role; nevertheless, at the level of the head of FAA/AST, the conflict does potentially arise.

Another idea would be to form a White House Commission to address safety and security issues and to define ways to advance the safe operation of the space tourism business. Just as there was a successful White House Commission that developed important new guidelines for aviation flight safety, a similar and parallel commission might be formed to carry out this task related to space tourism. This commission would be directed to complete its work within a specified time and be given a specific charge to carry out a number of tasks:

- It should devise ways to reduce personal spaceflight accidents and enhance flight safety by developing targeted and realistic quantitative and qualitative objectives and measures.
- It should develop a charge for the FAA (and NASA, as appropriate) in terms of creating standards for continuous safety improvement, and these goals should create targets based on performance against those standards.
- It should develop improved and more rigorous standards for the ultimate certification and licensing of spaceplanes (or their subsystems and/or their escape systems). These should also cover their operations. This could perhaps start with certification standards and best practices for spaceports and for training and simulation facilities.
- The Federal Aviation Rules for Private Spaceflight should be rewritten with statements in the form of performance-based regulations wherever possible.
- The FAA should develop better quantitative models and analytic techniques to assess spaceplane performance and to monitor safety enhancement processes. These should be based on industry best practices as new operational vehicles and safety and emergency escape systems come on line. Key to this process is systematic input from industry with regard to flight performance for many system and subsystem parameters.
- It should identify specific research topics for the Center of Excellence for Commercial Space Transportation in terms of safety-related development, and it should rank them in priority.
- The FAA Office of Commercial Space and the Department of Justice should work together to ensure that full "whistle-blower" protections are in place (including new legislation, if required), so that employees of the space tourism business can report safety infractions or risk factors of concern regarding safety violations or security

infractions to government officials without fear of retaliation or loss of employment. These employees include, but are not limited to, manufacturers and/or operators of spaceplanes; maintenance and other ground-based crew, owners, and operators of spaceports; and owners and operators of personal spaceflight training and simulator facilities. This could also include a safety and hazards reporting call-in line that is parallel to the NASA Safety Reporting System (NSRS).

The Private Spaceflight Federation is an important resource that should work with the White House Commission and the FAA to develop improved safety regulations, but a parallel entity representing space tourism passengers and crew is also needed. To this end, an entity representing spaceflight consumers should somehow provide input to the White House Commission's deliberations. This might be an offshoot of the Airline Passengers Association.

The Center of Excellence for Commercial Space Transportation, which was created in August 2010, is an important resource available for furthering safety research for the commercial space transportation industry. Member universities include the Florida Institute of Technology, Florida State University, the New Mexico Institute of Mining and Technology, New Mexico State University, Stanford University, the University of Central Florida, the University of Colorado at Boulder, the University of Florida, and the University of Texas Medical Branch at Galveston. Organizations such as the IAASS and the ISSF also lend support to these efforts. Specific research efforts in the safety arena should be given priority attention. [10]

Although the first commercial spaceplane flights will very likely be operated by U.S. firms or by a British company (such as Virgin Galactic) operating from within the United States, international regulations and licensing procedures may in time stem from other sources. Again, these include the ICAO, EASA, and other groups such as the IAASS, the ISSF, the Secure World Foundation, and the McGill Air and Space Law Institute. These entities and others may, over time, add important safety control capabilities, standards, and processes, but for the time being, the U.S. regulatory systems will lead the way. Let's hope they manage to get the right formula - which may actually be the simplest one. This strategy, as Dr. Nield explained to us, is to seek to hold the reins tight enough to achieve maximum safety, but not so tight as to stifle this new industry's innovation and rapid evolution. So far, most other countries seem willing to follow the U.S. model and are moving toward parallel or at least similar regulatory processes.

3. Who Are the Key Players in the Space Tourism Industry? And How Can One Best Follow What's Happening in This New Field?

To describe the current space tourism industry as chaotic, or somewhat like the Mad Hatter's Tea Party, would be to overstate the confusion and dramatic change now shaping this emerging field. However, it is not far short of the reality. Corporate reorganization, shifts in partnerships, mergers, or bankruptcies are weekly, if not daily, occurrences. Our research has come up with nearly 50 organizations that have been or are in the hunt. An even more compelling statistic is that over 25 companies have already folded or been absorbed by other companies. (To see the diversity just look at Appendix A.)

We can hope that "stability of design" will be found over time. These mature designs will most likely occur first in subsystems such as in propulsion, controls, life support systems, and environmental safety and escape systems. Some stability and standardization will ultimately be found in this new business, but this certainly does not describe the space tourism business as it exists today. As advocates such as Diamandis, Greason, Tai, and others contend, standardization today would inhibit and perhaps crush innovation. The evolution of this new technology and industry seems in many ways a historical replay of the early days of barnstorming pilots, where everyone was building their own plane, developing their own emergency escape system, and offering their own flight services to the public. Hundreds of garage mechanics and even bicycle manufacturers were trying to build a "better flying mousetrap." It was decades before the best of the airplane designs and the best of the aviation companies evolved. The same seems likely to be the case for the space tourism business. It may well be into the 2020s before this now "risky business" ultimately matures. In time, there may be a significant role for lighter-than-air vehicles, advanced ion engines, space elevators, satellite-based tether systems, or even technologies that have yet to be invented. For this reason, "type certification" of particular spaceplanes remains a long way away. In this respect, high-altitude, hypersonic corporate executive jets may in time partially lead the way because they will fly longer-duration flights and because the potential market for those that might purchase these planes is far larger. Their safety standards may also be higher.

It is projected that during the current decade or early in the next, a number of large aerospace companies such as Lockheed Martin, Boeing, and EADS-Airbus will develop and offer to the market such supersonic commercial jets with noise abatement systems.

One element of seeming stability for the new space tourism industry is the increasingly predominant CSF. Its membership represents a "who's who" of the current industry—at least from the United States—and it has been perhaps the main source of comments to the FAA with regard making rules that cover the operation of spaceplanes and the licensing of their use.

This group will undoubtedly continue to play a key role in advancing a space tourism business in the United States as well as in developing and amending FAA rules and regulations that govern safety and licensing practices. In terms of total wealth per individual member, it may well be one of the most exclusive clubs in the world. The CSF will clearly help the industry find stability, share best practices, and assist new entrants into the field to meet minimal standards for simulation, vehicle testing, and operational procedures. Its members will also provide the capital financing this new industry so desperately needs. The appointment of the very well respected former astronaut Michael Lopez-Alegria in some ways seemed to reflect the maturity of the CSF and its leading corporate members as this fledgling new industry finds its way forward.

One can only hope that, in time, a countervailing group to the CSF will represent the views of space tourism passengers—much as there is an Airline Passenger Association today. Also, professional publications that follow the details and key events of this new industry will probably evolve. Today, however, one must largely follow the development of private spaceflight through broader-brush media such as

Space.com, *Space News*, and *Aviation Week & Space Technology*, and by reading the many websites of the spaceplane and space tourism companies. A breath of fresh air that has now established itself within a little over a year is *Space Safety Magazine*, co-sponsored by the IAASS and the ISSF. One can also learn a good deal by reviewing the official websites of the FAA, EASA, ICAO , IAASS, and the Secure World Foundation.

4. Is the Space Tourism Business Commercially Viable or a Media-Hyped Anomaly That Will Soon Fade from the Scene?

This is the number one question that financial analysts, insurance companies, and commercial banks are asking, with a wide divergence of answers.

Almost two decades before the dot.com bubble burst, a cynical newspaper business reporter opened his remarks at a conference in Denver by saying, "The broadband Internet commercial e-tail business has rapidly grown from a zero-million-dollar activity to a zero-billion-dollar enterprise." Today the "e-tail" business has now grown into a truly viable enterprise and is a significant force in the global economy. However, 15 years ago, it was 99+ percent hype and less than 1 percent sales. Many feel that space tourism may suffer from the same start-up oversell. Serious studies have certainly been undertaken to try to understand the market for space tourism and to estimate the volume of demand and the scale of this enterprise as an on-going business. The optimistic market studies that Futron Inc. first conducted a decade ago have been replaced by the Tauri Group studies, which project much more conservative growth figures.

The ESA has commissioned a major study that projects slow initial growth. Altogether, tens of millions of dollars in advanced booking fees have now been deposited with Virgin Galactic, Space Adventures, and others to reserve seats on future flights. This sounds impressive, but the amount shrinks to insignificance when one starts to tally up the huge investment that has now been made in spaceports and spaceplanes. Indeed, perhaps $3 billion have been or will likely be spent on planes, test facilities, training centers, spaceports, and spaceplane launcher vehicles before commercial and still "experimental" flights take off.

This is a capital-intensive business, and a few billionaires have risked substantial investments in the hope that this is a venture that will "fly." (You can suppress a groan here, if you like.) A great deal of investment in spaceplanes, spaceport infrastructure, regulatory oversight, and highly trained personnel will all be required up front, well before serious revenues start flowing.

This industry is thus a highly vulnerable enterprise that requires a lot of money to be spent on insurance and risk management. A personal spaceflight launch insurance business could go south quickly if one of the commercial flights ends catastrophically at an early stage in the "take-off" of the industry. It is probably true that developing a successful insurance and risk management system for the space tourism business is more important than, say, developing a better rocket engine, or certainly more important than creating a highly functional spaceport. At this stage of the new and unproven industry's development, there are lots of imponderables. Just a few of these questions are as follows:

What is the return or "repeat" market? The suborbital flights take just a couple of hours or so from launch to landing, with only about 4 minutes of weightlessness and about 10 minutes of "black sky" sightseeing. Jeff Greason's plan for the XCOR-developed Lynx vehicle, which climbs to a "sub-suborbital" altitude of only 37 miles (or nearly 60 kilometers), will be a flight of about a half hour.

How many people will consider doing this sort of high-cost flight again? The answer is "no one knows." As one analyst of the space tourism ventures has suggested, this business will need to develop the "and then what?" or the "what thrill next?" market. The Space Adventures people are indeed working on a number of "accessories" that the company will offer to its clientele.

What will the prices and the pricing curve be over time? The price of a suborbital flight is currently pegged at around $200,000 to $250,000, (or about $100,000 for XCOR's lower-altitude ride), but these prices will inevitably decline over time. It is thought that people with over $1 million in assets and making $200,000 or so a year might be the market most likely to buy a seat on a spaceplane ride. Such people, who are in the upper 1 percent of the world's population in terms of wealth, however, are usually shrewd and intelligent buyers. In short, they got rich by investing wisely. If they are not among the first thousand to fly, it is just possible that they may wait until the prices fall quite a bit further. If the prices go down to, say, $50,000, there might be a new surge level of demand, and if the price falls to $20,000 or $25,000K, there might be yet another surge of potential consumers. No one yet knows what the price-versus-demand sensitivity ratios will be for space tourism flights. The demand curve for a "flight to nowhere" is hard for anyone to predict. Personally, we think that the low end of the curve, in terms of paying for a space tourism experience, may be the most successful—at least in the early years. By the "low end," we are thinking of IMAX movies, training to be an astronaut at a space center, a parabolic flight to achieve weightlessness, or a flight on a very high altitude jet like the Foxbat.

How many spaceplane operators can the market support? Sir Richard Branson seems to be making a pre-emptive coup by buying a fleet of spaceplanes from the SpaceShip Corporation, which he co-founded with Burt Rutan. But Rutan has said he plans to sell as many as 40 of these vehicles to space tourism operators and, as described in previous chapters, many dozens of others are planning to develop spaceplanes. Various projected numbers for these vehicles—between 40 and 100—that might fly, frankly, do not add up in terms of a viable business plan today. It does not seem that the market can sustain so many spaceplanes operating, at least in the next few years. Prices can remain high only if space rides remain a very limited commodity. The high cost of buying and operating a spaceplane would appear to require high prices for some time to come. Presumably over a million jet-setters who will shell out big bucks for first-class air fares are out there. However, the ultimate marketing answer remains elusive—only time will tell. Our best guess is that of the 40 or so spaceplane developers who say they will provide space tourism flights in coming years, only four or five will survive.

Will there be synergy between the corporate supersonic jet and spaceplanes to support space tourism? Major aerospace companies are designing and planning to deploy "quiet" 10- to 12-passenger supersonic corporate jets that will share a number of characteristics with spaceplanes. It is still too early to tell if the technology, the business plans, the manufacturing facilities, and the scale of production will have enough in common to help sustain both industries and to provide for their more effective regulation and control. This, however, could be a critical factor—especially in terms of sustaining a viable insurance market. The challenges of the supersonic jet market for executives are dramatically different from those of the spaceplane entrepreneurs. Currently, there are major safety, environmental, and regulatory obstacles. Challenges to be overcome include the prohibition against flights in the United States due to sonic booms, greenhouse gas contrails in the stratosphere, huge developmental and operating costs, and doubts about the size of the market. The plot line and the actors are certainly quite different as well. The spaceplane "heroes" are the small entrepreneurs and the daring and visionary space billionaires. The script for supersonic jets calls for the giant and anonymously "gray" aerospace companies like Lockheed, Boeing, and EADS-Airbus to develop these very fast and high-flying new craft.

Hollywood and the media thus find the spaceplane story a lot more compelling. Beyond the smaller supersonic business jets are the long-term and more grandiose plans like the European A2, which would fly 300 passengers, with hydrogen-fueled jet engines, to speeds of up to Mach 6. The synergy, or lack thereof, between space tourism and hypersonic transport is just one of the market enigmas that potential investors are left to ponder.

5. Does a Large Number of Spaceplane Flights Endanger the Ozone Layer and Present Other Environmental Dangers?

Although many diverse approaches are now underway, virtually all efforts involve flying some form of rocket into the upper atmosphere. As is known from the experience of the Concorde SST aircraft, sustained flights high into the stratosphere can be destructive to the ozone layer, and greenhouse gases were spewed out at an altitude where they do a lot of harm. The ozone layer shield in the stratosphere actually protects humans and flora and fauna from intense interstellar radiation. Those who believe that preserving humanity is actually desirable should recognize that aviation, rocket launches, and high-altitude craft have been and continue to be of environmental concern. Orbital debris, ultraviolet radiation, and "killer electrons" blitzes enabled by distortions in the geomagnetic flux are also of concern.

Past experience—with jet aviation, the Concorde, and so on—suggests that the space tourism business should also be subject to some form of environmental impact process and regulatory control. It also suggests that new technologies such as hydrogen propulsion systems, ion engines, space elevators, and tethers, which might be less destructive to the ozone layer and could reduce greenhouse gas emissions, should be developed as soon as practicable. It also might mean that space tourism fees—and perhaps even more likely, aviation tickets—could in the future be structured to include an environmental fee that would fund science or implement new environmental systems to replenish the environmental balance in the stratosphere.

Certainly, the environmental issues raised by the potential depletion of, or reduction in, the ozone layer as posed by spaceplanes and by supersonic corporate jets, are in some ways parallel, but they are largely different. Spaceplanes fly through the ozone layer, whereas hypersonic transports fly along the layer for sustained periods. In both cases, however, special "sustainability" fees might be imposed on these industries. Here, aviation should be expected to shoulder far more of the responsibility—to address methods to help restore the ozone layer and to reduce greenhouse gas emissions.

When one recalls that the atmosphere that saves us from extinction is proportionally thinner than the skin of an apple and that the ozone layer is much, much thinner than the "shine" on the peel, the extent of our vulnerability becomes quite clear. Currently, U.S. law requires an environmental impact statement to accompany every rocket launch. A key question is whether that requirement will change as the space tourism industry matures, and if so what the new regulatory process will be.

6. How Does the Entrepreneurial Approach to Developing New Space Systems Differ from That of NASA or the Other Space Agencies?

The short answer is that there are fundamental differences. Table

13.1 summarizes the approaches of large and "bureaucratic" governmental agencies versus agile and flexible entrepreneurial organizations.

Table 13.1 Worlds apart—the space agency approach versus the entrepreneurial approach.

Space Agency Approach	Private Sector Approach
Mission Objectives: Space agencies see their mission as experimental and to develop totally new space technologies and launcher systems.	**Mission Objectives:** The private sector is seeking to apply proven technology to a commercial business where safety and commercial confidence are crucial.
Technical Goals: Several space agencies are striving to develop systems to go to the moon and Mars and currently are seeking to develop systems capable of aero-braking and operating at speeds of up to Mach 30. Other parts of space agencies are trying to develop cost-effective, fuel-efficient supersonic jets, and the two programs are not at all closely linked.	**Technical Goals:** The private sector is largely planning suborbital space tourism missions that need operate only at speeds in the range of Mach 4 to 6. Thus, aero-braking and thermal protection systems requirements are less demanding.
Budget and Number of Employees: NASA operates with a budget on the order of $18 billion/year and has thousands of employees and tens of thousands of contractors working on a wide range of technologies and capabilities. Many other space agencies have budgets that are at least in the billion-dollar range and thousands of employees.	**Budget and Number of Employees:** Start-up spaceplane businesses and space tourism enterprises are very entrepreneurial and have just scores or hundreds employees. Just a few of the new space enterprises and newer space companies have a meaningful fraction of the total number of employees of a space agency. Every employee must thus be quite focused on design safety, safe operations, and well-defined objectives.
Regulatory Oversight: The space agencies are, in effect, in charge of their own safety and mission assurance. Their review panels are, for the most part, self-appointed, and GAO reviews can simply be noted or even ignored. Only after major accidents (such as the *Challenger* and the *Columbia*) has true independent review of safety been carried out.	**Regulatory Oversight:** Start-up spaceplane businesses and space tourism enterprises are subject to the requirements of the FAA for experimental licensing of vehicles, spaceports, and operations. They must meet the requirements of FAA rule making and ongoing FAA/AST oversight. The EASA, ESA, and the European Union are still trying to work out their own rules.
Scope of Activity: Many of the large space agencies are involved in a huge range of activities involving space sciences, space applications, operation of data relay satellites, longer-range R&D, the ISS, and new launch technology for lunar and Mars missions, educational programs, and so on.	**Scope of Activity:** Start-up spaceplane businesses and space tourism enterprises are tightly focused on developing reusable launch vehicles and operating them safely to fly space tourists into space. Everyone of a limited number of employees knows exactly his or her job, and any error is clearly identifiable and associated with an individual.
Historical Perspective and Work Ethic: NASA has been in existence for almost a half century and its vast centers, employees, and contractors act on the basis of a 50-year historical record. Politicians, to an extent, see NASA as a public works program.	**Scope of Activity:** Start-up spaceplane businesses and space tourism enterprises are tightly focused on developing reusable launch vehicles and operating them safely to fly space tourists into space. Everyone of a limited number of employees knows exactly his or her job, and any error is clearly identifiable and associated with an individual.

Space Agency Approach	Private Sector Approach
Incentives and Time Perspective: Space agencies have many thousands of employees and contractors working on literally hundreds of different programs. They have the margin created by many test flights and engine firings to achieve success. If a test fails, there are no dramatic consequences.	**Incentives and Time Perspective:** Everyone on an entrepreneurial team is working, often around the clock, to make their system safe and to make their enterprise succeed within tight deadlines figured in months or years, certainly not decades. Failures are taken very personally and can make or break events.

7. Will the Possibility of Increased Weaponization of Space Adversely Impact the Development of Space Tourism?

In recent years, more and more defense-related planning has involved the possible weaponization of outer space. Current international treaties and conventions, in theory, largely prohibit the deployment of weapons in earth orbit or on the moon. Nevertheless, there is a difficult and not well-understood distinction between defensive systems and offensive weapons. China's 2007 test of an anti-satellite missile system destroyed an LEO satellite and exploded thousands of new pieces of debris into earth orbit, intensifying concerns in this area. In early 2008, the United States followed by demonstrating its capability to destroy a "de-orbiting spy satellite." Fortunately, most of the debris in the U.S. event de-orbited in a very short time span because the missile strike was from the top down, not from the bottom upward.

As more and more satellites and space tourism facilities are deployed, these questions will become even more germane. An effort to further codify a ban on weapons in space, as well as being an aid in the reduction of orbital debris, would also be of great value to those operating space tourism businesses and would serve the interests of the peaceful uses of outer space and commercial space activities.

Some see the National Policy Directive, adopted by the Bush Administration in the fall of 2006, as protecting U.S. assets in space and discouraging attacks on U.S. satellites and private space initiatives. This has been replaced by the space policy of the Obama Administration, which seems to place more emphasis on space cooperation, the evolution of soft law such as a universal "Code of Conduct" for space activities, and joint space initiatives. New initiatives, such as those of the Secure World Foundation to develop and help implement new "rules of the road" in space and ESA's effort to get space-faring nations to sign on to a

"Code of Conduct," could prove very helpful to finding ways forward with respect to space weaponization.

8. How Is the Space Tourism Business Expected to Evolve?

Many have speculated that the space tourism business will need to offer a suite of new experiences to keep the supply of customers coming. Indeed, one needs only to examine the website of Space Adventures to see that a range of "space tourism" services with a broad scale of prices is already on offer. These range from about $5000 for a ride on the so-called "vomit comet" in a high parabola to experience a few seconds of weightlessness, all the way up to making the now-estimated $35 million cost of a trip to the ISS for an eight-day stay if the availability of Soyuz seats is resumed. Eric Anderson, president of Space Adventures, has indicated that his company will continue to push the envelope and ultimately will offer his clients the chance to engage in space walks for a mere $15 million add-on. And most recently, there has been talk of a $100 million customer willing to pay for a trip on a Soyuz vehicle around the moon.

The trick to attracting space tourism customers and keeping them in the "supply chain" is to offer new challenges, together with the incentive of fares that decline over time. Thus the "next new space thrill" may be critical to this becoming a viable market. Others suggest that if this technology is converted to actual destination travel for high-end executives, then this might represent the critical path to viability.

Indeed, some visionaries are thinking even further ahead than Space Adventures. For example, Bigelow, the hotel suite king and owner of Bigelow Aerospace of Las Vegas, Nevada, has committed his firm to deploying a space hotel. This is not a pipe dream. His prototype technology has already been demonstrated with two test "Genesis" space habitats in orbit, and they have sent back live video. In addition, Bigelow is not alone in his belief that viable space hotels can be developed sooner rather than later. IOS of Mojave, California, has sought to develop a one-and-a-half stage to orbit vehicle called the Neptune; this would convert the rocket's fuel tanks to an in-orbit habitat complete with windows to view the earth.

9. What Are the Near to Longer-Term Outcomes from Private Enterprise in Space?

Currently, the space tourism business is focused almost exclusively on providing a unique experience of space adventures within the context of a short, rather risky, flight that ends at the initial departure point. John Spencer, in his book *Space Tourism: Do You Want to Go?*, candidly observed: "We are in the experience business. Not the space business and not the launch business…" [11] However, Spencer clearly has a very short,term and, indeed, narrow-minded perspective of what space tourism portends. Clearly, Spencer would not be a strong advocate of having an "environmental" surcharge added to space tourism flights to help sustain the planet and perhaps clear the way for hypersonic transportation services. No one wanting to succeed in a highly profitable business wants to contend with a "downer" like paying extra to save earth for posterity.

Indeed, the next major space development will most likely be the development and flight of hypersonic craft for transcontinental transport. As noted in Chapter 9, the list of those currently involved in this development numbers nearly a dozen and is growing. There are certainly a number of projects in the United States and Europe, but there are also efforts in Russia, Japan, and India. In time, there will likely be a Chinese project as well. Current efforts include development of executive hypersonic transport as well as larger craft for public transport for as many as 300 people.

In the longer term, true space commercialization will occur if access to orbit becomes reliable and cost effective. Space manufacturing, space energy production, space mining, and even space hotels and space colonization will eventually be in our future. The first step along this path is, however, improved space transport. The necessary research and development to support entirely new systems for the safe, reliable, and cost-effective creation of Solar Power Satellites (SPS), space manufacturing, space resource extraction, colonies on the moon or Mars, or other longer-term commercial space activities could and should be conceived of now. Hydrogen engines that produce water vapor rather than the more egregious greenhouse gases should be another priority.

This does not ignore the fact that water vapor does indeed represent yet another greenhouse gas. Nevertheless, it is still less of a problem than carbon dioxide, and it is way, way less than nitrous oxide or methane, or solid fuels that create particulate wastes in the stratosphere.

We need to begin critical supporting efforts, and we need to do so now, if 21st-century space programs are truly to have meaning that goes beyond a trip to an amusement park in near space or what we call the protozone. Key infrastructure such as space elevators, advanced ion engines, mass-drivers, mag-lev devices on the surface of the moon, satellite tether lifting devices, and so on are the inevitable follow-ons to simple parabolic flights into space. Serious planning for these subsequent activities should not be "left to later," which is always the easy way out. In reality, however, is this an area in which the private sector can be expected to take the lead? Will this flirtation with space tourism and "Citizen Astronauts" be followed by another Kennedy-esque pronouncement demanding that America commit its resources to leading the world into further reaches of outer space? The jury is still out as to whether this is a first date, "hooking up,, or a long-term and meaningful marriage where governments and private enterprises work together to make something truly meaningful happen.

If this is seen through the eyes of a visionary such as a Diamandis or a Maryniak, the commercial space industry could be the moral and economic equivalent of the computer revolution of the 1980s. Innovation and initiative from the private sector could create new economic growth, fuel new clean energy systems, stimulate new clean aviation transport, and much more. This is where creative public and private leadership is both greatly needed and unfortunately still unlikely to occur. [12]

10. "Citizen Astronauts": Reality or Science Fiction?

The race is on to make "Citizen Astronauts" a reality. More than a dozen countries are involved in various aspects of this enterprise, and some 50 companies, at one time or another in the past decade, have felt that they might provide a viable answer—in terms of both technology and business models. Some envision future spaceports as not only places from which to ride into outer space but also as venues where individuals can shop, be entertained, and visit a space museum or IMAX theater. In short, they envision "excitement village." The thrilling aspect of the space tourism enterprise could be that this is a business where rocket scientists are teaming up with "imagineers." Diamandis, the XPRIZE founder, indeed wants to host annual rocket races in New Mexico—a sort of 3D Mach plus NASCAR race that grips the public imagination and spurs space commerce to new heights.

The background to all the previous chapters suggests there are many questions still to be resolved and puzzles to be solved in a host of areas. These include technology development, flight operations, safety regulation, risk management, business plans, marketing, technology transfer and intellectual property, and environmental issues coupled with new technology development and possibly "sustainability surcharges." Other key issues include the weaponization of space, future public and private uses of space vehicles, and the development of a host of new commercial space applications and industries. Finally, it raises questions about the relation between national and international governmental agencies and private enterprise in this amazing new commercial space business environment. Some of the new space entrepreneurs (a number of whom have billions of dollars at their personal disposal) have undertaken test flights, and many trials are ongoing right now. The results of these will decide whether their often-changing target dates will be met. When the FAA reported on its Commercial Space Conference, it emphasized that the leaders of this industry were saying, "No flights before their time." This is to say that the FAA and the space entrepreneurs saw themselves in accord when they said that there is no artificial timetable for commercial flights to begin. Testing and proof of concept must be complete before commercial passengers start streaming to the nearest spaceport.

Seldom in human history are the stars aligned so that so much could go wrong or so much could go right—depending on whether the right or wrong steps are taken. If private enterprise moves ahead with more sophisticated and mature ways to move people and facilities into space, very good things could happen. In time, we could find ways to generate low-cost and environmentally friendly electric power in space and relay it to where it is needed. We could find new ways to manufacture in orbit the drugs and materials that we need to meet many societal needs. We could develop ways to build colonies on the moon and, in the centuries ahead, even begin to terra-form Mars so that people could live there. We could find ways to prevent comets or meteorites from creating massive waves of destruction to our planet. We might even devise new tools to cope with problems related to global warming, meteorological and geological disasters, and genetic mutation. Commercial space could help to make education and healthcare easier to distribute and reach far more students at significantly lower costs.

This could be the high road to a better human civilization and a "survivable world" for our species. The same pathway might be taken to another future that is much less desirable. Here we might find accelerated global warming, the destruction of the ozone layer, the weaponization of space, and a runaway global population that is harmed rather than helped by humans' new-found access to space.

The choice is ours to make. Temperate choices and smart planning can help us arrive at a future that will sustain new generations yet to come and allow our potential to be realized. Unlike Mr. Spencer, in his book *Space Tourism: Do You Want to Go?*, we believe the future of commercial space travel is about much more than a cheap thrill (well, actually, a rather expensive thrill). The future of space tourism could actually turn out to be about the future of *Homo sapiens* and the survival of the human race. Let us hope we can make the right choices and exploit this clever new technology to create a better world and sustain the human race.

REFERENCES
[1] Space Tourism Society, http://spacetourismsociety.org.
[2] "Common Horizons," White Paper of the International Space University Southern Hemisphere Studies program, February 2013.
[3] Pelton, J.N., *Space Debris and Other Threats in Space*, Springer Press, NY, 2013.
[4] "Finding Doomsday Asteroids, *New York Times*, p. A22, April 3, 2007.
[5] Pelton, J.N., and Marshall, P., *Space Travel and Astronaut Safety*, AIAA, Reston, VA, 2009.
[6] Steve Levy interview of Larry Page, "On Going All Out," *Wired* magazine, pp. 68-69, February 2013.

[7] Nield, G. et al., "Certification Versus Licensing for Human Spaceflight in Commercial Space Transportation," *New Space*, Vol(1)1, pp. 46-50.

[8] MOU between the FAA and NASA for Achievement of Mutual Goals in Human Space Transportation, signed June 4, 2012, http://www.nasa.gov/pdf/660556main_NASA-FAA%20MOU%20-%20signed.pdf.

[9] Speech, and Questions and Answers by Patricia Grace Smith, Center for Strategic and International Studies, Washington D.C., March 25, 2007.

[10] Center of Excellence in Commercial Space Transportation, Announced by Department of Transportation Secretary La Hood, August 2013, http://www.faa.gov/about/office_org/headquarters_offices/ast/programs/center_of_excellence/.

[11] Spencer, S., *Space Tourism: Do You Want to Go?* Apogee Books, p. 38, 2004.

[12] Diamandis, P., and Kotler, S., *Abundance*, Free Press, NY, 2012.

Appendix A: Inventory of Private Space Companies around the World
(Past and Present)

http://rocketdungeon.blogspot.com/
http://www.spacefuture.com/vehicles/designs.shtml
http://www.hobbyspace.com/Links/RLV/RLVTable.html
And numerous other sources, as listed below.

Company	Rocket Launch Vehicle	Intended Markets	Capabilities	Launch Site
Advent Launch Services	Advent 1 stage (VTHL from ocean).	Sub-orbital. 300 kg to 100 km.	Full-scale liquid engine tests.	Ocean launch and landing.
Aera Space Tours/Sprague Corp.	Altairis (VTHL).	Sub-orbital. Space tourism.	Two-stage, RP-1/LOX propulsion. Seven passengers to 100 km sub-orbital flights in 2007.	U.S. Air Force Cape Canaveral Launch Facility; five-year agreement.
Airbus/EADS/ Astrium	EADS/ASTRIUM TDN (See ESA A-1 development) spaceplane (HTHL) capable of carrying 4-100 passengers in sub-orbital flight. http://www.airbusgroup.com/	Sub-orbital. Space tourism.	In development since 2006. In partnership with ESA.	Test flight of demonstrator in May 2014 off coast of Singapore.

LAUNCHING INTO COMMERCIAL SPACE

Company	Rocket Launch Vehicle	Intended Markets	Capabilities	Launch Site
Alliant-ATK (See new name—Orbital ATK)	Pathfinder ALV X-1 (VTVL). Sounding rocket that is no longer being upgraded as a launcher.	Discontinued.	Intended to be upgraded Alliant sounding rocket.	Planned Mid-Atlantic Spaceport, Wallops Island.
Andrews (known as Spaceflight Industries as of 2015)	Gryphon Aerospaceplane http://www.andrews-space.com/	Sub-orbital. Space tourism	6360 kg to 100 km and return. LOX/RP-1. Less than $1 million/flight (in design).	N.A.
Armadillo Aerospace (see successor organization, Exos Aerospace: http://www.exosaero.com/flight_profile)	Black Armadillo. Program is now in "hibernation." Exos formed by employees to pursue space launch capabilities.	Sub-orbital spaceflight. Micro gravity launch with SARGE system.	One stage. LOX/ethanol engine. (Limited capital investment.) Vertical takeoff and land (like Delta Clipper design). Micro gravity launches, unmanned.	White Sands, New Mexico.
Blue Origin	New Shepard (VTVL). http://www.blueorigin.com	Sub-orbital. Space tourism to 100 km.	Reusable launch vehicle. Hydrogen peroxide and kerosene fuel. Abort system.	Culberson County, Texas. HQ in Seattle, Washington.

Company	Rocket Launch Vehicle	Intended Markets	Capabilities	Launch Site
Boeing	NASA has selected Boeing's CST-100 spacecraft capsule to develop crew launch to the ISS and return to earth. Uses a human-rated Atlas and Boeing spacecraft capsule CST-100. http://www.boeing.com/space/crew-space-transportation-100-vehicle/	CST-100 for docking with ISS and capsule return.	Capsule capable of supporting five astronauts and cargo.	Kennedy Space Center.
Bristol SpacePlanes Ltd	Ascender (subscale flight models). Space Bus (concept only): 50 persons or 110 tons. Space Cab (concept only): Eight persons or 2 + 750kg.	Sub-orbital. Three people or 400 kg on space tourism flight.	Jet. Two turbofans to 8 km. RL-10 liquid rocket engine to 100 km.	United Kingdom.

LAUNCHING INTO COMMERCIAL SPACE

Company	Rocket Launch Vehicle	Intended Markets	Capabilities	Launch Site
C & Space (Rep. of Korea) and AirBoss Aerospace Inc. (AAI)	Proteus spaceplane (VTHL). http://www.hobbyspace.com/nucleus/index.php?itemid=207	Sub-orbital. Space tourism. Three crew members.	LOX/methane engines. ITAR approval pending.	To be decided.
DaVinci Program (now defunct)	Da Vinci (balloon launch and vertical landing). http://www.davinciproject.com/	Sub-orbital. Space tourism. Three crew members.	Balloon to 40,000 ft. Twin LOX/kerosene engines to 120,000 ft. Parachute landing.	Can be launched from any balloon launch site.
DTI Associates	Terrier-Orion (Terrier is surplus Navy missile motor, and Orion is surplus Army missile motor).	Sub-orbital. Cargo to LEO (290 kg to 190 km).	Motors and vehicle FAA/AST licensed.	Woomera, Australia.
Enterprise TALIS	The Talis Enterprise GmbH is a consortium of five companies to develop a spaceplane. Current status unclear (see S-3 spaceplane).	Sub-orbital: Black Sky prototype to be followed by Enterprise sub-orbital spaceplane with 130 km apogee. Pilot, passenger, and cargo.	Consortium of German and Swiss aerospace companies.	Any spaceport with approved runway.

INNOVATIONS IN SPACE TRAVEL

Company	Rocket Launch Vehicle	Intended Markets	Capabilities	Launch Site
Excalibur Almaz	Project to use Russian Almaz capsule for space missions. http://www.parabolicarc.com/tag/excalibur-almaz/	Status is unclear due to lawsuits.	Status unclear.	Unclear.
HARC Space (No longer active)	Balloon launch reusable vehicle.	Sub-orbital. Sounding and targeting vehicle to sub-orbital.	Balloon and liquid fuel rocket engines.	Can be launched by balloon at many sites.
IL Aerospace (Israel)	Balloon launch and then Negev vehicle to sub-orbital space. http://commercialspace.pbworks.com/f/ILAT.pdf	Sub-orbital. 10 km by balloon and then Negev rocket launch to 120 km.	Balloon and Negev solid fuel rocket with parachute to water landing.	Can be launched by balloon at many sites. Israel base.
Inter Orbital Systems (Mojave, California)	Sea Star (13 kg to LEO). Neptune (4500 kg to LEO). http://www.interorbital.com/	Sea Star. Microsat launch vehicle.	Stage and a half. Liquid bi-propellant rocket. FAA/AST licensed.	Off shore. Pacific Ocean. Los Angeles and Tonga.

Company	Rocket Launch Vehicle	Intended Markets	Capabilities	Launch Site
Japanese Aerospace Exploration Agency (JAXA)	HII transfer vehicle (unmanned but in time might be upgraded to manned and pressurized vehicle). Prototype spaceplanes at Mach 2 and Mach 4 (LOH_2 and LOX propulsion system).	Unmanned cargo resupply to the ISS. Launched on the HII vehicle.	Conceptual studies.	To be decided.
Japanese Rocket Society	Kankoh-Maru (latest version of earlier Phoenix design).	Orbital. Fifty passengers to 200 km LEO.	Single stage to orbit. Vertical takeoff.	No hardware designs.
JP Aerospace (Rancho Cordova, California)	Access to Orbit-Ascender Balloon System. This is a different system from that used by Bristol SpacePlanes. http://www.jpaerospace.com	Sub-orbital. High-altitude experiments or rocket launch.	Very high-altitude balloon. Can be used as launch platform to LEO using ion engines.	California sites.

Company	Rocket Launch Vehicle	Intended Markets	Capabilities	Launch Site
Kelly Space & Technology Inc (San Bernadino, California)	Spaceplane. http://www.kellyspace.com/	Sub-orbital. Crew and satellite and cargo launch.	Tow launch of reusable spaceplane.	San Bernadino Airport.
Lockheed Martin	X-33 unmanned spaceplane. Joint projects with UP Aerospace to fly model prototypes. Research on hypersonic spaceplane prototypes such as SR-72 and Falcon HTV-2. http://www.lockheedmartin.com/us/products/falcon-htv-2.html	Hypersonic transport and sub-orbital flights at Mach 6 speeds.	Horizontal takeoff and landing by unmanned test vehicles.	To be decided.
Masten Space	XA 1.0 (VTVL). XA 1.5 (VTVL). XA 2.0 (VTVL). http://www.masten.aero	Sub-orbital. XA 1.0: 100 kg to 100 km. XA 1.5: 200 kg to 500 km. XA 2.0: 2000 kg (five people) to 500 km.	Liquid reusable internalized engines.	To be decided.

LAUNCHING INTO COMMERCIAL SPACE

Company	Rocket Launch Vehicle	Intended Markets	Capabilities	Launch Site
Orbital ATK (formerly Orbital Sciences and Alliant-ATK)	Antares launcher and Cygnus capsule. http://en.wikipedia.org/wiki/Antares_(rocket)	Antares/Cygnus developed under NASA COTS program.	Resupply missions to ISS. Cygnus burns up on return flight.	Launched from Mid Atlantic Spaceport, Wallops Island.
PlanetSpace & Canadian Arrow (discontinued)	Silver Dart Spaceplane and Lifting Body (VTHL) (discontinued). _____ Canadian Arrow (discontinued)	Orbital. Crew of eight. _____ Sub-orbital. Crew of three.	First stage liquid propellant + OX. Second stage four JATO rockets-Abort.	Nova Scotia, DaVinci Spaceport.
Reaction Engines (U.K.)	Skylon single-stage-to-orbit vehicle powered by SABRE engine. http://www.reactionengines.co.uk/sabre_howworks.html	Orbital.	Scram jet liquid hydrogen and oxygen.	Will likely use new U.K. spaceport.

Company	Rocket Launch Vehicle	Intended Markets	Capabilities	Launch Site
S-3 Swiss Space Systems	S3's SOAR shuttle concept would be launched from a certified Airbus A300. http://spacecoalition.com/blog/swiss-soar-space-plane-project-expands-technical-partnerships	Launch small satellite to LEO and return spaceplane for horizontal landing.	Small spaceplane launched from high altitude by carrier vehicle. Designed to launch small satellites.	
Sierra Nevada	Main initiative is the Dream Chaser spaceplane, which became a Sierra Nevada activity through the acquisition of SpaceDev.	See SpaceDev.	See SpaceDev.	To be decided.
Space Adventures with Myasishchev Design Bureau & Federal Russian Space Agency	Explorer Spaceplane (C-21) and MX-55 high-altitude launcher plane (HTHL).	Sub-orbital. Space tourism.	Liquid fuel motors. Horizontal takeoff and horizontal landing (lifting body with parachute landing).	To operate from international spaceports including Dubai, Singapore, and the United States.

LAUNCHING INTO COMMERCIAL SPACE

Company	Rocket Launch Vehicle	Intended Markets	Capabilities	Launch Site
SpaceDev (now part of Sierra Nevada Corp. [California])	Dream Chaser (VLHL). This vehicle lost out in NASA development contracts to develop crewed vehicle to fly astronauts to the ISS.	Sub-orbital. Space tourism (one stage), six passengers. Orbital. Two stage manned access to ISS (now not likely).	Single hybrid engine. Neoprene and NO_2 for sub-orbit. Launch of spaceplane with three large hybrid boosters to reach LEO and the ISS.	To be decided.
Space Exploration Technologies (SpaceX)	Falcon 9, Dragon spaceplane. http://www.spacex.com/	Orbital. COTS to ISS.	Cluster of nine Merlin engines on Falcon 9.	Kwajalein Atoll launch complex.
Spacefleet (U.K.)	(HLHL) The EARL project is being developed by the Spacefleet Association. This is a reusable unmanned aerial vehicle. http://www.spacefleet.co.uk/Services.html	Low-cost sub-orbital launcher. Sub-orbital (100 km apogee) missions to have five minutes of weightlessness.	Small scale prototypes only at this stage.	Any spaceport with runway.

Company	Rocket Launch Vehicle	Intended Markets	Capabilities	Launch Site
Space Hab (contract agreement with NASA to support COTS)	Apex 1. Apex 2. Apex 3. http://astrotechcorp.com/about-us	Orbital. Launch to LEO. 300 kg (Apex 1) to 6000 kg (Apex 3). Apex 1 and 2 are unmanned. Apex 3 can be manned.	Open architecture to support different missions and NASA's COTS program.	To be decided.
Space Transport Corp. (Forks, Washington)	Rubicon 1 and 2 and N-SOLV now to be replaced by Spartan vehicle.	Sub-orbital. Two passengers to 80-100 km. Spartan can launch 5 kg to LEO.	Design, status of project, and financing are not clear.	To be decided.
Starchaser Industries (U.K. and Rocket City, New Mexico)	Thunderstar-Starchaser 5.	Sub-orbital. Space tourism. Launch to 60 km.	Bi-liquid. LOX and kerosene rockets. Parachute recovery.	To be decided.
Sub-Orbital Corp. and Myasishchev Design Bureau	M-55X and Cosmopolis XXI.	Sub-orbital. Two stage to 100 km. Pilot and two passengers. Space tourism.	First stage M-55X Geophysika. Second stage C-21 rocket-powered lifting body with parachute landing.	Flexible launch and takeoff sites.

LAUNCHING INTO COMMERCIAL SPACE

Company	Rocket Launch Vehicle	Intended Markets	Capabilities	Launch Site
TGV Rocket	Michelle B Rocket (Modular Incremental Compact High Energy Low Cost Launch Experiment).	Sub-orbital. Small crew or scientific instruments.	Single stage to orbit. Modular.	White Sands, New Mexico.
Transformation Space Corp. t/Space (Allied with Scaled Composites)	CXV (Crew Transfer Vehicle).	Orbital. Crew of four to LEO or the ISS & ISS re-supply missions.	Launches at high altitude from a large cargo carrier aircraft.	To be decided.
Triton Systems	Stellar-J (HTHL).	Orbital. 440 kg of cargo to LEO.	Launches via a cargo jet and LOX-kerosene.	To be decided.
UP Aerospace	Sl-1 Carrier Rocket-Space Loft XL.	Launch of small scientific packages of 50 kg in 220 km LEO.	Liquid-fueled rocket (licensed by FAA/AST).	Spaceport America, New Mexico.
Vela Technologies	Spacecruiser.	Sub-orbital. Spaceplane. Up to eight people.	Jet plus Propane/NO_2.	To be decided.

Company	Rocket Launch Vehicle	Intended Markets	Capabilities	Launch Site
Virgin Galactic. Now owns SpaceShip Corporation and has bought out Sierra Nevada (SpaceDev and Scaled Composites interests). (Mojave, California)	SS2-SS (HTHL) launched from White Knight carrier plane with horizontal landing at Spaceport America.	Sub-orbital. Space tourism. Seven people to 100 km.	Aluminum polyimide with NO_2 as oxidizer.	Mojave Airport, and Spaceport America in New Mexico.
XCOR Aerospace	Lynx (sub-orbital space) (HTHL). Improved Lynx (sub-orbital space) or launch small satellite to LEO (HTHL).	Sub-orbital. Space tourism and nanosatellite launch.	Isopropyl alcohol/LOX. Lynx is FAA/AST licensed for experimental launch.	White Sands, New Mexico. Operations to be moved to Texas.

APPENDIX B

MEMBERSHIP LIST OF THE COMMERCIAL SPACEFLIGHT FEDERATION (CSF) (As of April 2015)

EXECUTIVE MEMBERS	ASSOCIATE MEMBERS
Alaska Aerospace Corporation	ARES Corporation
Bigelow Aerospace	Arizona State University
Blue Origin	ASRC Federal
Excalibur Almaz	Barrios Technology
Jacksonville-Cecil Field Spaceport	BRPH
Masten Space Systems	Colorado Space Coalition
Midland International Air & Space Port	David Clark Company
Mojave Spaceport	ETC – NASTAR Center
Moon Express	Golden Spike Company
Orbital Outfitters	Griffin Communications
Planetary Resources	Heinlein Prize Trust
Sierra Nevada Corporation	Houston Airport System
Southwest Research Institute	IHA
Space Adventures	InterFlight Global
Space Florida	Jacobs Technology
Spaceport America	MDA Corporation
SpaceX	ORBITEC
Virgin Galactic	Paragon SDC
Virginia Commercial Space Flight Authority	Penn State Applied Research Laboratory
	Planet Labs
World View Enterprises	Planetary Resources
XCOR Aerospace	QinetiQ North America
	Qwaltec
	RS&H
	S3 USA Holdings
	Scaled Composites
	Space Coast Spaceflight Alliance
	Spaceflight Services
	Spaceport Sweden
	Waypoint 2 Space
	X PRIZE Foundation

Note: The CSF website lists its membership at http://www.commercialspaceflight.org/membership/member-organizations/. You will find a listing of the members and can click on these members' names to access their websites to find their most current information. It is significant to see substantial churn in its members since 2012, with almost 50 percent turnover with regard to organizations leaving as well as new members joining or moving up from

Associate Members to Executive Members. Some of the largest corporate members (such as the United Launch Alliance, Raytheon, AeroJet, and Pratt & Whitney Rocketdyne) have left in the past three years.

APPENDIX C

\multicolumn{2}{c}{Chronology of Events in the Evolution of Commercial Spaceflight (With Other Key Dates in Space Development)}	
Date	**Event**
1957	
October	Launch of Sputnik by the USSR—the world's first artificial satellite.
1958	
November	President Eisenhower created NASA, a reaction to the Soviet space program and the desire to launch a U.S. satellite.
1961	
April	Soviet cosmonaut Yuri Gagarin was the first man to orbit the earth.
May	First Mercury manned spaceflight of the U.S. space program.
1962	
February	John Glenn was the first U.S. astronaut to orbit the earth.
1965	
June	Gemini-4 astronauts undertook the first EVA (spacewalk).
1967	
June	Three astronauts were killed in an Apollo launch-pad accident.
Summer	Barron Hilton, president of Hilton Hotels, published a paper about space tourism: "Hotels in Space."
1968	
April	Premiere of "2001: A Space Odyssey," the Academy Award-winning movie by Stanley Kubrick and Arthur C. Clarke. The film included scenes of a manned space station and lifestyle in outer space.
1969	
July	First Apollo moon landing; Neil Armstrong was the first man on the moon.
1972	
December	Last manned space flight to the moon; Gene Cernan was the "last man on the moon."
1977	
May	Skylab was launched by NASA.
1981	
April	The Space Shuttle made its first flight.
1982	
Spring	British Aerospace started development of Horizontal Takeoff and Landing (HOTOL) spacecraft (canceled in 1986 due to lack of funding).
1984	
November	The first of a series of papers on SSTO and HTOL vehicles for

LAUNCHING INTO COMMERCIAL SPACE

	space tourism published by David Ashford of Bristol SpacePlanes (U.K.).
1985	
August	A study on space tourism by Society Expeditions of Seattle was presented to NASA and the L-5 Society Space Development Conference.
	The design of the "Phoenix" SSTO-VTOL passenger vehicle was published by Gary Hudson of Pacific American Launch Systems.
	NASA started design work on HL-20 astronaut escape vehicle (canceled in the early 90s after costs reached $2 billion).
1986	
January	The loss of the Space Shuttle *Challenger* with seven astronauts.
October	"Potential Economic Implications of the Development of Space Tourism" was presented at the IAF Congress, including an estimate of market.
1987	
October	The design of the passenger-carrying upper stage for German "Stenger" SSTO-HTOL vehicle was presented at the IAF Congress by Dietrich Koelle.
1988	
September	Space Shuttle flights resumed 2.5 years after the loss of *Challenger*.
1989	
October	The design for an orbital hotel was presented at the "Feasibility of Space Tourism" session at the IAF Congress by Shimizu Corporation, a major construction company.
1990	
	NASA funded the development of the Rockwell X-30 National Space Plane (NASP)—an SSTO spacecraft (work was terminated in 1993).
1991	
	McDonnell Douglas started work on the Delta Clipper DCX reusable launch vehicle (canceled in 1996 after eight test flights).
November	The International Space Conference of Pacific-basin Societies (ISCOPS) in Kyoto received papers on the Phoenix ("History of the Phoenix VTOL SSTO and Recent Developments in Single-Stage Launch Systems") and space tourism ("Benefits of commercial passenger space travel for society").
1992	
October	The IAF Congress considered "The Prospects for Space Tourism: Investigation on the Economic and Technological Feasibility of Commercial Passenger Transportation into Low Earth Orbit" by Sven Abitzsch and Fabian Eilingsfeld.
	Starchaser Industries was founded in the U.K. to develop Britain's role in the space industry.
1993	

April	The Japanese Rocket Society started a study program on the feasibility of space tourism and established its Transportation Research Committee to design a passenger launch vehicle.
July	Spacehab made the first flight on board the shuttle *Endeavour*.
1994	
March	The American Society of Civil Engineers (ASCE) "SPACE 94" conference in Albuquerque considered the feasibility of commercial "space business parks."
May	The International Symposium on Space Technology and Science (ISTS) in Yokohama received a paper from the JRS study program on Kankoh-Maru, the JRS passenger launch vehicle to carry 50 passengers to and from LEO.
	The Commercial Space Transportation Study Final Report included the first study by major U.S. aerospace companies of the potential market for space tourism. It concluded that space tourism wasn't feasible!
October	The design of Kankoh-Maru was presented at the annual IAF Congress, and a 1/20 scale model of Kankoh-Maru was displayed at Farnborough International Air Show.
1995	
Spring	NASA JSC started developing the X-38 Lifting Body, followed by a contract with Scaled Composites Inc. (the project was canceled in April 2002 due to budget pressures).
September	The Space Transportation Association (STA) in Washington D.C. started a study of space tourism with cooperation from NASA.
	Market research on the demand for space tourism in Canada, the United States, and Germany showed a huge potential market worldwide.
October	The U.S. Office of Commercial Space Transportation (OCST) formally moved into the FAA.
1996	
January	NASA started developing the X-43A and X-43C spaceplanes, a $250 million program (terminated in late 2004 after the third test flight).
March/April	The cover story of *Ad Astra*, the magazine of the U.S. National Space Society, was "Space Tourism"—for the first time.
May	The XPRIZE project launched at a gala dinner in St Louis. Speakers (including NASA Administrator Dan Goldin) linked the XPRIZE to space tourism.
	At the 20th ISTS in Japan, NASA's Barbara Stone presented "Space Tourism: The Making of a New Industry," which concluded, "Studies and surveys worldwide suggest that space tourism has the potential to be the next major space business."
June	The ASCE "SPACE 96" conference considered the legal issues that need to be resolved before private commercial facilities can be

	constructed in orbit.
July	NASA announced the award of a $900 million three-year contract to Lockheed Martin to build and fly the X-33 un-piloted, reusable rocket test-vehicle to speeds of Mach 15 (canceled in March 2001).
	The STA-NASA space tourism study steering group concluded that the obstacles facing the establishment of a space tourism industry could be overcome "within 15 years."
August	A contract was awarded to Orbital Sciences Corp. for developing the X-34 test reusable launch vehicle (RLV) (canceled in March 2001 after three test flights).
September	California Spaceport became the first commercial U.S. spaceport to be licensed by the FAA/AST.
November	The *Aerospace America* article on Kankoh-Maru ("Japan plans day trips to space") was the first mainstream aerospace journal coverage on the subject.
1997	
February	The IEEE Aerospace Conference at Snowmass, Colorado, featured three papers on space tourism.
March	The STA-NASA workshop in Washington D.C. considered a range of issues relating to establishing the space tourism business.
	The First International Symposium on Space Tourism was held in Bremen, Germany, and was organized by Space Tours GMBH.
April	*Aviation Week* published the article "Studies claim space tourism feasible" based on papers presented at the IEEE Aerospace Conference (February, above), the first time it had covered the subject.
May	The FAA/AST issued the spaceport operated by the Florida Space Authority a license to operate. (This was renewed for five years in 2002.)
July	The Cheap Access To Space (CATS) conference, held in Washington D.C., was jointly sponsored by NASA and the Space Frontier Foundation.
October	International Astronautical Federation (IAF) President Karl Doetsch referred to space tourism as "one of the only businesses which will enable the launch industry to grow significantly."
November	The "Space Tourism Society," based in LA and chaired by Buzz Aldrin, was announced.
December	The FAA/AST issued a launch site operator's license to the Mid-Atlantic Regional Spaceport (MARS).
1998	
January	The AIAA workshop in Banff, Canada, on international cooperation in space included space tourism as one of five themes and recommended that "in light of its great potential, public space travel should be viewed as the next large, new area of commercial space activity."

February	*Business Week*, *Fortune*, and *Popular Science* all published articles on the U.S. venture companies that were developing reusable launch vehicles: Kelly Space Technology, Kistler Aerospace, Pioneer Rocketplane, and Rotary Rocket.
March	A press conference on Capitol Hill announced the release of "General Public Space Travel and Tourism," the final report of a joint study by STA and NASA started in September 1995. NASA admitted that space tourism is both feasible and economically desirable.
	NASA Administrator Goldin in a speech at NASA's 40th-anniversary gala dinner said, "…in a few decades there will be a thriving tourist industry on the moon."
April	The first modules of the ISS were launched.
	The formation of Bigelow Aerospace Inc. was announced.
	Space 98, the biennial space conference of the ASCE in Albuquerque, had sessions on space tourism, space commercialization, space access, and space ports, among many others.
May	The XPRIZE Foundation announced the target of $10 million and launched the XPRIZE for the first commercial company to demonstrate a reusable passenger space vehicle.
	The FAA started a study for extending air traffic management upwards to include LEO.
September	*Space Policy Journal* published the article "Space Tourism: a response to continuing decay in the U.S. civil space financial support."
October	Papers on space tourism were featured at the European Space Agency (ESA) workshop on Space Exploration and Resources Exploitation.
1999	
July	NASA awarded a contract to Boeing Co. to develop the X-37 Approach and Landing Test Vehicle (ALTV), with a four-year program valued at $173 million.
2000	
January	Norman Augustine, ex-CEO of Lockheed Martin Corporation, predicted in *Aviation Week* that space tourism would become the main space activity.
	The formation of MirCorp, to commercialize the MIR space station, was announced.
March	Illustrations of the Japanese Kankoh-Maru and Bristol SpacePlanes' "Spacebus" appeared on the NASA website.
June	The ISTS Symposium in Japan included space tourism sessions on Universal Spacelines, the XPRIZE, airline operations, insurance, and certification of Kankoh-Maru for passenger carrying.
	MirCorp announced that the first fare-paying guest to visit Mir was

LAUNCHING INTO COMMERCIAL SPACE

	Dennis Tito, founder of Wilshire Associates.
	The 2nd annual conference of the Space Travel and Tourism Division of the Space Transportation Association took place in Washington D.C.
July-August	The International Space University (ISU) Summer Session included Design Project on Space Tourism (for the first time).
October	The first meeting on space tourism in France took place at the French space agency, CNES. There were presentations by CNES, ESA, Astrium, and ISU.
2001	
February	"The Prospects for Passenger Space Travel" conference in Washington D.C. was sponsored by the FAA.
March	NASA canceled the X-33 and X-34 RLV technology demonstrator programs after spending over $1 billion and encountering numerous technical problems.
April	Dennis Tito became the first paying space tourist, launching from Baikonur aboard a Russian Soyuz bound for the ISS. Tito returned safely after 128 orbits in 8 days.
June	Space Exploration Corp (SpaceX) announced a program to develop the Falcon series of two-stage launch vehicles.
	NASA funded research into whether U.S. citizens would like to take a trip into space. The survey confirmed the potentially huge market for space tourism and led to the conclusion that only space tourism offered a large enough market to enable RLVs to reduce the cost of getting to orbit.
	The first U.S. Congressional hearing on space tourism took place at the House subcommittee on Space and Aeronautics.
July	NASA released "General Public Space Travel and Tourism," a very positive report on the feasibility of space tourism originally produced by NASA in 1998.
	Rocketplane Inc. formed to develop plans for space vehicles.
August	MirCorp announced an agreement with the Russian government and RSC Energia to design, develop, launch, and operate the world's first private space station, Mini Station 1.
October	The 2nd IAF Congress in Toulouse, France, included several symposia dedicated to space tourism and other possible new space markets.
	XCOR Aerospace successfully completed the first phase of its flight test program for the EZ-Rocket, the world's first privately built rocket-powered airplane.
November	Space Adventures, Ltd., commissioned a market survey on space tourism. The market analysis stated that, at the price of $100,000, more than 10,000 people per year would purchase flights.
	XPRIZE competitor Starchaser Industries successfully launched its Nova single-seat sub-orbital rocket for the first time. The un-piloted

	test, from Morecambe Sands, England, reached 1688.8 meters (5541 feet).
	The Presidential Commission on the Future of the U.S. Aerospace Industry held its first hearing. Astronaut Buzz Aldrin, one of the Commissioners, argued the case for the importance of developing "high volume human space transportation."
2003	
February	The loss of the Space Shuttle *Columbia* with seven astronauts.
October	China launched Shenzhu 5 and astronaut Yang Li Wei to become the third nation to successfully launch a manned spaceflight.
2004	
January	President George W. Bush announced a new space vision "to explore the moon, Mars and beyond."
September	The SpaceShipOne team of Paul Allen and Burt Rutan won the Ansari XPRIZE for commercial spaceflight.
	Sir Richard Branson of Virgin Atlantic and Burt Rutan, who developed SpaceShipOne, announced the formation of Virgin Galactic to provide passenger flights into space "in 2.5 to 3 years."
December	Congress passed the Commercial Space Launch Amendments Act of 2004 (CSLAA), which makes the Department of Transportation and the FAA responsible for regulating human spaceflight.
2005	
January	Jeff Bezos announced plans to create a spaceport facility to support his Blue Origin launch operations of the New Shepard vehicle in Van Horn, Texas.
September	NASA announced broad design features for Project Constellation.
October	Interorbital Systems announced the first "Promotional Fare" spaceline ticket to Tim Reed of Gladstone, Missouri.
December	NASA issued an RFP to industry for commercial services to ferry supplies and astronauts to the ISS, the COTS scheme.
	Virgin Galactic reached a 20-year lease agreement for use of a New Mexico spaceport
2006	
January	The FAA published "Human Space Flight Requirements for Crew and Space Flight Participants; Proposed Rule" in the Federal Register. It contained recommended requirements for crew qualifications, training and notification, as well as training and informed consent requirements for spaceflight participants.
February	Rocketplane Inc. purchased Kistler Aerospace, forming

LAUNCHING INTO COMMERCIAL SPACE

	RpK.
	Space Adventures announced the Explorer and Xerus spaceplanes and spaceports in Singapore and the United Arab Emirates.
July	Bigelow Aerospace successfully launched Genesis-1, the prototype for an orbital hotel with a target date of 2015.
August	The FAA and the U.S. Air Force Space Command issued new common federal launch safety standards designed to create consistent, integrated space launch rules.
	SpaceX and Rocketplane Inc. were awarded a contract by NASA to provide the K-1 vehicle to ferry personnel and supplies to the ISS under the COTS scheme.
	PlanetSpace/Canadian Arrow announced a permanent spaceport facility in Cape Breton, Nova Scotia.
	The Personal Spaceflight Federation (PSF) was founded in Washington D.C.
September	Anousheh Ansari became the fourth paying passenger (and the first woman) to fly into orbit with Soyuz, spending two days on the ISS.
	SpaceDev Corporation announced plans for Dream Chaser, a passenger-carrying spacecraft based on NASA's HL-20 design.
	Virgin Galactic announced that it will cost $190,000 for flights with six passengers to reach a sub-orbital altitude of 140 kilometers; there were reports of "100s of reservations" being made, including celebrity names.
	The Space Shuttle flight program was resumed 3.5 years after the loss of *Columbia*, to continue construction of the ISS. A further 15 shuttle flights were planned.
2007	
April	Charles Simonyi from Los Angeles became the fifth fare-paying passenger on a Soyuz flight to the ISS.
June	Astrium announced the European space vehicle at the Paris Air Show.
	The NASA Space Shuttle *Atlantis* carried out a successful mission to the ISS.
July	Bigelow Aerospace launched the Genesis 2 inflatable module from Russia.
	Space Adventures offered a spacewalk on a future Soyuz mission for $15 million in addition to a $30 million flight to the ISS.
	Aerospace giant Northrop Grumman increased its stake in Scaled Composites from 40 percent to 100 percent.
	Space Adventures announced a circumlunar mission on a Soyuz spacecraft, with two places at "$100 million per

	couch."
	The first private space industry fatalities. Three were killed and three injured in an engine testing explosion at the Scaled Composites facility in the Mojave Desert.
	Armadillo Aerospace announced successful tethered tests at its New Mexico base.
September	Google announced the $30 million Lunar XPRIZE for the first private moon rover vehicle to send images and data to earth.
October	NASA dropped RpK from the COTS program for failing to meet financial goals.
	China launched its first lunar probe to begin a three-phase program to land an astronaut on the moon by 2020.
	John Carmack's Armadillo Aerospace narrowly failed in an attempt to win Northrop Grumman's $350,000 Lunar Lander challenge.
November	PlanetSpace announced its proposal to compete for the NASA COTS award.
December	Sir Richard Branson completed a two-day training program at the NASTAR Center in preparation for his inaugural flight on SS2.
2008	
January	Virgin Galactic unveiled new models of SS2 and WK2 in New York.
	NASA announced a $4.7 million contract to Zero Gravity Corporation to provide weightless flights to NASA-operated experiments and personnel.
February	The X PRIZE Foundation and Google announced the first 10 teams to register for the $30 million Google Lunar XPRIZE.
	Orbital Sciences Corp. won a $171 million NASA award to build and demonstrate a launch system capable of delivering cargo to the ISS under its $500 million COTS program (replacing RpK).
March	XCOR announced a two-seater spaceship smaller than a private jet to take people up for a 25-minute spaceflight.
June	The Personal Spaceflight Federation was renamed the Commercial Spaceflight Federation (CSF), which has over 50 Executive and Associate members.
October	Richard Garriott was the sixth passenger to fly to the ISS on board a Russian Soyuz spacecraft.
2009	
September	Guy Laliberte was the seventh passenger to fly to the ISS on a Russian Soyuz.
November	Masten Systems and Armadillo Aerospace won the Lunar

Launching into Commercial Space

	Lander prizes.
2010	
September	Boeing announced plans to develop a spacecraft to provide services with Space Adventures "by 2015."
December	The SpaceX Dragon capsule completed its first two-orbit flight.
2011	
July	The NASA Space Shuttle made its final flight.
October	Sir Richard Branson opened Spaceport America in New Mexico as the base for Virgin Galactic flights.
2012	
February	Congress extended FAA regulatory authority for commercial spaceflight through 2015.
April	Former NASA astronaut Michael Lopez-Alegria became president of the CSF.
May	SpaceX launched the Dragon space capsule to successfully complete the first commercial cargo supply mission to the ISS.
June	Excalibur Almaz announced plans for the first civilian spaceflight in 2015.
	China sent three astronauts on a successful mission to its orbiting laboratory.
October	SpaceX made its first "official" delivery mission to the ISS.
December	The Golden Spike project was announced, with commercial flights to the moon by 2020 being planned.
2013	
February	The first "Citizen Astronaut," Dennis Tito, announced plans to send a married couple on a 500-day voyage to Mars in 2018.
April	Orbital Sciences completed a successful test launch of its Antares rocket and Cygnus spacecraft.
	The first powered test flight of WK2 from Virgin Galactic's SS2.
	John Carmack announced that his Armadillo project was "in hibernation," but his employees continued the work by creating Exos Aerospace.
May	The U.N./ICAO symposium on "Emerging Space Activities and Civil Aviation—Challenges and Opportunities" took place in Montreal.
November	ISRO launched a Mars mission. It has been in Mars orbit since September 2014.
2014	
July	Eric Stallmer succeeded Michael Lopez-Alegria as president of the CSF.

September	NASA announced contracts worth $6.8 billion to Boeing and SpaceX to develop systems to fly crew to the ISS.
October	Orbital Sciences suffered a launch failure of its Cygnus spacecraft with supplies to the ISS.
	Alliant-ATK and Orbital Sciences combined to create Orbital-ATK.
	SS2 experienced a fatal crash when the flutter system to decelerate the spaceplane's speed deployed as SS2 departed from the White Knight carrier plane.
2015	
February	The ESA completed a successful test flight of its IXV Intermediate Experimental Vehicle.
Early in the year	ZerotoInfinity (of Spain) and World View (of the United States) High Altitude Capsules announced balloon ascents to 100,000 feet (over 30 kilometers).
March	The "Mars One" initiative was announced by CEO Bas Lansdorf.
April	The SpaceX resupply mission to the ISS.
May	Virgin Galactic Gateway Galley opened at Spaceport America in New Mexico.

APPENDIX D

Recent Commercial Space Launches Licensed by the U.S. FAA

Date	Payload	Vehicle	Company	Site
October 31, 2014	None	SpaceShipTwo	Scaled Composites	California
August 22, 2014	N/A	Falcon 9-R	Space Exploration Technologies Corporation	Texas
August 1, 2014	N/A	Falcon 9-R	Space Exploration Technologies Corporation	Texas
June 17, 2014	none	Falcon 9-R	Space Exploration Technologies Corporation	Texas
May 1, 2014	N/A	Falcon 9-R	Space Exploration Technologies Corporation	Texas
April 17, 2014	N/A	Falcon 9-R	Space Exploration Technologies Corporation	Texas
January 10, 2014	Flight PF03	SpaceShipTwo	Scaled Composites	California
October 7, 2013	750m no translation	Grasshopper	Space Exploration Technologies Corporation	Texas
September 5, 2013	Flight PF02	SpaceShipTwo	Scaled Composites	California
August 13, 2013	N/A	Grasshopper	Space Exploration Technologies Corporation	Texas
June 14, 2013	N/A	Grasshopper	Space Exploration Technologies Corporation	Texas
April 29, 2013	Flight PF01	SpaceShipTwo	Scaled Composites	California
April 19, 2013	N/A	Grasshopper	Space Exploration Technologies Corporation	Texas
March 7, 2013	N/A	Grasshopper	Space Exploration Technologies Corporation	Texas
December 17, 2012	N/A	Grasshopper	Space Exploration Technologies Corporation	Texas
November 1, 2012	No Payload	Grasshopper	Space Exploration Technologies Corporation	Texas
August 24, 2011	N/A	PM 2	Blue Origin	Texas
May 6, 2011	N/A	PM 2	Blue Origin	Texas
October 25, 2008	N/A	QUAD (Pixel)	Armadillo Aerospace	New Mexico
October 24, 2008	N/A	MOD-1	Armadillo Aerospace	New Mexico
October 24, 2008	N/A	MOD-1	Armadillo Aerospace	New Mexico
October 24, 2008	N/A	Ignignokt	Scott Zeeb d/b/a TrueZer0	New Mexico

October 24, 2008	N/A	MOD-1	Armadillo Aerospace	New Mexico
October 28, 2007	N/A	MOD-1	Armadillo Aerospace	New Mexico
October 28, 2007	N/A	MOD-1	Armadillo Aerospace	New Mexico
October 27, 2007	N/A	MOD-1	Armadillo Aerospace	New Mexico
October 27, 2007	N/A	MOD-1	Armadillo Aerospace	New Mexico
October 20, 2007	N/A	MOD-1	Armadillo Aerospace	Oklahoma
June 2, 2007	N/A	QUAD (Pixel)	Armadillo Aerospace	Oklahoma
June 2, 2007	N/A	QUAD (Pixel)	Armadillo Aerospace	Oklahoma
April 19, 2007	N/A	PM 1	Blue Origin	Texas
March 22, 2007	N/A	PM 1	Blue Origin	Texas
November 13, 2006	N/A	PM 1	Blue Origin	Texas

Source: http://www.faa.gov/data_research/commercial_space_data/launches/?type=Permitted

AUTHOR BIOGRAPHIES

 **Joseph N. Pelton** is the award-winning author of hundreds of articles and over 40 books in the fields of telecommunications, and space policy and systems, including the Pulitzer Prize-nominated book *Global Talk*. This book also won the Eugene Emme Literature award of the American Astronautics Society. Dr. Pelton is on the Executive Board of the International Association for the Advancement of Space (IAASS) and is associate editor of the *Journal of Space Safety Engineering*. He is the founder and vice chairman of the Arthur C. Clarke Foundation and founding president of the Society of Satellite Professionals International. He also served as dean and chairman of the Board of Trustees of the International Space University, director of the Interdisciplinary Telecommunications Program at the University of Colorado at Boulder, and director of Strategic Policy at Intelsat. He is the former executive editor of the *Journal of International Space Communications*. Dr. Pelton holds a Ph.D. from Georgetown University and is Director Emeritus of the Space and Advanced Communications Research Institute (SACRI) at George Washington University.

His awards include the Outstanding Educator Award from the International Communications Association, the H. Rex Lee Award for Public Service from the Public Service Satellite Consortium, and the ISCe Award for Outstanding Educational Achievement. He also received the 2001 Arthur C. Clarke Lifetime Achievement Award. He was elected to the International Academy of Astronautics, elected Associated Fellow of the AIAA, and elected Chairman of the Academic Committee of the IAASS, of which he is a fellow. He has made a number of media appearances on U.S. television and radio, BBC radio, and CBC of Canada.

Peter Marshall began as a journalist with BBC radio and television news, later becoming a news editor. In 1966, he moved to Visnews (then part-owned by the BBC and now Reuters-TV), a global TV news agency in London, where he became general manager. He was a pioneer in the use of satellites for TV news coverage and distribution and, in 1986, moved to Washington D.C. to create the Broadcast Services Division of the inter-governmental body INTELSAT. With the emergence of deregulation and competition, he returned to the private sector as president of Keystone Communications. This organization, now part of Globecast (a subsidiary of France Telecom), is the largest provider of global satellite broadcast services. Marshall is a past chairman of the Royal Television Society in the U.K. and past president of the Society of Satellite Professionals (SSPI) in the United States. In 2006, he was elected to the SSPI's "Hall of Fame" in recognition of his pioneering work for the industry. He continues to serve as a member of the board of the Arthur C. Clarke Foundation. Marshall is now a writer and consultant, based in his native U.K., and he has collaborated with Dr. Pelton on five books and major research projects, including *Communications Satellites—Global Change Agents* (published by Lawrence Erlbaum in 2006), *Space Exploration and Astronaut Safety* (published by AIAA in 2007), and *License to Orbit— the Future of Commercial Space Travel* (published by Apogee Books in 2011).